Fractional Dynamics and Control

Dumitru Baleanu • José António Tenreiro Machado
Albert C. J. Luo

Editors

Fractional Dynamics and Control

 Springer

Editors
Dumitru Baleanu
Mathematics and Computer
Sciences
Faculty of Art and Sciences
Cankaya University
Ankara
Turkey
dumitru@cankaya.edu.tr

José António Tenreiro Machado
Department of Electrical
Engineering
Polytechnic Institute of Porto
Institute of Engineering
Rua Dr António Bernadino
Almeida 431
4200-072 Porto
Portugal
jtm@isep.ipp.pt

Albert C. J. Luo
Mechanical and Industrial
Engineering
School of Engineering
Southern Illinois University
Room: EB2064
Edwardsville
USA
aluo@siue.edu

ISBN 978-1-4899-9252-9   ISBN 978-1-4614-0457-6 (eBook)
DOI 10.1007/978-1-4614-0457-6
Springer New York Dordrecht Heidelberg London

Printed on acid-free paper

Springer is part of Springer Science+Business Media (www.springer.com)

Preface

Fractional calculus is as old as the classical one and it started to be intensively studied and applied in many branches of science and engineering.

The fractional dynamics is growing faster during the few decades together with the fractional control theory which certified the fractional calculus as being a fundamental tool in describing the dynamics of complex systems.

This book is based on the 3rd conference on Nonlinear Science and Complexity (NSC), Ankara, Turkey, July 27–31, 2010. Due to the impact of topics on a very wide spectrum of problems in science and engineering, this conference provided a place to exchange recent developments, discoveries and progresses on nonlinear science and complexity. This conference is the continuation of the first 2006 Conference on Nonlinear Science and Complexity, Beijing, China, and the second 2008 conference on Nonlinear Science and Complexity held in Porto, Portugal.

This book it is entitled "Fractional Dynamics and Control" and it collects the recent development in nonlinear dynamics, nonlinear vibration and control. The aim of this book provides a fast exchange idea in nonlinear dynamical systems and control. One can learn the recent developments, including analytical, numerical and experimental results in such area.

The editors hope that this collection of papers may be fruitful for scholars, researchers and advanced technical members of industrial laboratory facilities, for developing new tools and products.

The editors thank to the Rector and to the President of the Board of Trustee of Cankaya University as well as to the Scientific and Technological Research Council of Turkey for the financial support needed to hold the discussions and debates.

Dumitru Baleanu
José António Tenreiro Machado
Albert C. J. Luo

Contents

Part IV   Fractional Order Modeling

Part I
Fractional Control

Chapter 1
A Formulation and Numerical Scheme for Fractional Optimal Control of Cylindrical Structures Subjected to General Initial Conditions

Md. Mehedi Hasan, X.W. Tangpong, and O.P. Agrawal

1 Introduction

Many processes in physics and engineering systems can be modeled more accurately by fractional derivatives (FDs) or fractional integrals (FIs) than traditional integer order derivatives or integrals [1]. Miller and Ross [2] mentioned that almost every field of science and engineering has the application of FDs. Applications include biomechanics [3], behaviors of viscoelastic materials [4–7], control [8–11], electrochemical processes [12, 13], dielectric polarization [14], colored noise [15] and chaos [16], etc. Some other applications of Fractional Calculus (FC) can be found in [1, 17–21]. Optimal control problems have found applications in many different fields, including engineering, science, and economics. As the demand for more accurate and high-precision systems increases, the demand to develop formulation and numerical scheme of Fractional Optimal Control Problems (FOCPs) also increases.

Optimal control problem requires the minimization of a functional over an admissible set of control functions subject to dynamic constraints on the state and control variables [22]. A FOCP is an optimal control problem in which either the performance index or the differential equations governing the dynamics of the system or both contain at least one fractional order derivative term [23]. The applications of FOCPs have been increasing in many fields, and therefore, there is a critical need for developing solution techniques for those problems. The formulation and solution scheme for FOCPs was first established by Agrawal [24] where he applied Fractional Variational Calculus (FVC) to deterministic and stochastic analysis of FOCPs. Later, Agrawal [25] presented a general formulation and solution scheme for FOCPs in the Riemann–Liouville (RL) sense that was based on variational virtual work coupled with the Lagrange multiplier technique. Since

M.M. Hasan • X.W. Tangpong (✉) • O.P. Agrawal
Department of Mechanical Engineering, North Dakota State University, Fargo, ND 58108, USA
e-mail: mmh11003@engr.uconn.edu; annie.tangpong@ndsu.edu; om@engr.siu.edu

D. Baleanu et al. (eds.), *Fractional Dynamics and Control*,
DOI 10.1007/978-1-4614-0457-6_1, © Springer Science+Business Media, LLC 2012

Caputo Fractional Derivative (CFD) seems more natural and allows incorporating the usual initial conditions, it becomes a popular choice for researchers. Agrawal and Tangpong and Agrawal [1, 23] formulated FOCPs in terms of CFDs instead of RL derivatives and an iterative numerical scheme was applied to solve the problem numerically. Using the definitions of the FOCPs, Frederico and Torres [26–28] formulated a Noether-type theorem in the general context of the fractional optimal control (FOC) in the sense of CFDs. Agrawal and Baleanu [29] used an approximated method to approximate the FDs of the system which led to a set of algebraic equations that can be solved by using numerical techniques. Baleanu et al. [30] proposed a different solution scheme, where a modified Grünwald–Letnikov (GL) definition was used to derive a central difference formula. Based on the expansion formula for FDs, Jelicic and Petrovacki [31] proposed a new solution scheme. Using the eigenfunction expansion-based scheme, Agrawal [32] solved the FOCP in a one-dimensional distributed system. Later Ozdemir et al. [33] formulated FOCPs of a two-dimensional distributed system by using the same eigenfunction expansion approach.

FOCPs of a distributed system have recently been presented in polar coordinates. Ozdemir et al. [34] presented a formulation of axial symmetric FOCPs where the FDs were expressed in terms of RL definition, and the GL definition was used to find the numerical solution. Based on the eigenfunction expansion scheme, FOCPs of a three-dimensional distributed system were investigated in cylindrical coordinate by Ozdemir et al. [35] where FDs were defined in the RL sense and the GL approximation scheme was used. Fractional diffusion problems were discussed in polar coordinates [36] and in cylinder and spherical coordinates [37, 38]; however, these works [36–38] did not discuss FOCPs.

In this chapter, we present a general formulation and numerical solution scheme for FOCPs in cylindrical coordinates and use a solid cylinder case and a hollow cylinder case as examples to demonstrate the method. FDs are defined in the Caputo sense and the separation of variable method is used to decouple the equations. The eigenfunction approach is adopted to eliminate the space parameter and it is indicated by the combination of state and control functions. For numerical solutions, the FD differential equations are converted into Volterra-type integral equations and the time domain is discretized into several segments. The formulation derived here is used to solve problems with different derivative orders and the calculation converges toward the analytical solution for integer order problems as the order approaches 1.

2   A General Formulation of a Fractional Optimal Control Problem

For the ease of understanding of the FOCPs problems that are to be discussed in Sect. 3, this section briefly describes the general formulation of FOCPs. A description in more details can be found in [23, 39]. The FOCP under consideration can be defined as follows. Find the optimal control $f(t)$ that minimizes the performance index

$$J(f) = \int_0^1 F(w, f, t)\,dt \tag{1.1}$$

subject to the dynamic constraints

$$_0^C D_t^\alpha w = G(w, f, t) \tag{1.2}$$

and the initial conditions

$$w(0) = w_0, \tag{1.3}$$

where $w(t)$ and $f(t)$ are the state and control variables, respectively, F and G are two arbitrary functions, and w_0 represents the initial condition of the state variable. Equation (1.1) may also include additional terms containing state variable at the end points. When $\alpha = 1$, the above problem reduces to a standard optimal control problem. The integration limits are taken from 0 to 1 for a normalized case. $_0^C D_t^\alpha$ denotes the left CFD of order α, and we consider α to be in the range of $(0, 1)$. The conditions considered here are for simplicity in discussions to follow. After combining (1.1) and (1.2) using a Lagrange multiplier technique, applying integration by parts and setting the coefficients of $\delta\lambda$, δw, and δf to zero, the following equations are obtained:

$$_0^C D_t^\alpha w = G(w, f, t), \tag{1.4}$$

$$_t^C D_1^\alpha \lambda = \frac{\partial F}{\partial w} + \left(\frac{\partial G}{\partial w}\right)^{\mathrm{T}} \lambda, \tag{1.5}$$

$$\frac{\partial F}{\partial f} + \left(\frac{\partial G}{\partial f}\right)^{\mathrm{T}} \lambda = 0, \tag{1.6}$$

$$w(0) = w_0 \text{ and } \lambda(1) = 0, \tag{1.7}$$

where λ is the Lagrange multiplier also known as the co-state or adjoint variable. $_t^C D_1^\alpha$ in (1.5) denotes the right CFD of order α. The details of the derivations of (1.4)–(1.7) are given in [24]. Equations (1.4)–(1.6) represent the Euler–Lagrange equations for the FOCP. These equations give the necessary conditions for the optimality of the FOCP considered here. Since (1.4) contains the LCFD and (1.5) contains the RCFD, the solution of such optimal control problems requires knowledge of not only forward derivatives but also that of backward derivatives to account for all end conditions. In classical optimal control theories, such issue is either not discussed or not clearly stated largely because the backward derivative of order 1 turns to be the negative of the forward derivative of order 1. In the limit of $\alpha \to 1$, (1.4)–(1.6) reduce to those obtained using the standard methods for classical optimal control problems.

3   Formulation of FOC of Cylinder Structures with Axial Symmetry

The FOCP in consideration is as follows: find the control $f(r,\theta,z,t)$ that minimizes the cost functional

$$J(f) = \frac{1}{2} \int_0^1 \int_0^L \int_0^{2\pi} \int_0^R [Q'w^2(r,z,\theta,t) + R'f^2(r,z,\theta,t)] r \, dr \, d\theta \, dz \, dt \qquad (1.8)$$

subject to the system dynamic constraints

$$\frac{\partial^\alpha w}{\partial t^\alpha} = \beta \left(\frac{\partial^2 w(r,z,\theta,t)}{\partial r^2} + \frac{1}{r} \frac{\partial w(r,z,\theta,t)}{\partial r} + \frac{1}{r^2} \frac{\partial^2 w(r,z,\theta,t)}{\partial \theta^2} + \frac{\partial^2 w(r,z,\theta,t)}{\partial z^2} \right)$$

$$+ f(r,z,\theta,t) \qquad (1.9)$$

For an axisymmetric case, there is no variations in θ, and therefore, (1.8) and (1.9) become

$$J(f) = \frac{1}{2} \int_0^1 \int_0^R \int_0^L [Q'w^2(r,z,t) + R'f^2(r,z,t)] r \, dr \, dz \, dt, \qquad (1.10)$$

$$\frac{\partial^\alpha w}{\partial t^\alpha} = \beta \left(\frac{\partial^2 w(r,z,t)}{\partial r^2} + \frac{1}{r} \frac{\partial w(r,z,t)}{\partial r} + \frac{\partial^2 w(r,z,t)}{\partial z^2} \right) + f(r,z,t) \qquad (1.11)$$

where $\frac{\partial^\alpha w}{\partial t^\alpha}$ is the partial Caputo derivative of order α and $0 < \alpha < 1$. Q' and R' are the two arbitrary functions that may depend on time. R and L and are respectively the cylinder's radius and length. For convenience, the upper limit of time t is taken as 1. The initial condition is represented by

$$w(r,z,0) = w_0(r,z). \qquad (1.12)$$

Two cases of cylindrical structures – solid and hollow – are discussed next.

3.1   Solid Cylinder

The boundary conditions for FOC of a solid cylinder are considered as

$$w(0,z,t) = w(R,z,t) = w(r,0,t) = w(r,L,t) = 0, \quad t > 0. \qquad (1.13)$$

The eigenfunction approach is used here to decouple the equations and the state and the control functions are found to be

$$w(r,z,t) = \sum_{i=1}^{n} \sum_{j=1}^{m} q_{ij}(t) J_0\left(u_j \frac{r}{R}\right) \sin\left(i\pi \frac{z}{L}\right), \tag{1.14}$$

$$f(r,z,t) = \sum_{i=1}^{n} \sum_{j=1}^{m} p_{ij}(t) J_0\left(u_j \frac{r}{R}\right) \sin\left(i\pi \frac{z}{L}\right), \tag{1.15}$$

where $J_0(u_j \frac{r}{R})$ and $\sin(i\pi \frac{z}{L})$ are the eigenfunctions in the radial direction and the axial direction respectively. The total numbers of eigenfunctions, n and m, are determined by convergence studies. J_0 is the zero-order Bessel function of the first kind and u_j are the roots of this Bessel function. $q_{ij}(t)$ and $p_{ij}(t)$ are the state and control eigencoordinates. Substituting (1.14) and (1.15) into (1.10), we obtain the cost function

$$J = \frac{R^2 L}{8} \int_0^1 \sum_{i=1}^{n} \sum_{j=1}^{m} J_1^2(u_j)[Q' q_{ij}^2(t) + R' p_{ij}^2(t)]dt. \tag{1.16}$$

By substituting (1.14) and (1.15) into (1.11) and equating the coefficients of $J_0(u_j \frac{r}{R}) \sin(i\pi \frac{z}{L})$, we obtain

$$_0^C D_t^\alpha q_{ij}(t) = -\beta \left(\left(\frac{u_j}{R}\right)^2 + \left(\frac{i\pi}{L}\right)^2 \right) q_{ij}(t) + p_{ij}(t). \tag{1.17}$$

From (1.4)–(1.7), (1.16) and (1.17), we further obtain

$$_t^C D_1^\alpha p_{ij}(t) = -\frac{Q'}{R'} q_{ij}(t) - \beta \left(\left(\frac{u_j}{R}\right)^2 + \left(\frac{i\pi}{L}\right)^2 \right) p_{ij}. \tag{1.18}$$

Substituting (1.14) into (1.12), and then multiplying the equation by $r J_0(u_j \frac{r}{R})$ $\sin(i\pi \frac{z}{L})$ on both sides and integrating it, we find the initial condition of the eigencoordinates

$$q_{ij}(0) = \frac{4 \int_0^L \int_0^R r w_0(r,z) J_0\left(u_j \frac{r}{R}\right) \sin\left(i\pi \frac{z}{L}\right) drdz}{R^2 L J_1^2(u_j)}, \tag{1.19}$$

where J_1 is the first-order Bessel function of the first kind. A numerical scheme that can be used to solve (1.17) and (1.18) is given in Sect. 4.

3.2   Hollow Cylinder

The boundary conditions of a hollow cylinder is considered as

$$w(a,z,t) = w(R,z,t) = 0, \quad t > 0. \tag{1.20}$$

Using the eigenfunction approach, the state function and the control function are found to be

$$w(r,z,t) = \sum_{i=1}^{n}\sum_{j=1}^{m} q_{ij}(t)u_0(\lambda_j r)\sin\left(i\pi\frac{z}{L}\right), \tag{1.21}$$

$$f(r,z,t) = \sum_{i=1}^{n}\sum_{j=1}^{m} p_{ij}(t)u_0(\lambda_j r)\sin\left(i\pi\frac{z}{L}\right), \tag{1.22}$$

where

$$u_0(\lambda_j r) = Y_0(\lambda_j a)J_0(\lambda_j r) - J_0(\lambda_j a)Y_0(\lambda_j r) \tag{1.23}$$

is the eigenfunction in the radial direction, and $\sin(i\pi\frac{z}{L})$ is the eigenfunction in the axial direction. J_0 and Y_0 are the zero-order Bessel function of the first kind and the second kind, respectively, and λ_j is the root of the characteristic equation for the eigenfunctions in the radial direction. Substituting (1.21) and (1.22) into (1.10), we obtain the cost function

$$J = \frac{L}{4}\int_0^1 \sum_{i=1}^{n}\sum_{j=1}^{m}(Q'q_{ij}{}^2 + R'p_{ij}{}^2)\left(\int_\alpha^R u_0^2(\lambda_j r)r dr\right)dt. \tag{1.24}$$

By substituting (1.21) and (1.22) into (1.11) and equating the coefficients of $u_0(\lambda_j r)\sin(i\pi\frac{z}{L})$, we obtain

$$_0^C D_t^\alpha q_{ij}(t) = -\beta\left(\left(\frac{\lambda_j}{R}\right)^2 + \left(\frac{i\pi}{L}\right)^2\right)q_{ij}(t) + p_{ij}(t). \tag{1.25}$$

From (1.4)–(1.7), (1.24) and (1.25), we obtain

$$_t^C D_1^\alpha p_{ij}(t) = -\frac{Q'}{R'}q_{ij}(t) - \beta\left(\left(\frac{\lambda_j}{R}\right)^2 + \left(\frac{i\pi}{L}\right)^2\right)p_{ij}. \tag{1.26}$$

Substituting (1.21) into (1.12), and then multiplying the equation by $ru_0(\lambda_j r)$ on both sides and integrate from a to R, we find the initial condition of the eigencoordinates

$$q_{ij}(0) = \frac{2\int_0^L\left(\int_0^R rw_0(r)u_0(\lambda_j r)dr\right)\sin\left(i\pi\frac{z}{L}\right)dz}{L\int_a^b ru_0^2(\lambda_j r)dr}. \tag{1.27}$$

A numerical scheme that can be used to solve (1.25) and (1.26) is presented in the following section.

4 Numerical Algorithm

For the ease of discussions to follow, this section briefly describes the numerical algorithm for the FOCPs, similar to that presented in [23, 25]. FOCPs expressed in a generic form are as follows.

$$_0^C D_t^\alpha w = -Aw + Bf, \tag{1.28}$$

$$_t^C D_1^\alpha f = -Cw - Df, \tag{1.29}$$

$$w(0) = w_0, \tag{1.30}$$

$$f(1) = 0. \tag{1.31}$$

Equations (1.28) and (1.29) can be expressed in the Volterra integral form as follows.

$$w(t) = w_0 + \frac{1}{\Gamma(\alpha)} \int_0^t (t-\tau)^{(\alpha-1)} (Bf(\tau) - Aw(\tau)) d\tau, \tag{1.32}$$

$$f(t) = -\frac{1}{\Gamma(\alpha)} \int_\tau^1 (\tau-t)^{(\alpha-1)} (Df(\tau) + Cw(\tau)) d\tau. \tag{1.33}$$

After discretizing the time domain into N segments and taking linear approximations of $w(t)$ and $f(t)$ between two successive temporal nodes, (1.32) reduces to

$$w(t_i) = w_0 - A \sum_{j=0}^{i} a_{ij} w(t_j) + B \sum_{j=0}^{i} a_{ij} f(t_j), \quad i = 1, 2, \ldots, N. \tag{1.34}$$

where the coefficients a_{ij} are defined as

$$a_{ij} = d_1 \begin{cases} (i-1)^\beta - i^\beta + \beta i^\alpha & \text{if } j = 0, \\ (k+1)^\beta + (k-1)^\beta - 2k^\beta & \text{if } 1 \le j \le i-1, \\ 1 & \text{if } j = i. \end{cases} \tag{1.35}$$

Here, $d_1 = \frac{h^\alpha}{\Gamma(\alpha+2)}$, $\beta = (\alpha+1)$ and $k = i-j$. Following the same approach, the value of $f(t)$ at node i becomes

$$f(t_i) = -C \sum_{j=1}^{N} b_{ij} w(t_j) - D \sum_{j=1}^{i} b_{ij} f(t_j), \quad i = 0, 1, \ldots, N-1, \tag{1.36}$$

where

$$b_{ij} = d_1 \begin{cases} 1 & \text{if } j = i, \\ (k+1)^\beta + (k-1)^\beta - 2k^\beta & \text{if } i+1 \le j \le N-1, \\ (M-1)^\beta - M^\beta + \beta M^\alpha & \text{if } j = N. \end{cases}$$

Here, $M = N - i$ and $k = j - i$. Equations (1.34) and (1.36) represent a set of $2N$ linear algebraic equations which can be solved using a standard linear algebraic equations solver.

5 Numerical Results and Discussions

5.1 Solid Cylinder

For FOC of a solid cylinder with axial symmetry, the initial condition was taken as

$$w_0(r,z) = r(r-R)\left[1 - \cos\left(2\pi\frac{z}{L}\right)\right]. \qquad (1.37)$$

For simplicity, we considered $Q' = R' = L = 1$ and $R = 1$. For simulation purposes, we discretized the spatial dimensions and the time domain into several segments and took different values of α. The number of eigenfunctions in the radial and axial directions were determined through convergence studies. It was found that the results converged with $m = n = 5$, where m is the number of eigenfunctions in the radial direction and n is the number of eigenfunctions in the axial direction. All simulation results presented in Figs. 1.1–1.8 were generated based on these values of m and n.

Figures 1.1 and 1.2 demonstrate the state and control variables as functions of time, and they both converge as the time steps are reduced. The convergence studies of the number of eigenfunctions and time steps need to be conducted before other parameter studies, and the convergence criterion varies with the specific problem. Figures 1.3 and 1.4 show changes of the state and control variables as functions of time for various orders of α and also compare the numerical results with the analytical results when $\alpha = 1$. In the limit of $\alpha = 1$, the numerical solutions recover

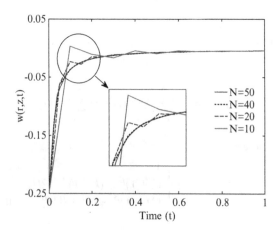

Fig. 1.1 Convergence of the state variable $w(r = 0.5, z = 0.25, t)$ for different number of time segments for $\alpha = 0.8$

Fig. 1.2 Convergence of the control variable $f(r = 0.5, z = 0.25, t)$ for different number of time segments for $\alpha = 0.8$

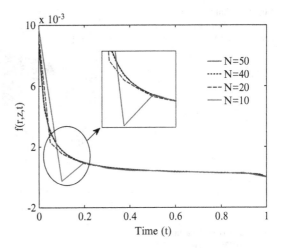

Fig. 1.3 State variable $w(r = 0.5, z = 0.25, t)$ for different values of α with $N = 50$

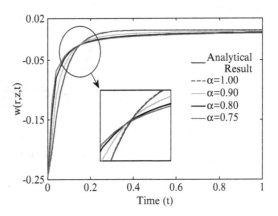

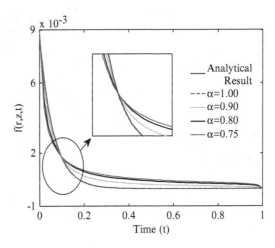

Fig. 1.4 Control variable $w(r = 0.5, z = 0.25, t)$ for different values of α with $N = 50$

Fig. 1.5 State variable $w(r, z = 0.25, t)$ for $N = 50$ and $\alpha = 0.90$

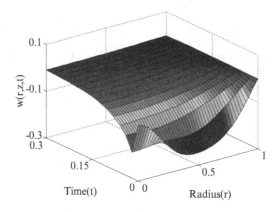

Fig. 1.6 Control variable $f(r, z = 0.25, t)$ for $N = 50$ and $\alpha = 0.90$

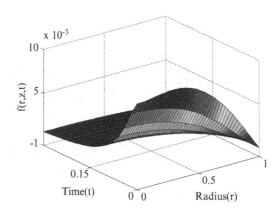

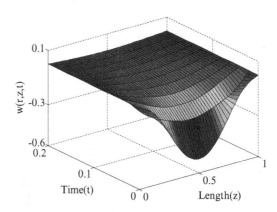

Fig. 1.7 State variable $w(r = 0.5, z, t)$ for $N = 50$ and $\alpha = 0.9$

Fig. 1.8 Control variable $f(r = 0.5, z, t)$ for $N = 50$ and $\alpha = 0.90$

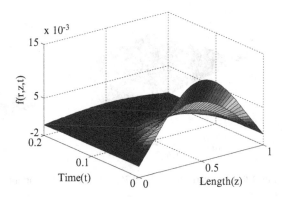

the analytical solutions of the integer order optimal control problem. The agreement of analytical results with the numerical results when $\alpha = 1$ validates the numerical algorithm. Figures 1.5 and 1.6 are the surface plots of the state and control variables in the radial direction. In both figures, the state and control variables initially have different values across the radial dimension due to the initial conditions; as the time progresses, each variable reaches the same value across the radius. The phenomenon shown in Fig. 1.5 is typical of a diffusion process. Figures 1.7 and 1.8 are the three-dimensional responses of the solid cylinder in longitudinal direction. Similar to the phenomena shown in Figs. 1.5 and 1.6, both the state and control variables approaches the same value across the length as the time progresses, representing a diffusion process. The dynamics constraint equation (1.11) becomes a heat diffusion equation when $\alpha = 1$; when $\alpha = 0.9$, the dynamics governed by (1.11) is close to a diffusion process, but not exactly the same as the integer order derivative case. For such dynamic problems, the fractional order differential equation can give more accurate results than the integer order differential equation.

5.2 Hollow Cylinder

For FOC of a hollow cylinder with axial symmetry, three cases of initial conditions have been considered.

Case 1:

$$w_0(r,z) = (r-a)(r-R)\sin\left(\pi\frac{z}{L}\right). \tag{1.38}$$

Case 2:

$$w_0(r,z) = (r-a)(r-R)\sin\left(3\pi\frac{z}{L}\right). \tag{1.39}$$

Case 3:

$$w_0(r,z) = (r-a)(r-R)\sin\left(\pi\frac{z}{L}\right) + (r-a)(r-R)\sin\left(3\pi\frac{z}{L}\right). \tag{1.40}$$

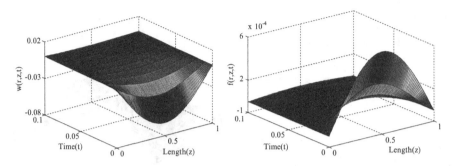

Fig. 1.9 State variable $w(r = 0.75, z, t)$ and control variable $f(r = 0.75, z, t)$ for $N = 100$ and $\alpha = 0.90$

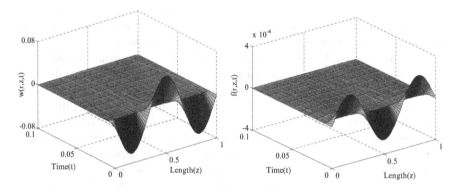

Fig. 1.10 State variable $w(r = 0.75, z, t)$ and control variable $f(r = 0.75, z, t)$ for $N = 100$ and $\alpha = 0.90$

The inner radius of the cylinder $\alpha = 0.5$, and $R = L = 1$. The initial condition of Case 3 is a linear combination of the initial conditions of Cases 1 and 2. Similar to the solid cylinder problem discussed in Sect. 5.1, convergence studies of the number of eigenfunctions in the radial and axial directions were first conducted for each case and all parameter studies were based on the values of $m = n = 5$, where m is the number of eigenfunctions in the radial direction and n is the number of eigenfunctions in the axial direction. Both the state variable and the control variable converge as the time step decreases, similar to the results depicted in Figs. 1.1 and 1.2. As the order of derivative α approaches 1, the results recover the analytical solutions, similar to the results shown in Figs. 1.3 and 1.4. Hence, the figures similar to Figs. 1.1–1.4 are not repeated here.

Figure 1.9 depicts responses of the state and the control as functions of time for Case 1, Fig. 1.10 presents the same responses for Case 2 and Fig. 1.11 for Case 3, all at the same location on the hollow cylinder. A notable observation has been made that the responses of Case 3 can be obtained by taking linear combination of the responses of Cases 1 and 2. This has an important implication in that for a

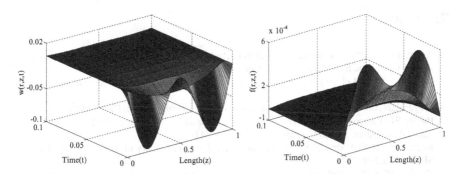

Fig. 1.11 State variable $w(r = 0.75, z, t)$ control variable $f(r = 0.75, z, t)$ for $N = 100$ and $\alpha = 0.90$

general form of initial condition, whether it is a periodic function or nonperiodic function, we can take Fourier transform of the function, find the response for each Fourier component using the method discussed above, and then apply superposition to obtain the response for the given initial condition. Hence, the method outlined in this chapter applies to FOCPs subjected to general form of initial conditions.

6   Conclusions

A general formulation and numerical scheme for FOC of a distributed system in cylindrical coordinate system are presented. Naturally, an axisymmetric problem arises for such a problem and a solid cylinder and a hollow cylinder, each with axial symmetry, are discussed as examples. Partial fractional time derivatives are defined in the Caputo sense and the performance index of the FOCP is defined as a function of both state and control variables. The separation of variable method and the eigenfunction approach is used to decouple the equations and define the problem in terms of the state and control variables. A few parameter studies are discussed including convergence of the state and control variables with respect to the number of segments in the time domain, and convergence of the number of eigenfunctions in the radial direction, as well as in the axial direction. The numerical results of the state and the control variables recover the analytical solutions as the order α approaches 1, and such agreement therefore validates the formulation and numerical scheme presented. When considering an initial condition, that is linear combination of two functions, the results also turn to be linear combination of the results of problems that consider the individual functions as initial conditions. Therefore, the method outlined in this chapter will apply to FOCPs subjected to general form of initial conditions.

Acknowledgment The authors Md. Hasan and X.W. Tangpong greatly acknowledge the support of ND EPSCoR, grant # FAR0017485.

References

1. Agrawal OP (2008a) A quadratic numerical scheme for fractional optimal control problems. ASME J Dynamic Syst, Measurement, Control 130(1):011010.1–011010.6
2. Miller KS, Ross B (1993) An introduction to the fractional calculus and fractional differential equations. Wiley, New York
3. Magin RL (2006) Fractional calculus in bioengineering. Begell House, Connecticut
4. Bagley RL, Calico RA (1991) Fractional order state equations for the control of viscoelastically damped structures. J Guid Control Dyn 14(2):304–311
5. Koeller RC (1984) Application of fractional calculus to the theory of viscoelasticity. J Appl Mech 51(2):299–307
6. Koeller RC (1986) Polynomial operators, Stieltjes convolution, and fractional calculus in hereditary mechanics. Acta Mechanica 58(3–4):251–264
7. Skaar SB, Michel AN, Miller RK (1988) Stability of viscoelastic control systems. IEEE Trans Automatic Control 33(4):348–357
8. Oustaloup A, Levron F, Mathieu B, Nanot FM (2000) Frequency-band complex noninteger differentiator: Characterization and synthesis. IEEE Trans Circ Syst I 47(1):25–39
9. Xue D, Chen YQ (2002) A comparative introduction of four fractional order controllers. In: Proceedings of the fourth IEEE world congress on intelligent control and automation (WCICA02), IEEE 4:3228–3235
10. Manabe S (2003) Early development of fractional order control. In: Proceedings of the ASME international design engineering technical conference, Chicago, IL, Paper No. DETC2003/VIB-48370
11. Monje CA, Calderón JA, Vinagre BM, Chen YQ and Feliu V (2004) On fractional PI^λ controllers: some tuning rules for robustness to plant uncertainties. Nonlinear Dyn 38(1–2):369–381
12. Ichise M, Nagayanagi Y, Kojima T (1971) An analog simulation of non-integer order transfer functions for analysis of electrode processes. J Electroanal Chem Interfacial Electrochem 33(2):253–265
13. Sun HH, Onaral B, Tsao Y (1984a) Application of positive reality principle to metal electrode linear polarization phenomena. IEEE Trans Biomed Eng 31(10):664–674
14. Sun HH, Abdelwahab AA, Onaral B (1984b) Linear approximation of transfer function with a pole of fractional power. IEEE Trans Automatic Control 29(5):441–444
15. Mandelbrot B (1967) Some noises with 1/f spectrum, a bridge between direct current and white noise. IEEE Trans Info Theory 13(2):289–298
16. Hartley TT, Lorenzo CF, Qammar HK (1995) Chaos in a fractional order Chua's system. IEEE Trans Circuits Syst: Part I: Fund Theory Appl 42(8):485–490
17. Hartley TT, Lorenzo CF (2002) Dynamics and control of initialized fractional-order systems. Nonlinear Dyn 29(1–4):201–233
18. Tricaud C, Chen YQ (2010a) An approximate method for numerically solving fractional order optimal control problems of general form. Comput Math Appl 59(5):1644–1655
19. Tricaud C, Chen YQ (2010b) Time-optimal control of systems with fractional dynamics. Int J Diff Equations. Article ID 461048. doi:10.1155/2010/461048
20. Machado JAT, Silva MF, Barbosa RF, et al (2010) Some applications of fractional calculus in engineering. Math Prob Eng: Article ID 639801, doi: 10.1155/2010/639801
21. Zamani M, Karimi-Ghartemani M, Sadati N, Parniani N (2007) FOPID controller design for robust performance using particle swarm optimization. J Fractional Calculus Appl Anal 10(2):169–188
22. Agrawal OP (1989) General formulation for the numerical solution of optimal control problems. Int J Control 50(2):627–638
23. Tangpong XW, Agrawal OP (2009) Fractional optimal control of a continuum system. ASME J Vibration Acoustics 131(2):021012

24. Agrawal OP (2004) A general formulation and solution scheme for fractional optimal control problems. Nonlinear Dyn 38(1–2):323–337
25. Agrawal OP (2006) A formulation and numerical scheme for fractional optimal control problems. J Vibration Control 14(9–10):1291–1299
26. Frederico GSF, Torres DFM (2006) Noethers theorem for fractional optimal control problems. In: Proceedings of the 2nd IFAC workshop on fractional differentiation and its applications, vol 2. Porto, Portugal, pp 142–147, July 19–21, 2006
27. Frederico GSF, Torres DFM (2008a) Fractional conservation laws in optimal control theory. Nonlinear Dyn 53(3):215–222
28. Frederico GSF, Torres DFM (2008b) Fractional optimal control in the sense of caputo and the fractional noethers theorem. Int Math Forum 3(10):479–493
29. Agrawal OP, Baleanu D (2007) A hamiltonian formulation and a direct numerical scheme for fractional optimal control problems. J Vibration Control 13(9–10):1269–1281
30. Baleanu D, Defterli O, Agrawal OP (2009) A central difference numerical scheme for fractional optimal control problems. J Vibration Control 15(4):583–597
31. Jelicic DZ, Petrovacki N (2009) Optimality conditions and a solution scheme for fractional optimal control problems. Struct Multidisciplinary Opt 38(6):571–581
32. Agrawal OP (2008b) Fractional optimal control of a distributed system using eigenfunctions. ASME J Comput Nonlinear Dyn 3(2):021204
33. Ozdemir N, Agrawal OP, Iskender BB, Karadeniz D (2009a) Fractional optimal control of a 2-dimensional distributed system using eigenfunctions. Nonlinear Dyn 55(3):251–260
34. Ozdemir N, Agrawal OP, Karadeniz D, Iskender BB (2009b) Fractional optimal control problem of an axis-symmetric diffusion-wave propagation. Physica Scripta T136:014024 (5pp)
35. Ozdemir N, Karadeniz D, Iskender BB (2009c) Fractional optimal control problem of a distributed system in cylindrical coordinates. Phys Lett A 373(2):221–226
36. Ozdemir N, Agrawal OP, Iskender BB, Karadeniz D (2009d) Analysis of an axis-symmetric fractional diffusion-wave problem. J Phys A: Math Theo 42:355208 (10pp)
37. Povstenko Y (2008) Time-fractional radial diffusion in a sphere. Nonlinear Dyn 53(1–2):55–65
38. Qi H, Liu J (2009) Time-fractional radial diffusion in hollow geometries. Meccanica, doi: 10.1007/s11012–009–9275–2
39. Hasan MM, Tangpong XW, Agrawal OP (2011) Fractional optimal control of distributed systems in spherical and cylindrical coordinates. J Vibration Control, in press

Chapter 2
Neural Network-Assisted PI$^\lambda$D$^\mu$ Control

Mehmet Önder Efe

1 Introduction

The need to handle the computational intensity of fractional order differintegration operators was an obstacle in between useful applications and theory. Rapid growth in the technology of fast computation platforms has made it possible to offer versatile design and simulation tools, from which the field of control engineering has benefited remarkably.

In [1–3], fundamental issues regarding the fractional calculus, fractional differential equations, and a viewpoint from the systems and control engineering are elaborated, and several exemplar cases are taken into consideration. One such application area focuses on PID control with derivative and integral actions having fractional orders, i.e., PI$^\lambda$D$^\mu$ control is implemented. In the literature, several applications of PI$^\lambda$D$^\mu$ controllers have been reported. The early notion of the scheme is reported by [3, 4]. In [5] and [6], tuning of the controller parameters is considered when the plant under control is a fractional order one. Ziegler–Nichols type tuning rules are derived in [7], and rules for industrial applications are designed in [8]. The application of fractional order PID controllers in chemical reaction systems is reported in [9], and the issues regarding the frequency domain are considered in [10]. Tuning based on genetic algorithms is considered in [11], where the best parameter configuration is coded appropriately and a search algorithm is executed to find a parameter set that meets the performance specifications. A similar approach exploiting the particle swarm optimization for finding a good set of gains and differintegration orders isin [12]. Clearly, the cited volume of works

M.Ö. Efe (✉)
Department of Pilotage, University of Turkish Aeronautical Association,
Akköprü, Ankara, Turkey
e-mail: onderefe@ieee.org

D. Baleanu et al. (eds.), *Fractional Dynamics and Control*,
DOI 10.1007/978-1-4614-0457-6_2, © Springer Science+Business Media, LLC 2012

demonstrates that the interest to PID control is growing also in the direction of fractional order versions. Unsurprisingly, the reason for this is the widespread use of the variants of PID controller and the confidence of the engineers in industry.

The idea of approximating the fractional order operators has been considered in [13], where a fractional order integrator is generalized by a neural network observing some history of the input and the output. The fundamental advancement introduced here is to generalize a PID controller using a neural structure with a similar network structure.

This chapter is organized as follows: Sect. 2 briefly gives the definitions of widely used fractional differintegration formulas and basics of fractional calculus; Sect. 3 describes the Levenberg–Marquardt training scheme and neural network structure; Sect. 4 presents a set of simulation studies, and the concluding remarks are given in Sect. 5 at the end of the chapter.

2 Fundamental Issues in Fractional Order Systems and Control

Let $\mathbf{D}^\beta$ denote the differintegration operator of order β, where $\beta \in \Re$. For positive values of β, the operator is a differentiator whereas the negative values of β correspond to integrators. This representation lets $\mathbf{D}^\beta$ to be a differintegration operator whose functionality depends upon the numerical value of β. With n being an integer and $n - 1 \leq \beta < n$, Riemann–Liouville definition of the β-fold fractional differintegration is defined by (2.1) where Caputo's definition for which is in (2.2).

$$\mathbf{D}^\beta f(t) = \frac{1}{\Gamma(n-\beta)} \left(\frac{\mathrm{d}}{\mathrm{d}t}\right)^n \int_0^t \frac{f(\tau)}{(t-\tau)^{\beta-n+1}} \mathrm{d}\tau \tag{2.1}$$

$$\mathbf{D}^\beta f(t) = \frac{1}{\Gamma(n-\beta)} \int_0^t \frac{f^{(n)}(\tau)}{(t-\tau)^{\beta-n+1}} \mathrm{d}\tau \tag{2.2}$$

where $\Gamma(\beta) = \int_0^\infty e^{-t} t^{\beta-1} \mathrm{d}t$ is the well-known Gamma function. In both definitions, we assumed the lower terminal zero and the integrals start from zero. Considering $a_k, b_k \in \Re$ and $\alpha_k, \beta_k \in \Re^+$, one can define the following differential equation:

$$(a_n \mathbf{D}^{\alpha_n} + a_{n-1} \mathbf{D}^{\alpha_{n-1}} + \cdots + a_0 \mathbf{D}^{\alpha_0}) y(t) = (b_m \mathbf{D}^{\beta_m} + b_{m-1} \mathbf{D}^{\beta_{m-1}} + \cdots + b_0 \mathbf{D}^{\beta_0}) u(t) \tag{2.3}$$

and with the assumption that all initial conditions are zero, obtain the transfer function given by (2.4).

$$\frac{Y(s)}{U(s)} = \frac{b_m s^{\beta_m} + b_{m-1} s^{\beta_{m-1}} + \cdots + b_0 s^{\beta_0}}{a_n s^{\alpha_n} + a_{n-1} s^{\alpha_{n-1}} + \cdots + a_0 s^{\alpha_0}} \tag{2.4}$$

Denoting frequency by ω and substituting $s = j\omega$ in (2.4), one can exploit the techniques of frequency domain. A significant difference in the Bode magnitude plot is to observe that the asymptotes can have any slope other that the integer multiples of 20 dB per decade, and this is a substantially important flexibility for modeling and identification research. When the state space models are taken into consideration, we have

$$\mathbf{D}^\beta \mathbf{x} = \mathbf{A}\mathbf{x} + \mathbf{B}u$$
$$y = \mathbf{C}\mathbf{x} + Du \tag{2.5}$$

and we obtain the transfer function via taking the Laplace transform in the usual sense, i.e.,

$$H(s) = \mathbf{C}\left(s^\beta \mathbf{I} - \mathbf{A}\right)^{-1}\mathbf{B} + D \tag{2.6}$$

For the state space representation in (2.5), if λ_i is an eigenvalue of the matrix $\mathbf{A}$, the condition

$$|\arg(\lambda_i)| > \beta\frac{\pi}{2} \tag{2.7}$$

is required for stability. It is possible to apply the same condition for the transfer function representation in (2.4), where λ_is denotes the roots of the expression in the denominator.

The implementation issues are closely related to the numerical realization of the operators defined in (2.1) and (2.2). There are several approaches in the literature and Crone is the most frequently used scheme in approximating the fractional order differintegration operators [1]. More explicitly, the algorithm determines a number of poles and zeros and approximates the magnitude plot over a predefined range of the frequency spectrum. In (2.8), the expression used in Crone approximation is given and the approximation accuracy is depicted for $N = 3$ and 9 in Fig. 2.1. According to the approximates shown, it is clearly seen that the accuracy is improved as N gets larger, yet the price paid for this is the complexity and the technique presented next is a remedy to handle the difficulties stemming from the implementation issues.

$$s^\beta \approx K\frac{\prod_{k=1}^{N} 1 + s/w_{pk}}{\prod_{k=1}^{N} 1 + s/w_{zk}} \tag{2.8}$$

The PI$^\lambda$D$^\mu$ controller with the operator described above has the transfer function given by (2.9), where $E(s)$ is the error entering the controller and $U(s)$ stands for the output.

$$\frac{U(s)}{E(s)} = K_p + \frac{K_i}{s^\lambda} + K_d s^\mu \tag{2.9}$$

In Fig. 2.2, it is illustrated that the classical PID controller variants correspond to a subset in the λ–μ coordinate system, and there are infinitely many parameter configurations that may lead to different performance indications.

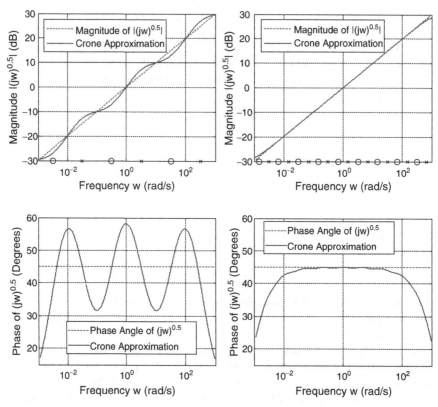

Fig. 2.1 Crone approximation to the operator $s^{0.5}$ with $\omega_{min} = 1e - 3\,\text{rad/s}$, $\omega_{max} = 1e + 3\,\text{rad/s}$. Left column: $N = 3$, Right column: $N = 9$

Fig. 2.2 Continuous values of the differintegration orders λ and μ enables to obtain infinitely many configurations of $PI^\lambda D^\mu$ controller where the variants of the classical PID controller correspond to a subset of the domain

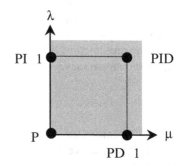

3 Neural Network-Based Modeling and Levenberg–Marquardt Training Scheme

In this work, we consider the feedforward neural network structure shown in Fig. 2.3, where there are m inputs, R neurons in the first hidden layer, and Q hidden layer in the second hidden layer. Since the neural structure is aimed to imitate a

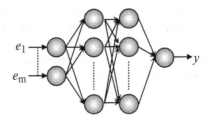

PI$^\lambda$D$^\mu$ controller, the model has a single output. The hidden layers have hyperbolic
tangent-type nonlinear activation while the output layer neuron is linear.

The powerful mapping capabilities of neural networks have made them useful
tools of modeling research especially when the entity to be used is in the form of
raw data. This particular property is mainly because of the fact that real systems
have many variables, the variables involved in the modeling process are typically
noisy, and the underlying physical phenomenon is sometimes nonlinear. Due to
the inextricably intertwined nature of the describing differential (or difference)
equations, which are not known precisely, it becomes a tedious task to see the
relationship between the variables involved. In such cases, black box models such as
neural networks, fuzzy logic, or the methods adapted from the artificial intelligence
come into the picture as tools representing the input/output behavior accurately.
In what follows, we describe briefly the Levenberg–Marquardt training scheme for
adjusting the parameters of a neural structure [14]. Since the algorithm is a soft
transition in between the Newton's method and the standard gradient descent, it
very quickly locates the global minimum (if achievable) of the cost hypersurface,
which is denoted by J in (2.10).

$$J = \frac{1}{2} \sum_{p=1}^{P} (d_p - y_p(e, \phi))^2 \tag{2.10}$$

where y_p denotes the response of the single output neural network, and d_p stands for
the corresponding target output. In (2.10), ϕ is the set of all adjustable parameters
of the neural structure (weights and the biases), and u is the vector of inputs which
are selected according to the following procedure:

$$\phi(t+1) = \phi(t) - \left(\mu I + \Phi(t)^\mathsf{T}\Phi(t)\right)^{-1} \Phi(t)^\mathsf{T} F(t) \tag{2.11}$$

where μ is the regularization parameter, $F(t) = [f_1 \, f_2 \ldots f_P]^T$ is the vector of errors
described as $f_i = d_i - y_i(e, \phi) \, i = 1, 2, \ldots, P$, where P is the number of training pairs
and Φ is the Jacobian given explicitly by (2.12)

$$\Phi = \begin{bmatrix} \dfrac{\partial f_1}{\partial \phi_1} & \dfrac{\partial f_1}{\partial \phi_2} & \cdots & \dfrac{\partial f_1}{\partial \phi_H} \\[2mm] \dfrac{\partial f_2}{\partial \phi_1} & \dfrac{\partial f_2}{\partial \phi_2} & \cdots & \dfrac{\partial f_2}{\partial \phi_H} \\[2mm] \vdots & \vdots & \ddots & \vdots \\[2mm] \dfrac{\partial f_P}{\partial \phi_1} & \dfrac{\partial f_P}{\partial \phi_2} & \cdots & \dfrac{\partial f_P}{\partial \phi_H} \end{bmatrix} \tag{2.12}$$

where there are H adjustable parameters within the vector ϕ. In the application of the tuning law in (2.11), if μ is large, the algorithm behaves more like the gradient descent; conversely, if μ is small, the prescribed updates are more like the Gauss–Newton updates. The algorithm removes the problem of rank deficiency in (2.11) and improves the performance of gradient descent significantly.

4   Simulation Studies

The first stage of emulating the response of a $PI^\lambda D^\mu$ controller is to select a representative set of inputs to be applied to the $PI^\lambda D^\mu$ controller and to collect the response. We have set $N = 9$ and follow the procedure described below.

> **For** $n = 1$ to #experiments
>   Set a random $K_p \in (0,2)$
>   Set a random $K_d \in (0,1)$
>   Set a random $K_i \in (0,1)$
>   Set a random $\mu \in (0,1)$
>   Set a random $\lambda \in (0,1)$
>   Apply $u(t)$ and obtain $y(t)$ for $t \in [0,3]$
>   Store $u(t), y(t), K_p, K_d, K_i, \mu, \lambda$
> **End**

A total of 200 experiments with step size 1 ms have been carried out to obtain the data to be used for training data. Once the set of all responses are collected, a matrix is formed, a generic row of which has the following structure:

$$[y(k), y(k-1), \cdots, y(k-d), K_p(k), K_d(k), K_i(k), \lambda(k), \mu(k)] \qquad (2.13)$$

where k is the time index indicating $y(k) = y(kT)$ and $T = 1$ ms, and there are $d + 6$ columns in each row and the delay depth d is a user-defined parameter. Denote the matrix, whose generic row is shown above, by Ω. In order to obtain the training data set, we downsample the matrix Ω by selecting the first row of every 100 consecutive row blocks. This significantly reduces the computational load of the training scheme, and according to the given procedure, 60,000 pairs of training data are generated and a neural network having $m = 16$ inputs is constructed. In Fig. 2.4, the evolution for the training data is shown with that obtained for the checking data, which is obtained by running 15 experiments and the same procedure of downsampling.

At 128th epoch, the best set network parameters is obtained, and after this time the checking error for the neural model starts increasing and the training scheme stops the parameter tuning when $J = 0.01778$. In what follows, we discuss the performance of the neural model as a $PI^\lambda D^\mu$ controller.

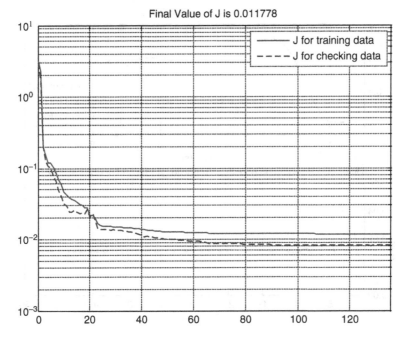

Fig. 2.4 Feedforward neural network structure with R neurons in the first, Q neurons in the second hidden layer

As an illustrative example, we consider the following control problem, which is simple yet our goal is to compare the responses of two controllers, namely, PI$^\lambda$D$^\mu$ controller and its neural network-based approximate. The plant dynamics is given below:

$$\frac{Y(s)}{U(s)} = \frac{1}{s(s+1)} \tag{2.14}$$

where Y is the plant output and U is the control input. We choose $K_p = 2.5$, $K_d = 0.9$, $K_i = 0.1$, $\mu = 0.02$, $\lambda = 0.7$ and apply a step command that rises when $t = 1$ s. The command signal, the response obtained with the PI$^\lambda$D$^\mu$ controller exploiting the above parameters, and the result obtained with the trained neural network emulator are shown on the top row of Fig. 2.5, where the response of PI$^\lambda$D$^\mu$ controller is obtained using the toolbox described in [15]. For a better comparison, the bottom row depicts the difference in between the plant responses obtained for both controllers individually. Clearly the results suggest that the neural network-based controller is able to imitate the PI$^\lambda$D$^\mu$ controller to a very good extent as the two responses are very close to each other.

A better comparison is to consider the control signals that are produced by the PI$^\lambda$D$^\mu$ controller (u_{FracPID}) and the neural network controller (u_{NNPID}). The

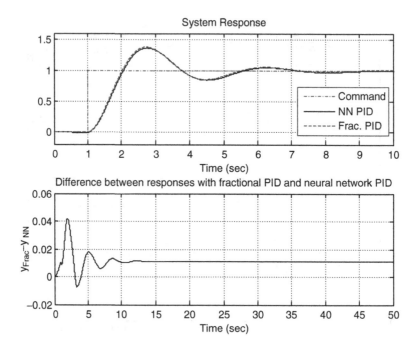

Fig. 2.5 For the first example, system response and the difference in between the two responses obtained with the $PI^\lambda D^\mu$ controller and its neural network-based substitute

results are seen in Fig. 2.6, where the two control signals are shown together on the top subplot, whereas the difference between them is illustrated in the bottom subplot. Clearly the two control signals are very close to each other; furthermore, the signal generated by the neural network is smoother than its alternative when $t = 1$. This particular example demonstrates that the neural network-based realization can be a good candidate for replacing the $PI^\lambda D^\mu$ controller.

Define the following relative error as given in (2.15), where T denotes the final time. For the results seen above, we obtain $e_{rel} = 0.1091$, which is an acceptably small value indicating the similarity of the two control signals seen in Fig. 2.6.

$$e_{rel} := \frac{\frac{1}{T}\int_0^T |u_{FracPID} - u_{NNPID}| \, dt}{\frac{1}{T}\int_0^T |u_{FracPID}| \, dt} \qquad (2.15)$$

In Table 2.1, we summarize a number of test cases with corresponding relative error values. The data presented in the table indicate that the proposed controller is able to perform well for a wide range of controller gains and for small values of λ and μ. However, for another control problem, the proposed scheme may perform

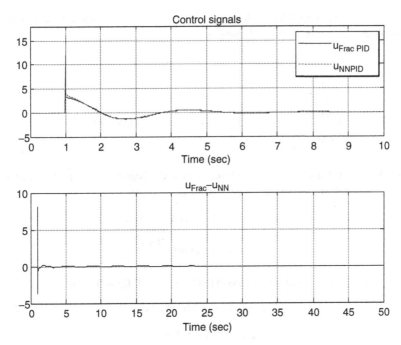

Fig. 2.6 The control signals generated by the PI$^\lambda$D$^\mu$ controller and its neural network-based substitute. The *bottom row* shows the difference between the two signals

Table 2.1 Performance of the proposed controller for a number of different parameter configurations

K_p	K_i	K_d	μ	λ	$e_{\text{rel.}}$
1.3000	0.9000	0.7000	0.0200	0.0900	0.0897
2.1000	0.9000	0.1000	0.0200	0.3900	0.0938
1.7000	0.7000	0.4000	0.0200	0.0900	0.0965
2.5000	0.7000	0.1000	0.0200	0.3900	0.0982
2.5000	0.7000	0.1000	0.0200	0.6900	0.0991
1.7000	0.3000	1.0000	0.0200	0.0900	0.1005
0.9000	0.9000	0.7000	0.0200	0.0900	0.1014
2.5000	0.9000	0.1000	0.0200	0.6900	0.1018
1.3000	0.7000	0.4000	0.0200	0.0900	0.1028
1.7000	0.1000	1.0000	0.0200	0.0900	0.1035
2.1000	0.9000	0.1000	0.0200	0.6900	0.1038
1.3000	0.7000	1.0000	0.0200	0.0900	0.1052
1.3000	0.9000	0.4000	0.0200	0.0900	0.1073
1.7000	0.5000	0.7000	0.0200	0.0900	0.1081
1.3000	0.9000	1.0000	0.0200	0.0900	0.1090
0.9000	0.9000	1.0000	0.0200	0.0900	0.1094
0.9000	0.7000	0.7000	0.0200	0.0900	0.1108
2.1000	0.7000	0.1000	0.0200	0.9900	0.1112
1.7000	0.5000	1.0000	0.0200	0.0900	0.1113

better for larger values of differintegration orders. To see this, as a second example, we consider the following plant dynamics:

$$x_1^{(0.1)} = x_2$$

$$x_1^{(0.4)} = x_3$$

$$x_3^{(0.8)} = f(x_1,x_2,x_3) + \Delta(x_1,x_2,x_3,t) + g(t)x_4 + \xi(t)$$

$$x_4^{(0.5)} = u \qquad (2.16)$$

where $\Delta(x_1,x_2,x_3)$ and $\xi(t)$ are uncertainties and disturbance terms that are not available to the designer. In the above equation, we have

$$f(x_1,x_2,x_3) = -0.5x_1 - 0.5x_2^3 - 0.5x_3|x_3| \qquad (2.17)$$

$$g(t) = 1 + 0.1\sin\left(\frac{\pi t}{3}\right) \qquad (2.18)$$

$$\Delta(x_1,x_2,x_3,t) = (-0.05 + 0.25\sin(5\pi t))x_1 + (-0.03 + 0.3\cos(5\pi t))x_2^3$$

$$+ (-0.05 + 0.25\sin(7\pi t))x_3|x_3| \qquad (2.19)$$

$$\xi(t) = 0.2\sin(4\pi t) \qquad (2.20)$$

The plant considered is a nonlinear one having four states, disturbance terms, and uncertainties. The time-varying gain multiplying the state x_4 in (2.14) makes the problem further complicated, and we compare the neural network substitute of the $PI^\lambda D^\mu$ controller given by

$$\frac{U(s)}{E(s)} = 2 + \frac{0.7}{s^{0.9}} + 0.6s^{0.75} \qquad (2.21)$$

The results are illustrated in Figs. 2.7 and 2.8. The responses of the system for both controllers are depicted in Fig. 2.7, where we see that the two responses are very close to each other. The similarity in the fluctuations around the setpoint is another result to emphasize. The outputs of the controllers are analyzed in Fig. 2.8, where we see that the $PI^\lambda D^\mu$ controller generates a very large magnitude spike when the step change in the command signal occurs, whereas the neural network-based substitute produces a smoother control signal, and this is reflected as a slight difference in between the plant responses to controllers being compared. The two controllers produce similar signals when the plant output is forced to lie around unity, which is seen in the middle subplot of Fig. 2.8, and the difference between the two control signals is seen to be bounded by 0.05 during this period. The value of e_{rel} for this case is equal to 20.3283, which seems large but noticing the peak in the top subplot of Fig. 2.8; this could be seen tolerable as the $PI^\lambda D^\mu$ controller requests high magnitude control signals when there is a step change in the command.

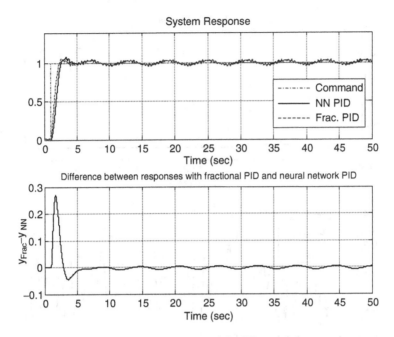

Fig. 2.7 For the second example, system response and the difference in between the two responses obtained with the PI$^\lambda$D$^\mu$ controller and its neural network-based substitute

A last issue to consider here is the possibility of increasing the performance obtained by the chosen neural network structure, which is 16-25-10-1. One can argue that the neural network could be realized as a single hidden layer one, or with two hidden layers with less number of neurons in each. In obtaining the neural model, whose results are discussed, many trials have been performed, and it is seen that the approximation performance could be increased if there are more neurons in the hidden layers. In a similar fashion, a better map could be constructed if earlier values of the incoming error signal are taken into consideration. This enlarges the network size and makes it more intense computationally to train the model. Depending on the problem in hand, the goal of this chapter is to demonstrate that a fractional order PI$^\lambda$D$^\mu$ control could be replicated to a certain extent using neural network models, and the findings of the chapter support these claims thoroughly.

5   Conclusions

This chapter discusses the use of standard neural network models for imitating the behavior of a PI$^\lambda$D$^\mu$ controller, whose parameters are provided explicitly as the inputs to the neural network. The motivation in focusing this has been the difficulty of realizing fractional order controllers requiring high orders of approximation for

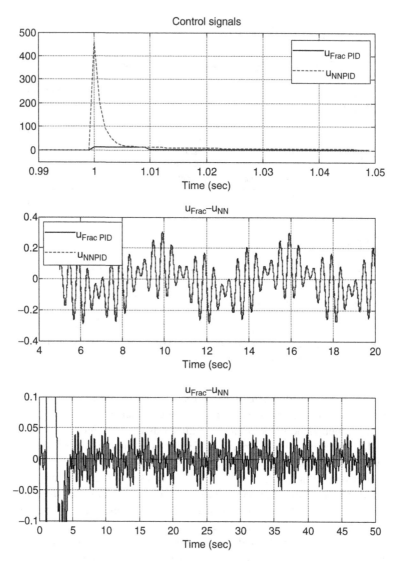

Fig. 2.8 The control signals generated by the PI$^\lambda$D$^\mu$ controller and its neural network-based substitute. The *top row* illustrates the two signals when the step change occurs. The *middle row* depicts the closeness of the two signals when $t > 5$ s, and the *bottom row* shows the difference in between the two signals

accuracy. The method followed here is to collect a set of data and to optimize the set of parameters to obtain an emulator of the PI$^\lambda$D$^\mu$ controller. Aside from the parameters of the PI$^\lambda$D$^\mu$ controller, the neural model observed some history of the input and outputs a value approximating the response of the PI$^\lambda$D$^\mu$ controller. Several exemplar cases are presented, and it is seen that the use of neural network models is a practical alternative in realizing the PI$^\lambda$D$^\mu$ controllers. Furthermore, the developed neural model allows modifying the controller parameters online as those parameters are supplied as eternal inputs to the network.

Acknowledgment This work is supported in part by Turkish Scientific Council (TÜBİTAK) Contract 107E137.

References

1. Das S (2008) Functional fractional calculus for system identification and controls, 1st edn. Springer, New York
2. Oldham KB, Spanier J (1974) The fractional calculus. Academic, London
3. Podlubny I (1998) Fractional differential equations, 1st edn. Elsevier Science & Technology Books, Amsterdam
4. Podlubny I (1999) Fractional-order systems and (PID mu)-D-lambda-controllers. IEEE Trans Automatic Control 44(1):208–214
5. Zhao C, Xue D, Chen Y-Q (2005) A fractional order PID tuning algorithm for a class of fractional order plants. Proceedings of the IEEE international conference on mechatronics & automation, Niagara Falls, Canada
6. Caponetto R, Fortuna L, Porto D (2002) Parameter tuning of a non integer order PID controller. In: Proceedings of the fifteenth international symposium on mathematical theory of networks and systems, Notre Dame, Indiana
7. Valerio D, Sa Da Costa L (2006) Tuning of fractional PID controllers with Ziegler–Nichols-type rules. Signal Process 86:2771–2784
8. Monje CA, Vinagre BM, Feliu V, Chen Y-Q (2006) Tuning and auto-tuning of fractional order controllers for industry applications. Control Eng Pract 16:798–812
9. Leu JF, Tsay SY, Hwang C (2002) Design of optimal fractional-order PID controllers. J Chinese Institute Chem Eng 33(2):193–202
10. Vinagre BM, Podlubny I, Dorcak L, Feliu V (2000) On fractional PID controllers: a frequency domain approach. In: IFAC workshop on digital control. past, present and future of PID control. Terrasa, Spain, pp 53–58
11. Cao J-Y, Liang J, Cao B-G (2005) Optimization of fractional order pid controllers based on genetic algorithms. In: Proceedings of the fourth international conference on machine learning and cybernetics, Guangzhou, 18–21 August
12. Maiti D, Biswas S, Konar A (2008) Design of a fractional order PID controller using particle swarm optimization technique. In: 2nd national conference on recent trends in information systems (ReTIS-08), February 7–9, Kolkata, India
13. Abbisso S, Caponetto R, Diamante O, Fortuna L, Porto D (2001) Non-integer order integration by using neural networks. The 2001 IEEE International Symposium on Circuits and Systems (ISCAS 2001), 6–9 May, vol. 2, pp 688–691
14. Hagan MT, Menhaj MB (1994) Training feedforward networks with the Marquardt algorithm. IEEE Trans Neural Network 5(6):989–993
15. Valerio D (2005) Ninteger v.2.3 fractional control toolbox for MatLab

Chapter 3
Application of Backstepping Control Technique to Fractional Order Dynamic Systems

Mehmet Önder Efe

1 Introduction

Recently there has been a dramatic increase in the number of research outcomes regarding the theory and applications of fractional order systems and control [4, 12, 14]. Despite the emergence of the theory dates back to a letter from Leibniz to L'Hôpital in 1695, asking the possible consequences of choosing a derivative of order $1/2$, the theory has been stipulated and with the advances in the computational facilities, many important tools of classical control have been reformulated for (or adapted to) fractional order case, such as PID controllers [15,20], stability considerations [2,3,8,9], Kalman filtering [18], state space models and approaches [4, 13, 17], root locus technique [11], applications involved with the partial differential equations [10, 16], discrete time issues [4, 12, 14, 18], and so on. A system to be identified can well be approximated by an integer order model or it can be approximated by a much simpler model that is a fractional order one. Having the necessary techniques and tools for such cases becomes a critical issue and with this motivation in mind, this chapter focuses on adapting the backstepping control technique for fractional order plant dynamics.

Backstepping technique has been a frequently used nonlinear control technique that is based on the definition of a set of intermediate variables and the procedure of ensuring the negativity of Lyapunov functions that add up to build a common control Lyapunov function for the overall system. Due to this nature, the backstepping technique is applicable to a particular – yet wide – class of systems, which includes most mechanical systems, biochemical processes, etc. The technique has

M.Ö. Efe (✉)
Department of Pilotage, University of Turkish Aeronautical Association,
Akköprü, Ankara, Turkey
e-mail: onderefe@ieee.org

D. Baleanu et al. (eds.), *Fractional Dynamics and Control*,
DOI 10.1007/978-1-4614-0457-6_3, © Springer Science+Business Media, LLC 2012

successfully been implemented in the field of robotics to as one of the state variables is of type position and the other is of type velocity [1,5–7].

Although the tools and approaches of fractional order mathematics and backstepping control are not new, implementation of backstepping control for fractional order system dynamics is new. The reason is the definition of derivative that is generalized by Leibniz rule. The rule, which also generalizes the integer order cases, yields infinitely many terms for the product, and it becomes difficult to figure out stability by choosing a square-type Lyapunov function and obtaining its time derivative. This chapter discusses a remedy to this within the context of backstepping control method. The contribution of the current study is to extend the backstepping technique to fractional order plants.

This chapter is organized as follows: Sect. 2 briefly gives the definitions of widely used fractional differintegration formulas and basics of fractional calculus, Sect. 3 describes the backstepping technique for fractional order plant dynamics, Sect. 4 presents a set of simulation studies covering a second-order linear system with known dynamics, and a third-order nonlinear system having uncertainties and disturbances, and the concluding remarks are given in Sect. 5 at the end of the chapter.

2 Fractional Order Differintegration Operators

Let $\mathbf{D}^\beta$ denote the differintegration operator of order β, where $\beta \in \mathfrak{R}$. For positive values of β, the operator is a differentiator, whereas the negative values of β correspond to integrators. This representation lets $\mathbf{D}^\beta$ to be a differintegration operator whose functionality depends upon the numerical value of β. With n being an integer and $n - 1 \leq \beta < n$, Riemann–Liouville definition of the β-fold fractional differintegration is defined by (3.1) where Caputo's definition for which is in (3.2).

$$\mathbf{D}^\beta f(t) = \frac{1}{\Gamma(n-\beta)} \left(\frac{\mathrm{d}}{\mathrm{d}t}\right)^n \int_0^t \frac{f(\tau)}{(t-\tau)^{\beta-n+1}} \mathrm{d}\tau \tag{3.1}$$

$$\mathbf{D}^\beta f(t) = \frac{1}{\Gamma(n-\beta)} \int_0^t \frac{f^{(n)}(\tau)}{(t-\tau)^{\beta-n+1}} \mathrm{d}\tau \tag{3.2}$$

where $\Gamma(\beta) = \int_0^\infty e^{-t} t^{\beta-1} \mathrm{d}t$ is the well-known Gamma function. In both definitions, we assumed the lower terminal zero and the integrals start from zero. Considering $a_k, b_k \in \mathfrak{R}$ and $\alpha_k, \beta_k \in \mathfrak{R}^+$, one can define the following differential equation

$$(a_n \mathbf{D}^{\alpha_n} + a_{n-1} \mathbf{D}^{\alpha_{n-1}} + \cdots + a_0) y(t)$$

$$= (b_m \mathbf{D}^{\beta_m} + b_{m-1} \mathbf{D}^{\beta_{m-1}} + \cdots + b_0) u(t) \tag{3.3}$$

and with the assumption that all initial conditions are zero, obtain the transfer function given by (3.4).

$$\frac{Y(s)}{U(s)} = \frac{b_m s^{\beta_m} + b_{m-1} s^{\beta_{m-1}} + \cdots + b_0}{a_n s^{\alpha_n} + a_{n-1} s^{\alpha_{n-1}} + \cdots + a_0} \tag{3.4}$$

Denoting frequency by ω and substituting $s = j\omega$ in (3.4), one can exploit the techniques of frequency domain. A significant difference in the Bode magnitude plot is to observe that the asymptotes can have any slope other that the integer multiples of 20 dB/decade and this is a substantially important flexibility for modeling and identification research. When it comes to consider state space models, one can define

$$\mathbf{D}^{\beta} \mathbf{x} = \mathbf{A} \mathbf{x} + \mathbf{B} u$$

$$y = \mathbf{C} \mathbf{x} + D u \tag{3.5}$$

and obtain the transfer function via taking the Laplace transform in the usual sense, i.e.,

$$H(s) = \mathbf{C} \left(s^{\beta} \mathbf{I} - \mathbf{A} \right)^{-1} \mathbf{B} + D \tag{3.6}$$

For the state space representation in (3.5), if λ_i is an eigenvalue of the matrix $\mathbf{A}$, the condition

$$|\arg(\lambda_i)| > \beta \frac{\pi}{2} \tag{3.7}$$

is required for stability. It is possible to apply the same condition for the transfer function representation in (3.4), where λ_is denote the roots of the expression in the denominator.

The implementation issues are tightly related to the numerical realization of the operators defined in (3.1) and (3.2). There are several approaches in the literature and Crone is the most frequently used scheme in approximating the fractional order differintegration operators [4]. More explicitly, the algorithm determines a number of poles and zeros and approximates the magnitude plot over a predefined range of the frequency spectrum. In (3.8), the expression used in Crone approximation is given and the approximation accuracy is depicted for $N = 9$ in Fig. 3.1 and for $N = 40$ in Fig. 3.2. According to the shown approximates, it is clearly seen that the accuracy is improved as N gets larger, yet the price paid for this is the complexity.

$$s^{\beta} \approx K \frac{\prod_{k=1}^{N} 1 + s/w_{p_k}}{\prod_{k=1}^{N} 1 + s/w_{z_k}} \tag{3.8}$$

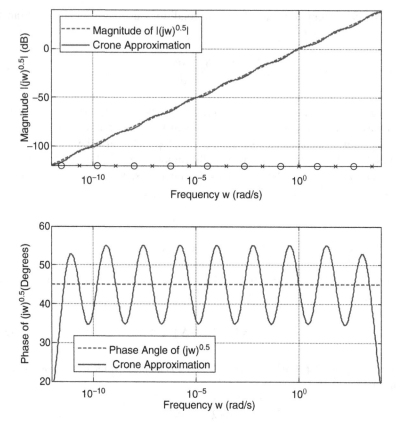

Fig. 3.1 Crone approximation to the operator $s^{0.5}$ with $\omega_{min} = 1e - 12\,\mathrm{rad/s}$, $\omega_{max} = 1e + 4\,\mathrm{rad/s}$, $N = 9$

3 Backstepping Control Technique for Fractional Order Plant Dynamics

Denote the β-fold differintegration operator $\mathbf{D}^{\beta}x$ by $x^{(\beta)}$ and consider the system

$$x_1^{(\beta_1)} = x_2$$
$$x_2^{(\beta_2)} = f(x_1, x_2) + g(x_1, x_2)u \qquad (3.9)$$

where x_1 and x_2 are the state variables, $0 < \beta_1, \beta_2 < 1$ are positive fractional differentiation orders, $f(x_1, x_2)$ and $g(x_1, x_2)$ are known and smooth functions of the state variables and $g(x_1, x_2) \neq 0$. Define the following intermediate variables of backstepping design.

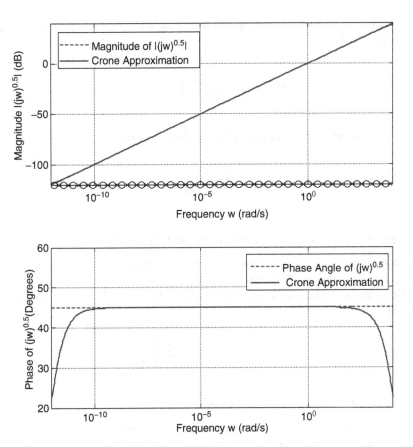

Fig. 3.2 Crone approximation to the operator $s^{0.5}$ with $\omega_{min} = 1e - 12 \, \mathrm{rad/s}$, $\omega_{max} = 1e + 4 \mathrm{rad/s}$, $N = 35$

$$z_1 := x_1 - r_1 - A_1$$
$$z_2 := x_2 - r_2 - A_2 \qquad (3.10)$$

where $A_1 = 0$ and $r_1^{(\beta_1)} = r_2$.

Theorem 3.1. *Let z be the variable of interest and choose the Lyapunov function given by (3.11).*

$$V = \frac{1}{2}z^2 \qquad (3.11)$$

If $zz^{(\beta)} < 0$ if $0 < \beta < 1$ is maintained then $z\dot{z} < 0$ is satisfied.

Proof. Consider the Riemann–Liouville definition, which is rewritten for the given conditions in (3.12).

$$zz^{(\beta)} = \frac{z}{\Gamma(1-\beta)} \frac{d}{dt} \int_0^t \frac{z(\tau)}{(t-\tau)^\beta} d\tau \qquad (3.12)$$

If $zz^{(\beta)} < 0$ is satisfied, then the variable z and the integral

$$\frac{d}{dt} \int_0^t \frac{z(\tau)}{(t-\tau)^\beta} d\tau \qquad (3.13)$$

are opposite signed, i.e., $\int_0^t \frac{z(\tau)}{(t-\tau)^\beta} d\tau$ is monotonically decreasing for positive z, and monotonically increasing for negative z. Since the denominator of the integrand is always positive, this can only arise if $z\dot{z} < 0$ is satisfied.

Considering the Caputo's definition in (3.14), having $zz^{(\beta)} < 0$ can arise when $z\dot{z} < 0$.

$$zz^{(\beta)} = \frac{z}{\Gamma(1-\beta)} \int_0^t \frac{\dot{z}(\tau)}{(t-\tau)^\beta} d\tau \qquad (3.14)$$

This proves that forcing $zz^{(\beta)} < 0$ implies $z\dot{z} < 0$. $\qquad\qquad\square$

Now we will formulate the backstepping control technique for the plant described by (3.9) by repetitively checking the quantities $z_1 z_1^{(\beta_1)}$ and $z_1 z_1^{(\beta_1)} + z_2 z_2^{(\beta_2)}$ as explained below.

Step 1: Check $z_1 z_1^{(\beta_1)}$

$$\begin{aligned}
z_1 z_1^{(\beta_1)} &= z_1 \left(x_1^{(\beta_1)} - r_1^{(\beta_1)} \right) \\
&= z_1 (x_2 - r_2) \\
&= z_1 (z_2 + r_2 + A_2 - r_2) \\
&= z_1 (z_2 + A_2)
\end{aligned} \qquad (3.15)$$

Step 2: With $k_1 > 0$, choose $A_2 = -k_1 z_1$, this would let us have

$$z_1 z_1^{(\beta_1)} = -k_1 z_1^2 + z_1 z_2 \qquad (3.16)$$

Step 3: Check $z_1 z_1^{(\beta_1)} + z_2 z_2^{(\beta_2)}$.

$$
\begin{aligned}
z_1 z_1^{(\beta_1)} + z_2 z_2^{(\beta_2)} &= -k_1 z_1^2 + z_1 z_2 + z_2 \left(x_2^{(\beta_2)} - r_2^{(\beta_2)} - A_2^{(\beta_2)} \right) \\
&= -k_1 z_1^2 + z_2 \left(x_2^{(\beta_2)} - r_2^{(\beta_2)} - A_2^{(\beta_2)} + z_1 \right) \\
&= -k_1 z_1^2 + z_2 \left(f + gu - r_2^{(\beta_2)} - A_2^{(\beta_2)} + z_1 \right)
\end{aligned}
$$
(3.17)

Step 4: Force $z_1 z_1^{(\beta_1)} + z_2 z_2^{(\beta_2)} = -k_1 z_1^2 - k_2 z_2^2$, $k_2 > 0$, this requires

$$
f + gu - r_2^{(\beta_2)} - A_2^{(\beta_2)} + z_1 := -k_2 z_2
$$
(3.18)

Step 5: Solve for u

$$
u = -\frac{1}{g(x_1, x_2)} \left(f(x_1, x_2) - r_2^{(\beta_2)} + k_1 z_1^{(\beta_2)} + z_1 + k_2 z_2 \right)
$$
(3.19)

It is possible to generalize the above procedure for higher order systems of the form

$$
x_i^{(\beta_i)} = x_{i+1}, \ i = 1, 2, \cdots, q - 1
$$
$$
x_q^{(\beta_q)} = f(x_1, x_2, \cdots, x_q) + g(x_1, x_2, \cdots, x_q) u
$$
(3.20)

and the control law

$$
u = -\frac{1}{g} \left(f - r_q^{(\beta_q)} - A_q^{(\beta_q)} + z_{q-1} + k_q z_q \right)
$$
(3.21)

where $k_q > 0$ and

$$
A_1 = 0, z_0 = 0
$$
(3.22)

$$
A_{i+1} = -k_i z_i + A_i^{(\beta_i)} - z_{i-1}, i = 1, 2, q - 1
$$
(3.23)

and the result of applying the control law in (3.21) is given below.

$$
\sum_{i=1}^{q} z_i z_i^{(\beta_i)} = -\sum_{i=1}^{q} k_i z_i^2
$$
(3.24)

According to the aforementioned theorem, ensuring the negativeness of the right-hand side of (3.24) is equivalent to ensuring the negativity of $\sum_{i=1}^{q} z_i \dot{z}_i$, and the trajectories in the coordinate system spanned by $z_1, \ldots, z_q$ converge the origin.

4 Simulation Studies

In this section, we consider two sets of simulations to justify the claims. The first system is linear and a second-order one with all necessary parameters are known perfectly. The system is given by (3.25).

$$\begin{pmatrix} x_1^{(0.7)} \\ x_2^{(0.6)} \end{pmatrix} = \begin{pmatrix} 0 & 1 \\ -2 & -3 \end{pmatrix} \begin{pmatrix} x_1 \\ x_2 \end{pmatrix} + \begin{pmatrix} 0 \\ 1 \end{pmatrix} u \tag{3.25}$$

The system is desired to track a sinusoidal profile for a period of 50 s, and then the following of a pulse-like command is claimed. The results are illustrated in Figs. 3.3 and 3.4.

According to the presented results, precise tracking of the command signals is achieved with $N = 35$ term approximation for the fractional order differentiation operators. The numerical realization has been performed in Matlab environment with Ninteger toolbox [19]. The results seen in Fig. 3.3 have been obtained with $k_1 = k_2 = 10$, and those in Fig. 3.4 are obtained with $k_1 = k_2 = 0.1$. The former case reveals better tracking performance while the latter produces smother control signals, and the comparison guides the designer for setting the best parameter values for the design expectations.

In the second set of simulations, a third-order system dynamics with several uncertainty terms is considered. The system dynamics is given by (3.26).

$$x_1^{(0.7)} = x_2$$

$$x_1^{(0.6)} = x_3$$

$$x_3^{(0.5)} = f(x_1, x_2, x_3) + \Delta(x_1, x_2, x_3, t) + g(t)u + \xi(t) \tag{3.26}$$

where $\Delta(x_1, x_2, x_3)$ and $\xi(t)$ are uncertainties and disturbance terms that are not available to the designer. In the above equation, we have

$$f(x_1, x_2, x_3) = -0.5x_1 - 0.5x_2^3 - 0.5x_3|x_3| \tag{3.27}$$

$$g(t) = 1 + 0.1\sin\left(\frac{\pi t}{3}\right) \tag{3.28}$$

$$\Delta(x_1, x_2, x_3, t) = (-0.05 + 0.25\sin(5\pi t))x_1 + (-0.03 + 0.3\cos(5\pi t))x_2^3$$
$$+ (-0.05 + 0.25\sin(7\pi t))x_3|x_3| \tag{3.29}$$

$$\xi(t) = 0.2\sin(4\pi t) \tag{3.30}$$

The results of the simulations are shown in Fig. 3.5, where it is seen that the reference signal for the first state variable is followed very precisely when $k_1 = k_2 = 10$ and $N = 35$. Regarding the second state variable, due to the sharp changes

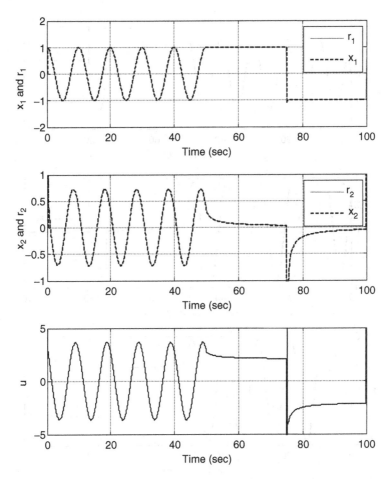

Fig. 3.3 Simulation results for the system described by (3.24). $k_1 = k_2 = 10, N = 35$

in the reference signal, several instantaneous peaks are visible. The effect of the disturbances and approximation errors are seen as a slight degradation in the tracking performance of the third state variable. The last row in Fig. 3.5 shows the control signal that yields the shown tracking performances. Clearly the control signal has very sharp responses when there are sudden changes in the command signal. In Fig. 3.6, the approximation parameter is reduced to $N = 9$ and the simulations were repeated. Apparently in this case the state tracking performance even for the second state is visibly degraded, and we conclude that the numerical issues in implementing the fractional order differintegration operators influence the performance significantly.

Since the reference signal contains instantaneous changes, the responses are affected at these instants. In order to clarify this situation, we study the second example once again but in this time, we choose the reference signal as a filtered

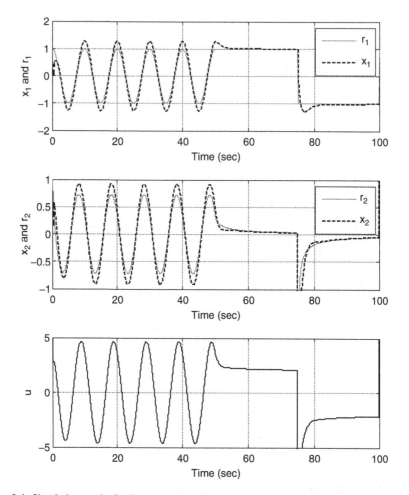

Fig. 3.4 Simulation results for the system described by (3.24). $k_1 = k_2 = 0.1, N = 35$

version of the reference signal considered in the previous cases. More explicitly, we choose

$$R_1(s) = \frac{1}{(s+1)^6}C(s) \tag{3.31}$$

where $C(s)$ is the command signal used so far and $R_1(s)$ is the Laplace transform of $r_1(t)$. The results are shown in Fig. 3.7, where it is seen that both the trajectory tracking performance and the control signal smoothness are very good provided that the smoothness of the command signal is assured.

The presented results demonstrate that the backstepping design can be adapted for fractional order plant dynamics and the use of better approximations for fractional order operators can lead to improved performance indications.

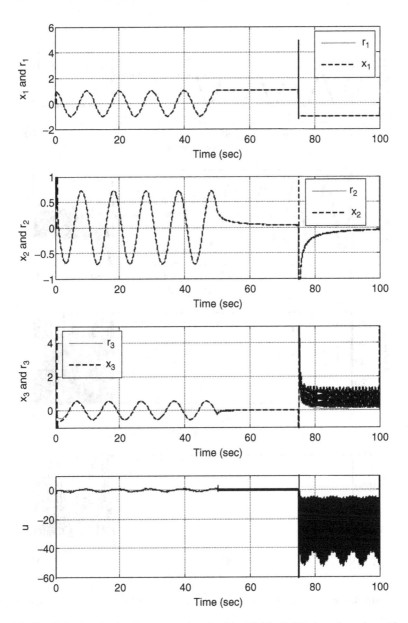

Fig. 3.5 Simulation results for the system described by (3.26)–(3.30), $k_1 = k_2 = k_3 = 10$ and $N = 35$

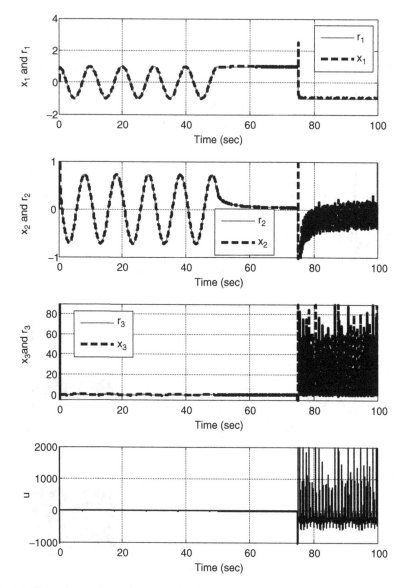

Fig. 3.6 Simulation results for the system described by (3.26)–(3.30), $k_1 = k_2 = k_3 = 10$ and $N = 9$

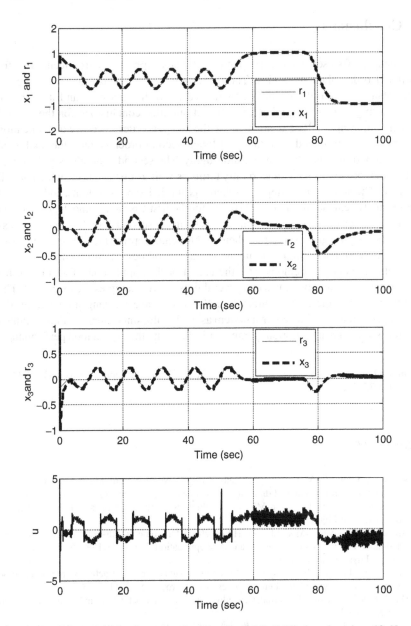

Fig. 3.7 Simulation results for the system described by (3.26)–(3.30), $k_1 = k_2 = k_3 = 10$, $N = 35$ and the reference signal is a filtered one as described by (3.31)

5 Conclusions

This chapter focuses on the adaptation of backstepping control technique for fractional order plant dynamics. The derivation of the control law for a second order plant is given, the result is generalized for q-th order case and it is shown that ensuring $zz^{(\beta)} < 0$ implies $z\dot{z} < 0$ and stability conclusions for the control laws maintaining $zz^{(\beta)} < 0$ are tied to the integer order case. Two application examples are scrutinized. The first is a linear second order system, the analytical details embodying which is known thoroughly. The second example is a nonlinear system that possesses some uncertainty terms as well as disturbances, which are all bounded. The adapted backstepping scheme is applied to both systems, and it is seen that the analytical claims are met perfectly for the first case and some degradation in the performance due to the uncertainties is seen in the second case. If the smoothness of the command signal is assured, then a significant improvement in the trajectory tracking performance and the command signal smoothness is observed.

Briefly, the chapter demonstrates the use of backstepping control technique for fractional order plant dynamics and several illustrative examples are discussed. The results show that the design parameters N and k_is have a strong influence on the overall performance of the control system as well as the smoothness of the command signal is seen to be an important parameter influencing the closed loop performance.

Acknowledgment This work is supported in part by Turkish Scientific Council (TÜBİTAK) Contract 107E137.

References

1. Adigbli T, Grand C, Mouret JB, Doncieux S (2007) Nonlinear attitude and position control of a microquadrotor using sliding mode and backstepping techniques. 3rd US-European Competition and Workshop on Micro Air Vehicle Systems & European Micro Air Vehicle Conference and Flight Competition, pp 1–9
2. Ahmed E, El-Sayed AMA, El-Saka HAA (2006) On some Routh–Hurwitz conditions for fractional order differential equations and their applications in Lorenz, Rössler, Chua and Chen systems. Phys Lett A 358:1–4
3. Chen Y-Q, Ahna H-S, Podlubny I (2006) Robust stability check of fractional order linear time invariant systems with interval uncertainties. Signal Process 86:2611–2618
4. Das S (2008) Functional fractional calculus for system identification and controls, 1st edn. Springer, New York
5. Hua C, Liu PX, Guan X (2009) Backstepping control for nonlinear systems with time delays and applications to chemical reactor systems. IEEE Trans Industrial Electronics 56(9):3723–3732
6. Krstic M, Kanellakopoulos I, Kokotovic P (1995) Nonlinear and adaptive control design. Wiley, New York
7. Madani T, Benallegue A (2006) Backstepping sliding mode control applied to a miniature quadrotor flying robot. In: Proceedings of the 32nd Annual Conference on IEEE Industrial Electronics, November 6–10, Paris, France, pp 700–705

8. Matignon D (1998) Stability properties for generalized fractional differential systems. In: ESAIM Proceedings, Fractional Differential Systems, Models, Methods and Applications, vol 5, pp 145–158
9. Matignon D (1996) Stability results for fractional differential equations with applications to control processing. Comput Eng Syst Appl 963–968
10. Meerschaert MM, Tadjeran C (2006) Finite difference approximations for two-sided space-fractional partial differential equations. Appl Numerical Math 56:80–90
11. Merrikh-Bayat F, Afshar M (2008) Extending the root-locus method to fractional order systems. J Appl Math (Article ID 528934)
12. Oldham KB, Spanier J (1974) The fractional calculus. Academic, New York
13. Ortigueira MD (2000) Introduction to fractional linear systems. Part 1: continuous time case. IEE Proc Vis Image Signal Process 147(1):62–70
14. Podlubny I (1998) Fractional differential equations, 1st edn. Elsevier Science & Technology Books, Amsterdam
15. Podlubny I (1999) Fractional-order systems and (PID mu)-D-lambda-controllers. IEEE Trans Automatic Control 44(1):208–214
16. Podlubny I, Chechkin A, Skovranek T, Chen Y-Q, Vinagre BM (2009) Matrix approach to discrete fractional calculus II: partial fractional differential equations. J Comput Phys 228:3137–3153
17. Raynaud H-F, Zerganoh A (2000) State-space representation for fractional order controllers. Automatica 36(7):1017–1021
18. Sierociuk D, Dzielinski AD (2006) Fractional Kalman filter algorithm for the states, parameters and order of fractional system estimation. Int J Appl Math Comput Sci 16(1):129–140
19. Valerio, D. (2005) Ninteger v.2.3 fractional control toolbox for MatLab
20. Zhao C, Xue D, Chen Y-Q (2005) A fractional order PID tuning algorithm for a class of fractional order plants. In: Proceedings of the IEEE Int. Conf. on Mechatronics & Automation, Niagara Falls, Canada

Chapter 4
Parameter Tuning of a Fractional-Order PI Controller Using the ITAE Criteria

Badreddine Boudjehem and Djalil Boudjehem

1 Introduction

Recently the concept of fractional calculus are widely introduced in many areas in science and engineering. In control systems, this concept are successfully used to construct fractional order controllers. As a result, the closed loop control system performances are improved in comparison with classical controllers.

In podlubny [11] proposed a generalization of the PID controller namely fractional PID ($PI^\lambda D^\mu$) where λ and μ are the order of integration and derivation respectively that can be real numbers. In comparison with classical PID these controllers have two extra parameters. Therefore classical design method may not be applied directly to adjust all fractional controller parameters.

Several research works have proposed new design techniques and tuning rules, for fractional controllers. Some of them are based on an extension of the classical control theory. In Valerio and da costa [16] a tuning method for fractional PID controller based on Ziegler–Nichols-type rules was proposed. Monje et al. [6] present a frequency domain approach based on the expected crossover frequency and phase margin. A state-space tuning method based on pole placement was also used (see [3]). Recent tuning method based on quantitative feedback theory (QFT) are presented in [9].

B. Boudjehem (✉)
Département de Génie Electrique, Université de Skikda, Route El-Hadeik,
BP 26, Skikda 21000, Algérie

Laboratoire d'Automatique et Informatique de Guelma (LAIG), Université de Guelma,
BP 401, Guelma 24000, Algérie
e-mail: b_boudjehem@yahoo.fr

D. Boudjehem
LAIG, Université de Guelma, BP 401, Guelma 24000, Algérie
e-mail: dj_boudjehem@yahoo.fr

D. Baleanu et al. (eds.), *Fractional Dynamics and Control*,
DOI 10.1007/978-1-4614-0457-6_4, © Springer Science+Business Media, LLC 2012

Many methods for control design are based on optimization techniques. The common approach is to minimize a performance index [1]. An optimization approach was proposed in [6], for the PI fractional controller tuning. A nonlinear functional minimization subject to some given nonlinear constraints are solved using matlab minimization function. An intelligent optimization method for designing fractional order PID controller based on particle swarm optimization (PSO) was presented (see [2]). In Leu et al. [4], an optimal fractional order PID controller based on specified gain and phase margins with a minimum ISE criterion has been designed by using a differential evolutionary algorithm. Tuning fractional PID controller based on ITAE criterion by using Particle Swarm Optimization has been also presented in [12]. In Tavazoei [15] the infiniteness and finiteness of different performance indices in class of fractional-order systems have been presented.

In this paper we propose a tuning method for fractional PI controller based on minimizing integral time absolute error (ITAE) by means of diffusive representation. The feedback control system is implemented in Matlab/Simulink. Simulation results show the effectiveness of the proposed design method in comparison with classical PI controller. The paper is organized as follows. Section 2 gives an overview on fractional order controllers and the diffusive representation. Section 3 presents the design method procedure. An illustrative example is given in Sect. 4 to demonstrate the effectiveness of the proposed method. Finally conclusions are stated in Sect. 5.

2   Fractional Order Operators and Controllers

2.1   Fractional Order Operator

There are several different definitions of fractional operators (see [10] and [5]). One of the most used definition of the fractional integration is the Riemann–Liouville definition:

$$D^{-\alpha}f(t) = \frac{1}{\Gamma(\alpha)} \int_0^t (t-\tau)^{\alpha-1} f(\tau) d\tau \qquad (4.1)$$

while the fractional derivative definition is

$$D^{\beta}f(t) = D^m[D^{-\gamma}f(t)], \qquad (4.2)$$

where

$$\Gamma(\alpha) = \int_0^{\infty} x^{\alpha-1} e^{-x} dx \qquad (4.3)$$

is the Gamma function, α is the order of the integration, m is an integer number and $\gamma = m - \beta$.

The Laplace transform method is a powerful tool in the frequency domain for both the system analysis and the controller synthesis. The Laplace transform of the fractional integral given by Riemann–Liouville, under zero initial conditions for order α is:

$$L\left(D^{-\alpha}f(t)\right) = s^{-\alpha}F(s),\tag{4.4}$$

where $F(s)$ is the normal Laplace transformation $f(t)$.

2.2 Fractional PID Controller ($PI^{\lambda}D^{\mu}$)

Fractional $PI^{\lambda}D^{\mu}$ controller is a system described by a fractional differential equation:

$$K_p\left(y(t) + \frac{1}{T_i}D^{-\lambda}y(t) + T_dD^{\mu}y(t)\right) = e(t),\tag{4.5}$$

where D is the derivative operation, K_p is the proportional gain, T_i is the integration constant, T_d is the derivative constant, λ is the integration order and μ is the derivative order. The Laplace transform of (4.5), lead to the following transfer function:

$$C(s) = K_p\left(1 + \frac{1}{T_i}s^{-\lambda} + T_ds^{\mu}\right).\tag{4.6}$$

Taking $\mu = 0$ and/or $T_d = 0$ we obtain a fractional PI. We note that if $\mu = 1$ and $\lambda = 1$, we obtain a classical PID controller.

2.3 Diffusive Representation of Fractional Operators

There are several approaches that have been used to implement fractional order integration (see [7] and [14]). An alternative is to use the so-called "Diffusive approach" (see [8]).

The diffusive realization of the pseudo differential operator H, with impulse response h, $u \to g = H\left(d/dt\right)u$ is defined by the dynamic input–output system:

$$\begin{cases} \partial_t\varphi(\xi,t) = -\xi\varphi(\xi,t) + u(t), \\ g(t) = \int_0^{\infty}\mu(\xi)\varphi(\xi,t)d\xi \\ \varphi(\xi,0) = 0, \quad \xi \succ 0. \end{cases}\tag{4.7}$$

The system (4.7) is the diffusive realization of H.

The impulse response $h(t)$ is expressed from $h(t)$ by:

$$h(t) = \int_0^{+\infty} e^{-\xi t} \mu(\xi) d\xi \tag{4.8}$$

So the diffusive symbol is also given by: $\mu = L^{-1} h$.

The transfer function of the operator H is given by:

$$H(s) = \int_{-\infty}^{+\infty} \frac{\mu(\xi)}{s+\xi} d\xi. \tag{4.9}$$

We thus have the three equivalent representations:

Diffusive rep. $\overset{L}{\to}$	Convolution rep. $\overset{L}{\to}$	Symbol
μ	$h(t)$	$H(s)$
$\mu \# v$	$h(t)*r(t)$	$H(s).R(s)$

In the particular case of fractional integrators

$$H\left(\frac{d}{dt}\right) = \left(\frac{d}{dt}\right)^{-\alpha}, \quad 0 < \alpha < 1. \tag{4.10}$$

The diffusive symbol is expressed as (see [13]): $\mu(\xi) = \frac{\sin(\pi\alpha)}{\pi} \frac{1}{\xi^\alpha}, x > 0$ where a is the order of integration.

The numerical approximation of the fractional order system based on diffusive representation is simple and present more advantages.

3   The Proposed Design Method

Let us consider the feedback control system depicted in Fig. 4.1. Where $G(s)$ is the controlled system transfer function and $C(s)$ is the controller transfer function. The controller used is a fractional PI controller with transfer function given by (4.11):

$$C(s) = K_p \left(1 + \frac{1}{T_i s^\lambda}\right). \tag{4.11}$$

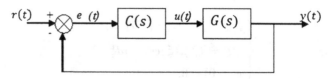

Fig. 4.1 Feedback control system

These controllers have three unknown parameters K_p, T_i, and λ, that must be determined to achieve the desired specifications. Using (4.7), the diffusive realization of the controller with input e and output u is given by

$$\begin{cases} \partial_t \varphi(\xi,t) = -\xi \varphi(\xi,t) + e(t), \\ u(t) = \int_0^\infty v(\xi) \varphi(\xi,t)\, d\xi, \end{cases} \tag{4.12}$$

where the diffusive symbol is:

$$v(\xi) = K_p \left(\delta(\xi) + \frac{1}{T_i} \frac{\sin(\pi\lambda)}{\pi} \frac{1}{\xi^\lambda} \right). \tag{4.13}$$

Thus the transfer function of the fractional controller PI^λ by means of diffusive representation are

$$C(s) = \int_{-\infty}^{+\infty} \frac{v(\xi)}{s+\xi}\, d\xi, \tag{4.14}$$

where v is defined by (4.13).

Our objectives are to adjust the three fractional PI parameters (K_p, T_i, and λ) that minimizing is ITAE defined by the objective function:

$$J_{\text{ITAE}}(K_p, T_i, \lambda) = \int_0^T t|e(t)|\, dt, \tag{4.15}$$

where t is the time and $e(t)$ is the error step set-point change.

The procedures to determine the PI^λ fractional controller parameters are summarized in the following:

1. Implement the feedback control system in Matlab/Simulink including diffusive realization of the fractional PI controller through Simulink model
2. Calculate the ITAE
3. Use a function of Matlab optimization toolbox to minimize the objective function J. The initial controller parameters is set to be those determined by one of existing tuning rules

4 An Example of Application

In this section, an example of application is given to illustrate the proposed method. Consider the forth order system:

$$G(s) = \frac{k}{s(s+1)(s+2)(s+3)}. \tag{4.16}$$

To illustrate the robustness to parameter variations, we consider only that the gain can be changed with a variation range of $K \in [1, 1.8]$. For the simulation the function of matlab optimization tool box are used to minimize the objective function J_{ITAE}.

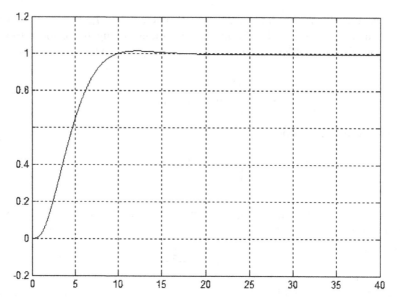

Fig. 4.2 Step response of the controlled system with fractional PI controller

The stability margin based Ziegler–Nichols is used for determine the initial parameters K_p and T_i whereas the order $\lambda = 1$. The optimized parameters of the fractional PI controller are: $K_p = 5.64$, $T_i = 76.32$, and $\lambda = 1.12$.

Therefore, the transfer function of the fractional controller is:

$$C(s) = 5.64 \left(1 + \frac{1}{76.32 s^{1.12}} \right). \tag{4.17}$$

Figure 4.2 presents the step response of the controlled system with fractional PI controller. This figure shows clearly that the overshoot and set time are acceptable.

Figure 4.3 illustrates the step responses with fractional controller for different values of K. This figure shows that the fractional controller designed by the proposed method permits to have a time responses with slightly iso-overshoot for different values of gain K.

In order to prove the efficiency of the proposed method we compare our results with those obtained using classical PI controller, the same procedure are applied to tune classical PI controller Therefore the optimal transfer function of classical PI controller is:

$$C(s) = 2.0512 \left(1 + \frac{1}{3.0681e6s} \right). \tag{4.18}$$

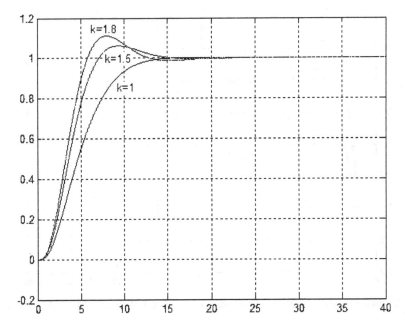

Fig. 4.3 Step response of the controlled system with fractional PI controller for different values of K

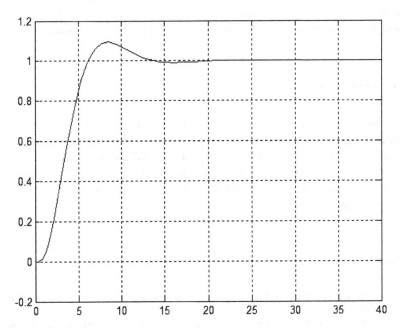

Fig. 4.4 Step response of the controlled system with classical PI controller

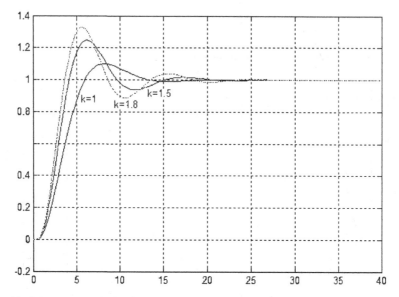

Fig. 4.5 Step response of the controlled system with classical PI controller for different values of K

Figure 4.4 presents the step response of the controlled system with classical PI controller. Figure 4.5 illustrates the step responses with classical controller for different values of K. This figure shows that the responses with this controller present different overshoot. So that the system is not robust to gain variations.

5 Conclusions

An optimal design method for fractional PI controller has been presented. The method is based on using diffusive representation of fractional operator. The optimal settings have been obtained by minimizing ITAE using Matlab optimization toolbox. The simulation results have shown the effectiveness of the proposed method in comparison with classical PI controller. In addition the system obtain is robust to gain variations.

References

1. Aström K, Hägglund T (1995) PID controllers: Theory, design, and tuning. Instrument Society of America, North Carolina
2. Cao J, Cao BG (2006) Design of fractional order controller based on particle swarm optimisation. Int J Contr Autom Syst 4:775–781

3. Dorcak L, Petras I, Kostial I, Terpak J (2001) State-space controller desing for the fractional-order regulated system. International carpathian control conference, Krynica, Poland
4. Leu JF, Tsay SY, Hwang C (2002) Design of optimal fractional-order PID controllers. J Chin Inst Chem Eng 33:193–202
5. Miler S (1993) Introduction to the fractional calculus and fractional differential equations. Wiley, NewYork
6. Monje CA, Calderón AJ, Vinagre BM et al (2004) On fractional PI^λ controllers: Some tuning rules for robustness to plant uncertainties. Nonlinear Dynam 8:369–381
7. Point T, Trigeassou JC (2002) A method for modelling and simulation of fractional systems. Signal Process 83:2319–2333
8. Montseny G (2004) Simple approach to approximation and dynamical realization of pseudodifferential time operators such as fractional ones. IEEE Trans Circ Syst II 51:613–618
9. Natarj PSV, Tharewal S (2007) On fractional-order QFT controllers. J Dyn Syst Meas Contr 129:212–218
10. Oldham K, Spanier J (1974) The fractional calculus: Theory and application of differentiation and integration to arbitrary order. Wiley, NewYork
11. Podlubny I (1999) Fractional-order systems and PID-controllers. IEEE Trans Automat Contr 44:208–213
12. Maiti D, Acharya AA, Chakraborty M, konar A, Janarthanan R (2008) Tuning PID and $PI^\lambda D^\mu$ controller using the Integral Time Absolute Error criteria. International Conference on Information and Automation for Sustainability ICIAF8, Colombo, Srilanka
13. Laudebat L, Bidan P, Montseny G (2004) Modeling and optimal identification of pseudodifferential dynamics by means of diffusive representation-Part I: Modeling. IEEE Trans Circ Syst 51:1801–1813
14. Oustaloup A, Levron F, Mathieu B, Nanot F (2000) Frequency-band complex non integer differentiator: Characterization and synthesis. IEEE Trans Circ and Syst I 47:25–39
15. Tavazoei MS (2010) Notes on integral performance indices in fractional-order control systems. J Proc Contr 20:285–291
16. Valério D, Sá da Costa J (2006) Tuning of fractional PID controllers with ziegler-nichols-type rules. Signal Process 86:2771–2784

Chapter 5
A Fractional Model Predictive Control for Fractional Order Systems

Djalil Boudjehem and Badreddine Boudjehem

1 Introduction

Fractional calculus allows a more compact representation and problem solution for many systems. The idea of fractional integrals and derivatives has been known since the development of regular calculus. Probably the first physical system to be widely recognized as one demonstrating fractional behavior is the semi-infinite lossy (RC) transmission line (see [5]). Another equivalent system is the diffusion of heat into semi-infinite solid (see [8]). Other systems that are known to display fractional order dynamics are viscoelasticity, colored noise, electrode–electrolyte polarization, dielectric polarization, boundary layer effects in ducts, and electromagnetic waves. Because many systems are known to display fractional order dynamics, they can't be controlled the same way as those which doesn't. Unfortunately, these systems had been considered to be similar to systems with integer order dynamics for a long time. However, in the last decade we noticed the born of the fractional control that deals with those specific systems. The significance of fractional control system is that it is the generalization of the classical integer order control theory, which could lead to a more adequate modeling and more robust control performance.

Predictive control is a family of control techniques that optimize a given criterion by using a model to predict system evolution and compute a sequence of future control actions. Predictive control accepts a variety of models, objective functions,

D. Boudjehem (✉)
Laboratory of Automatics and Informatics of Guelma (LAIG), Department of Electrical Engineering, University of Guelma BP 401, Guelma 24000, Algeria
e-mail: dj_boudjehem@yahoo.fr

B. Boudjehem
Department of Electrotechnique, University of Skikda, Roud ElHadeik BP26 Skikda 21000, Algeria
e-mail: b_boudjehem@yahoo.fr

D. Baleanu et al. (eds.), *Fractional Dynamics and Control*,
DOI 10.1007/978-1-4614-0457-6_5, © Springer Science+Business Media, LLC 2012

and constraints, providing flexibility in handling a wide range of operating criteria present in industrial processes, See [4,6,7]. A variety of processes can be controlled using MPC [3].

Generally, in MPC linear models are used to predict the system dynamics, even though the dynamics of the closed-loop system is nonlinear or displays fractional order dynamics.

This paper focuses on the use of fractional order system description to model fractional order dynamics in model predictive control (MPC) to construct a fractional order MPC. The fractional order model could model various real materials more adequately than integer order ones and provide a more adequate description of many actual dynamical processes, which will improve the MPC performances and lead to a more robust control performance, See [1,2]. The results in this paper show that the use of fractional order models not only give better results than the use of integer order models, but also it performs better with the presence of noise which, can be interpreted as a kind of robustness.

The paper is organized as follows. In Sect. 1 we present some theoretical aspects of fractional order systems and fractional order approximation. In Sect. 2. the basic concept of MPC is introduced. An outline of the fractional model MPC schemes is also presented and simulation results are discussed in Sect. 3.

2   Fractional Order Systems

A fractional order system is that system described by the following fractional order differential equation

$$a_n D^{\alpha_n} f(x) + a_{n-1} D^{\alpha_{n-1}} f(x) + a_{n-2} D^{\alpha_{n-2}} f(x) + \cdots$$
$$= b_n D^{\beta_n} f(x) + b_{n-1} D^{\beta_{n-1}} f(x) + b_{n-2} D^{\beta_{n-2}} f(x) + \cdots, \qquad (5.1)$$

where $D^{\alpha_n} =_0 D_t^{\alpha_n}$, is called the fractional derivative of order α_n with respect to variable t and with the starting point $t = 0$.

In fractional calculus, the fractional derivative is defined due to Riemann and Liouville fractional integral version given by (5.2), (See [9, 12]):

$$_0D_x^{-v} f(x) = \int_0^x \frac{(x-t)^{v-1}}{(v-1)!} f(t) dt \qquad (5.2)$$

Or

$$_0D_x^{-v} f(x) = \frac{1}{\Gamma(v)} \int_0^x (x-t)^{v-1} f(t) dt, \qquad (5.3)$$

where $\Gamma(v)$ is the Euler's *Gamma* function defined by:
$\Gamma(v) = (v-1)!$, with the property: $\Gamma(v+1) = v\Gamma(v)$.

The fractional derivation is then defined by:

$$D^v f(x) = \frac{1}{\Gamma(n-v)} \left(\frac{d}{dx}\right)^n \int_0^x (x-t)^{n-v-1} f(t) dt, \qquad (5.4)$$

where n is an integer number defined as $n-1 < v < n$.

The Laplace transform of a fractional derivative can be calculated easily by applying the regular Laplace operator on (5.3):

$$\mathcal{L}[D^v f(x)] = s^v F(s) - \sum_{k=0}^{n-1} s^{n-k-1} D^{k-n+v} f(0) \quad \text{if Re}(v) > 0 \qquad (5.5)$$

and n is the integer number defined before. The summation in the right hand side of (5.5) will be equal to zero if $\text{Re}(v) \le 0$.

Approximation of the Fractional Order Transfer Function

For simulations and implementations, we need to approximate the fractional order transfer functions of powers of $v \in \mathbb{R}$ by the usual integer order $n \in \mathbb{Z}$ transfer functions with a similar behavior. The integer transfer function may then have to include an infinite number of poles and zeros. But it is always possible to get good approximations. In fact, there are a number of approximations that exist and makes use of a recursive distribution of poles and zeros. See [13, 14]. The most common approximation used is that proposed by [14]:

$$s^v \approx C \prod_{k=-N}^{N} \frac{1 + (s/\omega_{z_k})}{1 + (s/\omega_{p_k})}, \quad v > 0 \qquad (5.6)$$

Where ω_l, ω_h are the lower and higher frequency approximation interval. This means that the approximation is valid in that frequency interval. The gain C has the role of approximation tuning, so it is adjusted until both sides of (5.6) will have unit gain at 1 rad/s. The approximation limits N is chosen before hand, and the good performance of the approximation strongly depends thereon. Low values result in simpler approximations, but also cause the appearance of a ripple in both gain and phase behaviors; such a ripple may be practically eliminated increasing N, but the approximation will be computationally heavier. Frequencies of poles and zeros in (5.6) are given by:

$$\alpha = (\omega_h/\omega_l)^{v/N}, \qquad (5.7)$$

$$\eta = (\omega_h/\omega_l)^{(1-v)/N}, \qquad (5.8)$$

$$\omega_{z_0} = \omega_0 \sqrt{\eta}, \qquad (5.9)$$

$$\omega_{p_k} = \omega_{z_k} \alpha, \tag{5.10}$$

$$\omega_{z_{k+1}} = \omega_{p_k} \eta. \tag{5.11}$$

In general, it is usual to split fractional powers of s like this:

$$s^\nu = s^n s^\delta, \quad \nu = n + \delta, \tag{5.12}$$

where n is an integer number defined as: $n < \nu < n + 1$, thus the values of δ will be compromised between 0 and 1. In this manner we need only to approximate the latter term.

3   The MPC

The MPC problem is formulated as solving on-line, a finite horizon open-loop optimal, control problem subject to system dynamics and constraints involving states and controls. However, the success of the MPC strategy depends critically on the choice of the model, See [3, 4, 7].

3.1   The MPC Principle

For a given plant, at instant k, the reference trajectory $r(k)$ is defined to be the ideal trajectory along which the plant should return to the set point trajectory $w(k)$. It is frequently assumed that the reference trajectory approaches the set point exponentially from the current output value $y(k)$, with the time constant of the exponential, which we shall denote T_{ref}, defining the speed of response [4].

The current error between the output and the set point is then defined to be:

$$\text{err}(k) = w(k) - y(k), \tag{5.13}$$

then the reference trajectory is chosen such that the error i steps later, if the output followed it exactly, would be

$$\text{err}(k + i) = \exp^{-iT_s/T_{\text{ref}}} \text{err}(k), \tag{5.14}$$

where T_s is the sampling interval. That is the reference trajectory is defined to be:

$$r(k + i|k) = w(k + i) - \text{err}(k + i). \tag{5.15}$$

The notation $r(k + i|k)$ indicates that the reference trajectory depends on the conditions at time k, in general, See [7, 11].

A predictive controller has an internal model which is used to predict the behavior of the plant, starting at the current time k, over a future prediction horizon H_p. This predicted behavior depends on the assumed input trajectory $\hat{u}$ that is to be applied over the prediction horizon. The idea is to select that input which promises the best predicted behavior. The notation $\hat{u}$ rather than u here indicates that at time k we only have a prediction of what the input at time $k + i$ may be; the actual input at that time, $u(k + i)$, will probably be different from $\hat{u}(k + i|k)$. Note that we assume having the output measurement $y(k)$ available when deciding the value of the input $u(k)$. We should also notice that the model output $y(k)$ depends only on the past inputs $u(k - 1), u(k - 2), \ldots$, not including the present input $u(k)$ [7].

Once a future input trajectory has been chosen, only the first element of that trajectory is applied as the input signal to the plant. That is, we set $u(k) = \hat{u}(k|k)$, where $u(k)$ denotes the actual signal applied. Then the whole cycle of output measurement, prediction, and input trajectory determination is repeated, one sampling interval later. Since the prediction horizon remains of the same length as before, but slides along by one sampling interval at each step, this way of controlling a plant is often called a receding horizon strategy.

3.2   Computing the Optimal Control Signal

To compute the control signal $\hat{u}(k)$ that will be applied at time instant k, we should solve an optimization problem where, the receding horizon cost function that will be minimized is defined by the following equation:

$$J = \sum_{i \in P} R \times [r(k+i|k) - y(k+i|k)]^2$$
$$+ \sum_{H_u} Q \times [\Delta u(k+i-1)]^2, \tag{5.16}$$

where P denotes the set of indices i which correspond to coincidence points and H_u is the control horizon, R and Q are weighting matrices. In the simplest case these matrices are set to identity matrix.

Conceptually, the internal model can first be used to predict the free responses $\hat{y}_f(i + H_p|k)$ of the plant, which are the responses that would be obtained at the coincidence points if the future input trajectory remained at the latest value $u(k-1)$. Depending on the form of the model, these values will be obtained, and if a step or pulse response is available as the model, then all the available past inputs are needed.

Now let $S(H_p)$ be the response of the model to a unit step input, H_p steps after the unit step is applied. The predicted output at time $k + H_p$ is as follows:

$$\hat{y}(k+Hp|k) = \hat{y}_f(k+Hp|k) + S(H_p)\Delta\hat{u}(k|k), \tag{5.17}$$

where

$$\Delta\hat{u}(k|k) = \hat{u}(k|k) - u(k-1) \tag{5.18}$$

is the change from the current input $u(k-1)$ to the predicted input $\hat{u}(k|k)$. We want
to achieve

$$\hat{y}(k+Hp|k) = r(k+Hp|k) \qquad (5.19)$$

Hence, the optimal change of input can be calculated easily by solving the
minimization of the cost function given by (5.16).

4 Fractional Model for Predictive Control

Since fractional order models describes fractional systems better than integer order
models do, we propose to use a fractional order model rather than an integer order
one as in the used conventional MPC.

Figure 5.1 shows the MPC scheme using the fractional order model. The
implementation of this idea means that (5.16) have to be in the following form:

$$J = \sum_{i \in P}[r(k+i) - y_{\text{frac}}(k+i|k)]^2$$
$$+ \sum_{H_u}[\Delta u(k+i-1)]^2, \qquad (5.20)$$

where y_{frac} is the fractional order system output. Due to the approximation calculation of y_{frac} the determination of the control signal $u(k)$ or its increment $\Delta u(k)$
will be quit difficult, and for this reason the matrices R and Q are set to identity.
Therefore, depending on the number of poles and zeros used for the fractional
system approximation, we can adjust the control system for better improvement.
Consequently, a better fractional system dynamic approximation will be achieved.

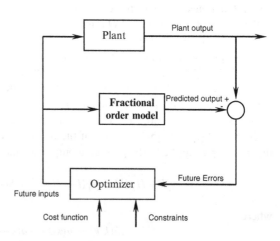

Fig. 5.1 General structure of
the fractional MPC

We will show in the next section how can we improve the performances of the predictive control, and maintain the system stability in the case of changing the prediction horizon by the use of a fractional order model rather than integer order model in the predictive control.

4.1 Simulation Results

In this section we consider the following non commensurate fractional order plant given by Podlubny ([9]):

$$G(s) = \frac{1}{0.8s^{2.2} + 0.5s^{0.9} + 1}.$$ (5.21)

To implement the fractional system model we used the approximation given by (5.6)–(5.11).

For the comparison purpose, we will use the integer order model proposed by Podlubny ([10]) given by (5.22), which represent the nearest integer order model to the system defined by (5.21). Since the fractional orders are approximated to those nearest integer ones this model will be used to implement the integer order MPC:

$$G(s) = \frac{1}{0.8s^2 + 0.5s + 1}.$$ (5.22)

The performance of the results is then asserted and compared to the performance obtained with integer order model.

Figure 5.2 shows the step response of the fractional order system.

Figures 5.3 and 5.4 show the effect of the prediction horizon on the closed loop control system using both fractional order and integer order MPC respectively. In Fig. 5.3 the step responses shows an increase in the maximum overshoot as the prediction horizon increases. Unfortunately, this not the case for the integer order MPC as shown in Fig. 5.4. Since integer order MPC was completely unstable for a prediction horizon $H_p = 6$ the prediction horizon started from $H_p = 8$.

Hence, for the comparison reasons we used the prediction horizon $H_p = 12$ with both MPC's.

Figure 5.5 shows the controlled plant output to a square input signal using both fractional order and conventional MPC. Notice that both plant outputs reaches the correct set-point. However, the use of the fractional order MPC leads to better improvements of the control of the fractional order system compared to the use of the integer order MPC.

Figure 5.6 shows the control input signal using both fractional model and integer order MPC. To see the effect of the noise presence, we add a noise of a standard deviation of 0.1 which switches from $+0.1$ to -0.1 halfway as shown in Fig. 5.9. Then, the resulting noisy input signal is shown in Fig. 5.8. Figure 5.7 shows the

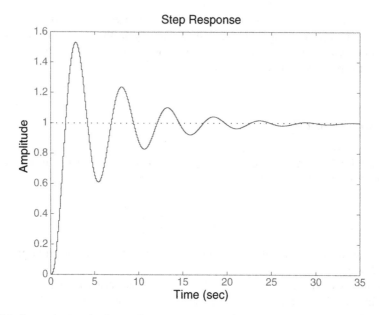

Fig. 5.2 Step response of the plant

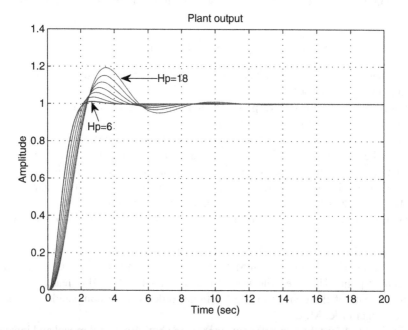

Fig. 5.3 The effect of the prediction horizon on the fractional model MPC

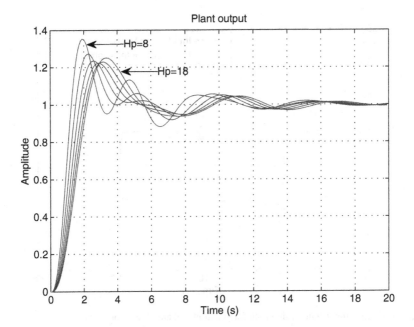

Fig. 5.4 The effect of the prediction horizon on the integer model MPC

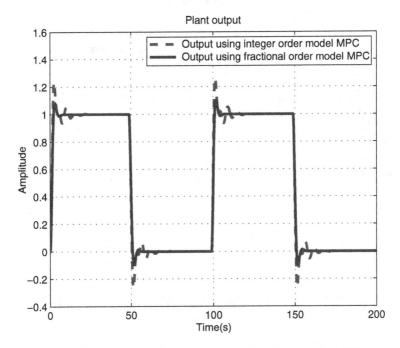

Fig. 5.5 Controlled plant output using both fractional model and integer order MPC

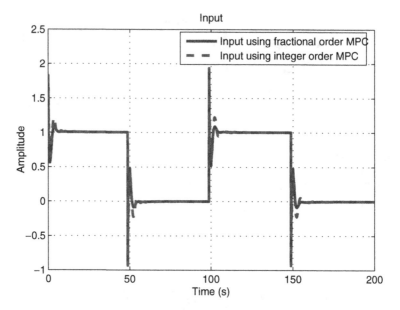

Fig. 5.6 Control input signal using both fractional model and integer order MPC

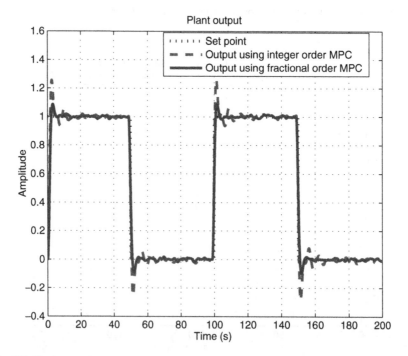

Fig. 5.7 The input noise

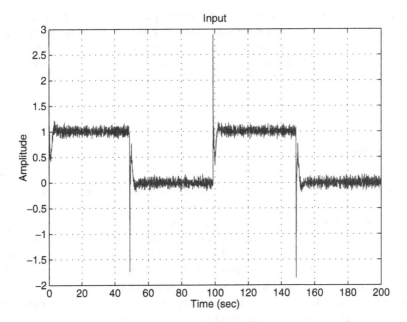

Fig. 5.8 The noisy input using fractional order MPC

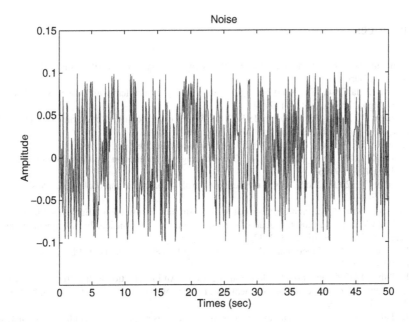

Fig. 5.9 The system output with noisy input

plant output with a noise presence. Notice that both plant outputs reaches the correct set-point despite the unknown input noise. However, the use of the fractional order MPC leads to better improvements (maintain the maximum overshoot level as in the first case. See Fig. 5.5) of the control of the fractional order system compared to the use of the integer order MPC, where an increase in the maximum overshoot is remarkable.

5 Conclusion

In this paper, a fractional order MPC is proposed for fractional order systems control. The benefits of fractional order models for real dynamical objects and processes become more and more obvious. Through the fractional order and integer order dynamical models, the proposed fractional order MPC has been presented. The simulation results illustrate that the use of fractional order models to control systems that present fractional dynamic behaviors, to construct a fractional model MPC achieves better control performances compared to those of the conventional MPC. This approach allows an efficient formulation of MPC while guaranteeing stability and performance of the closed-loop control system.

References

1. Shantanu D (2008) Functional fractional calculus for system identification and controls. Springer, Berlin
2. Valério D, da Costa JSá (2006) Tuning of fractional PID controllers with Ziegler-Nichols-type rules. Signal Process 86:2771–2784
3. Guzmán JL, Berenguel M, Dormido S (2005) Interactive teaching of constrained generalized predictive control. IEEE Contr Syst Mag 25:52–66
4. Camacho EF, Bordóns C (2004) Model predictive control. Springer, Berlin
5. Clarke T, Achar BNN, Hanneken JW (2004) MittagLeffler functions and transmission lines, J Mol Liq 114:159–163
6. Rossiter JA (2003) Model-based predictive control: A practical approach. CRC Press, Boca Raton, FL
7. Maciejowski J (2002) Predictive control with constraints. Prentice-Hall, Englewood Cliffs, NJ
8. Kulish VV, Lage JL (2000) Fractional-diffusion solutions for transient local temperature and heat flux. Trans ASME 122:372–376
9. Podlubny I (1999) Fractional-order systems and $PI^{\Lambda} D^{\mu}$ controllers. IEEE Trans Automat Contr 44:208–214
10. Podlubny I (1999) Fractional differential equations: An introduction to fractional derivatives, fractional differential equations to methods of their solution and some of their applications. Academic Press, San Diego
11. Richalet J (1993) Industrial applications of model based predictive control. Automatica 29:1251–1274
12. Miller KS, Ross B (1993) An introduction to the fractional calculus and fractional differential equations. Wiley, New York

13. Chareff A, Sun HH, Tsao YY, Onaral B (1992) Fractional system as represented by singularity function. IEEE Trans Automat Contr 37:1465–1470
14. Oustaloup A (1991) La commande CRONE: commande robuste d´ordre non entier. Hermès, Paris

Chapter 6
A Note on the Sequential Linear Fractional Dynamical Systems from the Control System Viewpoint and L^2 -Theory

Abolhassan Razminia, Vahid Johari Majd, and Ahmad Feiz Dizaji

1 Introduction

Fractional calculus is a generalization of ordinary differentiation and integration to an arbitrary real-valued order. This subject is as old as the ordinary differential calculus and goes back to the times when Leibniz and Newton invented differential calculus. The problem raised by Leibniz in a letter dated 30 September 1695 for a fractional derivative that has become an ongoing topic for more than hundreds of years [1, 2].

This subject has attracted the attention of researchers from different fields in the recent years. While it was developed by mathematicians few hundred years ago, efforts on its usage in practical applications have been made only recently. It is known that many real systems are fractional in nature; thus, it is more effective to model them by means of fractional order than integer order systems. Applications such as modeling of damping behavior of viscoelastic materials [3], cell diffusion processes [4], transmission of signals through strong magnetic fields [5], and finance systems [6] are some examples. Moreover, fractional order dynamic systems have been used in both design and implementation of control systems. Studies have shown that a fractional order controller can provide better performance than an integer order one and can lead to more robust control performance [7].

On the other hand, it is has been reported that specific fractional differential equations with order less than three may exhibit complex dynamical evolution, even chaotic dynamics [8, 9]; however, this is not the case for ordinary differential equations due to the famous Poincaré–Bendixson theorem [10].

A. Razminia • V.J. Majd (✉)
Intelligent Control Systems Laboratory, School of Electrical and Computer Engineering, Tarbiat Modares University, Tehran, Iran
e-mail: a.razminia@gmail.com; majd@modares.ac.ir

A.F. Dizaji
Department of Engineering Science, School of Engineering, University of Tehran, Tehran, Iran
e-mail: dizaji@ut.ac.ir

D. Baleanu et al. (eds.), *Fractional Dynamics and Control*,
DOI 10.1007/978-1-4614-0457-6_6, © Springer Science+Business Media, LLC 2012

Here L^2 space as a special case of L^p space is considered. The L^p spaces are function spaces defined using natural generalizations of p-norms for finite-dimensional vector spaces. They are sometimes called Lebesgue spaces. One of the main topics in L^p space is approximation theory [11], which is used here for deriving a relationship between an arbitrary function from L^2 space and the output of a linear time invariant fractional control system. The dynamics of the control system are stated in the sequential format. Sequential fractional operator is a special case among other fractional operators which is discussed in the next section.

Due to the computing demands of the fractional differentiation and integration and the need for approximating some complicated functions which are used in physical process, we present a novel method in this chapter by which any arbitrary function from L^2 space can be approximated by the output of a linear time invariant time fractional control system (LTIFCS). Using a rigorous proof, we provide the construction procedure of LTIFCS whose output approximated the given function in the sense of L^2-norm.

This chapter is organized as follows: after this introduction in Sect. 1, a historical review is presented in Sect. 2. Some basic concepts and mathematical preliminaries are summarized in Sect. 3. Section 4 studies the main result which is stated in a theorem form. Finally conclusions in Sect. 5 close the chapter.

2 Historical Review

How well can functions be approximated in a finite interval by the output of a controllable and observable SISO fractional linear time-invariant system? This question was first considered in the framework of constructing the inverse of a linear system in the integer order system basin [12]. Here we want to extend this problem for the first time, in a more comprehensive field, i.e., fractional order systems. Although in [13] Unser's method has been extended for fractional systems, the method presented in this chapter is completely different and has a more general form.

Over the past few decades, the theory of linear control theory has centered on stability properties that only really make sense on an infinite interval. However, there are many problems in which the only thing of interest is a finite interval. Indeed, this work began with questions of constructing flyable trajectories for linear systems [14]. Since the appearance of those initial papers, there has been a growing interest on the construction of curve approximator using linear control theory, which has developed into the theory of interpolating and smoothing splines.

The theory of interpolating splines began with the paper of Schoenberg [15] in 1946 and has begun to be widely used since 1960s with the availability of digital computers. The theory of smoothing splines began with the work of [16]. A more theoretic problem can be found in [14]. A body of literature has developed around the concept of control theoretic splines. The initial work was about flight control application [11].

In the area of smoothing splines, Martin and colleagues have developed the theory of smoothing splines based on linear control theory [17]. Some modifications in a more general prospective under weaker conditions and more general spaces, e.g., Banach spaces, of the problem studied in [17] were treated in [17] and [18] recently. These works were motivated by the trajectory planning problem to avoid high accelerations that were observed when it was required to be exactly at point at a given time. The constraint of interpolation was relaxed to approximation, and a simple optimal control problem yielded generalized smoothing splines. This technique was shown to be adaptable to a number of situations not easily handled with traditional polynomial smoothing spline techniques.

The question in the background of all of the control theoretic splines was about its convergence. This problem was specifically discussed in [19]. Reference [20] also gives a fairly comprehensive theory of convergence of splines in a L^2 sense. However, it was necessary to impose some rather restrictive conditions on the linear system, which although were natural from an approximation view point, they were not natural from the viewpoint of control theory. Beside these restrictions and developments, we want to extend these tasks to a more general form, i.e., fractional order control systems. The new theory might have an important impact on the interpolation and spline theory. Moreover, this methodology can be used fairly in the signal processing and system identification.

3   Preliminaries

Fractional calculus as an extension to ordinary calculus possesses definitions that stem from the definitions existing for ordinary derivatives. Some of the current definitions for fractional derivatives are described in [21]. The Riemann–Liouville definition is the simplest definition to use. Based on this definition, the qth order fractional integral of $f \in L^1[0,t]$ and the terminal value a is given by [22]:

$$J_a^q f(t) = f_q(t) = \frac{1}{\Gamma(q)} \int_a^t (t-s)^{q-1} f(s) ds; \quad q \in R^+ \tag{6.1}$$

where the Euler–Gamma function is defined as follows:

$$\Gamma(q) = \int_0^\infty e^{-q} u^{q-1} du; \quad q > 0 \tag{6.2}$$

Another definition for the fractional derivative is the Caputo's definition [22]:

$$D_*^\alpha f(t) = J_a^{m-\alpha} D^m f(t) = \begin{cases} \frac{1}{\Gamma(m-\alpha)} \int_a^t \frac{f^{(m)}(\tau)}{(t-\tau)^{\alpha+1-m}} d\tau & m-1 < \alpha < m \\ \frac{d^m}{dt^m} f(t) & \alpha = m \end{cases} \tag{6.3}$$

The following property holds for the Laplace transform of the Caputo derivative [23]:

$$L\{D_*^{\alpha} f(t)\} = s^{\alpha} L\{f(t)\} - \sum_{k=0}^{m-1} s^{\alpha-k-1} f^{(k)}(0) \tag{6.4}$$

It can be seen that in the Caputo derivative, one needs the integer order derivatives of the initial conditions while in the Riemann–Liouville derivative, the fractional derivative of the initial conditions is used. Therefore, it is more suitable to use Caputo definition.

Miller and Ross in [24] introduced a so-called sequential fractional derivative in the following way: MR^D

$$MR^{D^{\alpha}} = D^{\alpha}, \, 0 < \alpha \leq 1$$
$$MR^{D^{k\alpha}} = MR^{D^{\alpha}} MR^{D^{(k-1)\alpha}}, \, k = 2,3,\ldots \tag{6.5}$$

where D^{α} is the Caputo fractional derivative. Next, we define a sequential fractional differential equation in an operator format as follows:

$$\left({}_{MR}D^{n\alpha} + a_{n-1} {}_{MR}D^{(n-1)\alpha} + \cdots + a_1 {}_{MR}D^{\alpha} + a_0 \right) y(t) = f(t) \tag{6.6}$$

without loss of generality we restrict $0 < \alpha < 1$.

As in the usual case, it is easy to obtain the connection between (6.6) and the corresponding system of linear fractional differential equations. If we apply the following change of variables to (6.6):

$$x_1(t) = y(t), D^{\alpha} x_j(t) = x_{j+1}(t), \quad (j = 1,2,\ldots,n-1) \tag{6.7}$$

where we have used the Caputo fractional derivative. Using the above notation we have:

$$D^{\alpha} x_1 = x_2,$$
$$D^{\alpha} x_2 = x_3,$$
$$\vdots$$
$$D^{\alpha} x_{n-1} = x_n,$$
$$D^{\alpha} x_n = -\sum_{k=0}^{n-1} a_k x_k + f(t) \tag{6.8}$$

Or in a matrix form with a companion matrix A,

$$D^{\alpha} x(t) = Ax(t) + f(t) \tag{6.9}$$

This is a standard linear fractional differential equation (LFDE).

Definition 6.1 ([25]). Let $1 \leq p < \infty$ and (S, Σ, μ) be a measure space. Consider the set of all measurable functions from S to C (or R) whose absolute value raised to the p-th power has finite integral, or equivalently, that:

$$\|f\|_p := \left(\int_S |f|^p d\mu \right)^{1/p} < \infty \tag{6.10}$$

In this chapter, we consider a special case with $p = 2$ which is the most important case in the L^p spaces. Consider the dynamical system:

$$D^{\alpha} x(t) = Ax(t) + bu(t)$$
$$y = Cx \tag{6.11}$$

Definition 6.2 ([26]). The system (6.11) is observable on $[t_0, t_f]$ iff $x(t)$ for $t \in [t_0, t_f]$ can be deduced from knowledge of $y(t)$ and $u(t)$ on $[t_0, t_f]$.

Definition 6.3 ([26]). System (6.11) is controllable iff for any (x_0, x_1) there exists a control $u(t)$ defined on $[t_0, t_f]$ which drives the initial state $x(t_0) = x_0$ to the final state $x(t_f) = x_1$.

4   Main Results

As in the integer order case, let x_h and x_p be the homogeneous and particular solutions of (6.9), respectively.

Proposition 6.1. *For the LFDE (6.9), we have the general solution as follows:*

$$x(t) = x_h(t) + x_p(t) \tag{6.12}$$

Proof. Inserting x_h and x_p in (6.9) we have:

$$D^{\alpha} x_h = Ax_h,$$
$$D^{\alpha} x_p = Ax_p + f,$$
$$\Leftrightarrow D^{\alpha} x = D^{\alpha} (x_h + x_p) = D^{\alpha} x_h + D^{\alpha} x_p = Ax_h + (Ax_p + f)$$
$$= A(x_h + x_p) + f = Ax + f. \tag{6.13}$$

So based on this proposition it is sufficient to find the homogeneous and particular solutions separately. Using a similar method in finding the homogeneous solution for integer order differential equations, we can find the homogeneous solution for LFDE as:

$$x_h(t) = \sum_{k=0}^{\infty} \frac{A^k (t)^{(1+k)\alpha - 1}}{\Gamma[(k+1)\alpha]} M \tag{6.14}$$

To show that this can be the homogenous solution of (6.9), based on the differinte-grability of x_h it is easy to differentiate (6.14) [27]:

$$D^\alpha \left(\sum_{k=0}^\infty \frac{A^k (t)^{(1+k)\alpha-1}}{\Gamma[(k+1)\alpha]} M \right) = \sum_{k=0}^\infty A^k D^\alpha \frac{(t)^{(1+k)\alpha-1}}{\Gamma[(k+1)\alpha]} M$$

$$= \sum_{k=0}^\infty A^k \frac{(t)^{k\alpha-1}\Gamma((1+k)\alpha)}{\Gamma[(k+1)\alpha]\Gamma(k\alpha)} M = A. \sum_{j=0}^\infty \frac{A^j (t)^{(1+j)\alpha-1}}{\Gamma[(j+1)\alpha]} M \qquad (6.15)$$

where M is a constant matrix. Also one can show that the particular solution is as follows [28]:

$$x_p = \int_0^t \sum_{k=0}^\infty \frac{A^k (t-\zeta)^{(1+k)\alpha-1}}{\Gamma[(k+1)\alpha]} f(\zeta) d\zeta$$

$$= \sum_{k=0}^\infty \frac{A^k}{\Gamma[(k+1)\alpha]} \int_0^t (t-\zeta)^{(k+1)\alpha-1} f(\zeta) d\zeta \qquad (6.16)$$

Therefore, the general solution is:

$$x(t) = x_h(t) + x_p(t) = \sum_{k=0}^\infty \frac{A^k (t)^{(1+k)\alpha-1}}{\Gamma[(k+1)\alpha]} M$$

$$+ \sum_{k=0}^\infty \frac{A^k}{\Gamma[(k+1)\alpha]} \int_0^t (t-\zeta)^{(k+1)\alpha-1} f(\zeta) d\zeta \qquad (6.17)$$

Now we return to a control system problem. Consider a fractional control system with zero initial conditions and companion matrix which is not a restrictive assumption:

$$D^\alpha x(t) = Ax(t) + bu(t)$$

$$y = Cx \qquad (6.18)$$

According to the previously developed theory, we can write:

$$y = Cx = C. \sum_{k=0}^\infty \frac{A^k}{\Gamma[(k+1)\alpha]} \int_0^t (t-\zeta)^{(k+1)\alpha-1} bu(\zeta) d\zeta$$

$$= \sum_{k=0}^\infty \frac{CA^k}{\Gamma[(k+1)\alpha]} \int_0^t (t-\zeta)^{(k+1)\alpha-1} bu(\zeta) d\zeta \qquad (6.19)$$

Now we present the main result of the chapter:

Theorem 6.1. *Consider the system (6.18) which is controllable and observable. Defining a linear operator as follows:*

$$\Theta(u)(t) = \int_0^t C \sum_{k=0}^{\infty} \frac{A^k (t-\zeta)^{(1+k)\alpha-1}}{\Gamma[(k+1)\alpha]} bu(\zeta) d\zeta \tag{6.20}$$

Therefore, for every $m(t) \in L^2[0,T]$ there exists a $u \in L^2[0,T]$ such that:

$$\inf_{u \in L^2[0,T]} \int_0^T (m(t) - y(t))^2 dt = 0 \tag{6.21}$$

Proof. It is enough to establish that the range space of the $\Theta(u)$ is dense in $L^2[0,T]$. Suppose that $\Theta(u)$ is not dense in $L^2[0,T]$. Thus there exists a nonzero function $v \in L^2[0,T]$ such that $\langle \Theta(u), v \rangle = 0$ for all $u \in L^2[0,T]$. Thus:

$$\langle \Theta(u), u \rangle = \int_0^T v(t) \int_0^T C(t-\zeta)^{\alpha-1} \sum_{k=0}^{\infty} \frac{A^k (t-\zeta)^{k\alpha}}{\Gamma[(k+1)\alpha]} bu(\zeta) d\zeta dt$$

$$= \sum_{k=0}^{\infty} \frac{CA^k b}{\Gamma[(k+1)\alpha]} \int_0^T v(t) \int_0^T (t-\zeta)^{(k+1)\alpha-1} u(\zeta) d\zeta dt$$

$$= \sum_{k=0}^{\infty} \frac{CA^k b}{\Gamma[(k+1)\alpha]} \int_0^T \int_\zeta^T (t-\zeta)^{(k+1)\alpha-1} u(\zeta) d\zeta v(t) dt\, u(\zeta)$$

$$= \langle u, \Theta^*(v) \rangle \tag{6.22}$$

where

$$\Theta^*(v)(\zeta) = \sum_{k=0}^{\infty} \frac{CA^k b}{\Gamma[(k+1)\alpha]} \int_\zeta^T (t-\zeta)^{(k+1)\alpha-1} v(t) dt$$

$$= \int_\zeta^T \sum_{k=0}^{\infty} \frac{CA^k b}{\Gamma[(k+1)\alpha]} (t-\zeta)^{(k+1)\alpha-1} v(t) dt \tag{6.23}$$

Therefore, it is concluded that $\Theta^*(v)(\zeta) = 0$. Now we use the fact that $\Theta(u)$ is onto if $\Theta^*(v)$ is one-to-one [29]. It suffices to show that $v = 0$ almost everywhere. Let

$$\wp(\zeta) = \int_\zeta^T \sum_{k=0}^{\infty} \frac{A^k b}{\Gamma[(k+1)\alpha]} (t-\zeta)^{(k+1)\alpha-1} v(t) dt \tag{6.24}$$

Using the following important facts [30]

1. $D_\zeta^\alpha \zeta^l = \dfrac{\Gamma(l+1)}{\Gamma(l+1-\alpha)} \zeta^{l-\alpha}$

2. $D_\zeta^\alpha \displaystyle\int_0^\zeta K(\zeta,\tau)\mathrm{d}\tau = \int_0^\zeta D_\zeta^\alpha K(\zeta,\tau)\mathrm{d}\tau + \lim_{\tau\to\zeta-0} D_\zeta^{\alpha-1} K(\zeta,\tau)$ (6.25)

We deduce that

$$D_\zeta^\alpha \wp(\zeta) = D_\zeta^\alpha \int_\zeta^T \sum_{k=0}^\infty \frac{A^k b}{\Gamma[(k+1)\alpha]} (t-\zeta)^{(k+1)\alpha-1} v(t)\mathrm{d}t$$

$$= \left\{ \int_\zeta^T \sum_{k=0}^\infty \frac{A^k b}{\Gamma[(k+1)\alpha]} D_\zeta^\alpha (t-\zeta)^{(k+1)\alpha-1} v(t)\mathrm{d}t + bv(\zeta) \right\}$$

$$= -A \int_\zeta^T \sum_{j=0}^\infty \frac{A^j b}{\Gamma[(k+1)\alpha]} (t-\zeta)^{(j+1)\alpha-1} v(t)\mathrm{d}t - bv(\zeta)$$

$$= -A\wp(\zeta) - bv(\zeta)$$ (6.26)

or briefly

$$D_\zeta^\alpha \wp(\zeta) = -A\wp(\zeta) - bv(\zeta), \quad \wp(T) = 0$$ (6.27)

Hence, from $\Theta^*(v)(\zeta) = 0$ and using the controllability condition of the original system (A,b), it is obtained that for an initial value $\wp_0$ we have

$$\sum_{k=0}^\infty \frac{(-A)^k \zeta^{(k+1)\alpha-1}}{\Gamma((k+1)\alpha)} \wp_0 - \sum_{k=0}^\infty \frac{(-A)^k}{\Gamma((k+1)\alpha)} \int_0^\zeta (\zeta-\tau)^{(k+1)\alpha-1} bv(\tau)\mathrm{d}\tau = 0$$

$$\Rightarrow C\wp(\zeta) = 0$$ (6.28)

Now introducing the vector valued function $\wp(\zeta)$ as follows, we can manipulate the relations easily:

$$C = [c_1 \quad \cdots \quad c_n], \quad \wp(\zeta) = [\wp_1(\zeta) \quad \wp_2(\zeta) \quad \cdots \quad \wp_n(\zeta)]$$ (6.29)

Using these notations, we have

$$\sum_{i=0}^n c_i \wp_i(\zeta) = 0$$ (6.30)

since (A,b) and (A,C) are controllable and observable, respectively. Thus, we can write:

$$\sum_{i=0}^{n} c_i D_\zeta^\alpha \wp_i(\zeta) = 0 \tag{6.31}$$

Introducing the following state variables:

$$D_\zeta^\alpha \wp_1 = -\wp_{i+1}, \quad i = 1,2,\ldots,n-1 \tag{6.32}$$

we have:

$$D_\zeta^\alpha \begin{bmatrix} \wp_1 \\ \wp_2 \\ \vdots \\ \wp_n \end{bmatrix} = \begin{bmatrix} 0 & -1 & 0 & \cdots & 0 \\ 0 & 0 & -1 & 0 & \vdots \\ \vdots & \ddots & \ddots & -1 & 0 \\ 0 & 0 & \cdots & \cdots & -1 \\ \frac{c_1}{c_n} & \frac{c_1}{c_n} & \cdots & \cdots & \frac{c_1}{c_n} \end{bmatrix} \begin{bmatrix} \wp_1 \\ \wp_2 \\ \vdots \\ \wp_n \end{bmatrix} = Q \begin{bmatrix} \wp_1 \\ \wp_2 \\ \vdots \\ \wp_n \end{bmatrix} \tag{6.33}$$

Using (6.17), we have:

$$\wp(\zeta) = \sum_{k=0}^{\infty} \frac{(Q)^k \zeta^{(k+1)\alpha-1}}{\Gamma((k+1)\alpha)} \wp(0) \tag{6.34}$$

Since $\wp(T) = 0$ we conclude that $\wp(0) = 0$ and consequently $\wp(\zeta) = 0$. This establishment is based on the assumption that $c_n = 0$, but this approach can be easily extended to more general case. Therefore, we have:

$$\sum_{k=0}^{\infty} \frac{(-A)^k \zeta^{(k+1)\alpha-1}}{\Gamma((k+1)\alpha)} \wp_0 - \sum_{k=0}^{\infty} \frac{(-A)^k}{\Gamma((k+1)\alpha)} \int_0^\zeta (\zeta-\tau)^{(k+1)\alpha-1} bv(\tau) d\tau = 0 \tag{6.35}$$

and hence that

$$\wp_0 - \sum_{k=0}^{\infty} \frac{(A)^k}{\Gamma((k+1)\alpha)} \int_0^\zeta (\tau)^{(k+1)\alpha-1} bv(\tau) d\tau = 0 \tag{6.36}$$

Differintegrating of (6.36) once we have:

$$D_\zeta^\alpha \left(\wp_0 - \sum_{k=0}^{\infty} \frac{(A)^k}{\Gamma((k+1)\alpha)} \int_0^\zeta (\tau)^{(k+1)\alpha-1} bv(\tau) d\tau \right)$$

$$= D_\zeta^\alpha \wp_0 - \sum_{k=0}^{\infty} \frac{(A)^k}{\Gamma((k+1)\alpha)} D_\zeta^\alpha \int_0^\zeta (\tau)^{(k+1)\alpha-1} bv(\tau) d\tau$$

$$= 0 - \lim_{\tau \to \zeta - 0} D_\zeta^\alpha (\tau)^{(k+1)\alpha - 1} bv(\tau) = 0$$

$$\Rightarrow v(\zeta) = 0 \; a.e. \tag{6.37}$$

Hence, this shows that Θ^* is one-to-one and hence that Θ is onto a dense subspace of $L^2[0, T]$.   □

This idea indicates an important point in control theory. Indeed given a function in L^2 space, one can propose a fractional order transfer function which can approximate that function in term of L^2 norm. One of the main advantages of this idea is that we can analyze a very complex function in L^2 space by transferring it to a fractional order transfer function space which may exhibit a more general transfer function than the usual ones. By transferring it to the second space, we can develop an analysis using well-known existing methods such as frequency methods, root-locus methods, and state space techniques, which are well established and easy to use.

5 Conclusions

In this chapter, considering a fractional differential equation system with constant matrices and using some facts from functional analysis, we proved that any arbitrary function belonging to L^2 can be approximated by an output of a linear time-invariant fractional differential control system. More precisely, we gave a proof that every L^2 function can be represented as the L^2 limit of functions that are the outputs of fractional linear control systems. Our analysis was developed in the Hilbert space time-invariant fractional differential control system.

References

1. Cafagna D (2007) Fractional calculus: a mathematical tool from the past for the present engineer. IEEE Trans Indus Electron:35–40
2. Gao X (2006) Chaos in fractional order autonomous system and its control. IEEE:2577–2581
3. Rossikhin YA, Shitikova MV (2006) Analysis of damped vibrations of linear viscoelastic plates with damping modeled with fractional derivatives. Signal Process 86(10):2703–2711
4. Langlands TAM (2006) Solution of a modified fractional diffusion equation. Phys A Stat Mech Appl 367:136–144
5. Tenreiro Machado JA, Jesus IS, Galhano A, Cunha JB (2006) Fractional order electromagnetics. Signal Process 86(10):2637–2644
6. Laskin N (2000) Fractional market dynamics. Phys A 287:482–492
7. Podlubny I (1999) Fractional order systems and $PI^\lambda D^\mu$ -controllers. IEEE Trans Autom Control 44(1):208–214
8. Baranyi P, Yam Y (2006) Case study of the TP-model transformation in the control of a complex dynamic model with structural nonlinearity. IEEE Trans Ind Electron 53:895–904

9. Baleanu D (2006) Fractional Hamiltonian analysis of irregular systems. Signal Process 86(10):2632–2636
10. Razminia A, Majd VJ, Baleanu D (2011) Chaotic incommensurate fractional order Rossler system: active control and synchronization. Advances in Difference Equation 15:1–12
11. Sun S, Egerstedt M, Martin C (2000) Control theoretic smoothing splines. IEEE Trans Automat Control 45(12);2271–2279
12. Brockett RW, Mesarovic MD (1965) The reproducibility of multivariable systems. J Math Anal Appl 11:548–563
13. Panda R, Dash M (2006) Fractional generalized splines and signal processing. Signal Process 86(9):2340–2350
14. Zhang Z, Tomlinson J, Enqvist P, Martin C (1995) Linear control theory, splines and approximation. In: Lund J (eds) Computation and control III. Birkhäuser, Boston, pp 269–288
15. Schoenberg I (1946) Contributions to the problem of approximation of equidistant data by analytic functions. Quart Appl Math 4:45–49
16. Wahba G (1990) Spline models for observational data. CBMS–NSF Regional Conf. Ser. in Appl. Math., SIAM, Philadelphia
17. Jonsson U, Martin C (2007) Approximation with the output of linear control systems. J Math Anal Appl 329:798–821
18. Shapiro JH (2008) On the outputs of linear control systems. J Math Anal Appl 340:116–125
19. Egerstedt M, Martin C (2003) Statistical estimates for generalized splines. ESAIM Control Optim Calc Var 9:553–562 (electronic)
20. Zhou Y, Egerstedt M, Martin C (2001) Optimal approximation of functions. Commun Inf Syst 1(1):101–112
21. Kilbas AA, Srivastava HM, Trujillo JJ (2006) Theory and applications of fractional differential equations. Elsevier, Amsterdam
22. Loverro A (2004) Fractional calculus: history, definitions and applications for the engineer. University of Notre Dame, Notre Dame
23. Podlubny I (2005) Fractional derivatives: history, theory, application. Utah State University, Utah
24. Miller KS, Ross B (1993) An introduction to the fractional calculus and fractional differential equations. Wiley, New York
25. Adams RA, Fournier JF (2003) Sobolev spaces, 2nd edn. Academic, New York
26. Matignon D, Novel BD Some results on controllability and observability of finite dimensional fractional differential equations
27. Oldham KB, Spanier J (1974) "The fractional calculus", mathematics in science and engineering. Academic, New York
28. Bonilla B, Rivero M, Trujillo JJ (2007) On systems of linear fractional differential equations with constant coefficients. Appl Math Comput 187:68–78
29. Folland GB (1999) Real analysis, modern techniques and their applications. Wiley, New York
30. Podlubny I (1999) "Fractional differential equations", mathematics in science and engineering V198. Acadamic, New York

Chapter 7
Stabilization of Fractional Order Unified Chaotic Systems via Linear State Feedback Controller

E.G. Razmjou, A. Ranjbar, Z. Rahmani, R. Ghaderi, and S. Momani

1 Introduction

The real world sometime possesses a fractional order dynamic [17]. Accordingly, fractional order controllers such as CRONE [13], TID [7], fractional PID controller [14], and lead-lag compensator [16] have been implemented to improve the performance and robustness of some closed loop control systems. An application of fractional algebra is the modeling of the fractional order chaotic systems. This kind of modeling provides more accuracy, less complexity as well as the possibility to increase the stability region [17].

Chaos, as an application of the fractional order modeling, is a very interesting nonlinear phenomenon. High sensitivity to initial conditions is a main characteristic of chaotic systems. Therefore, these systems are found to be difficult for synchronization or control [5]. Due to the complexity of these systems, control and stabilization task of chaotic nonlinear systems have been one of the arising interests in the control engineering area. In the past decade, great efforts have been devoted toward the chaos control, including stabilization of unstable equilibrium points, and more generally, unstable periodic solutions. Particularly, in case of chaos suppression of known chaotic systems, some useful methods have been developed. These include time delay feedback control [15], bang–bang control [19], optimal control [8], intelligent control [22], adaptive control [23], etc.

A unified chaotic system is a chaotic system that depends on a parameter, e.g., $\alpha \in [0,1]$. If $0 \leq \alpha < 0.8$, the unified chaotic system is reduced to the

E.G. Razmjou • A. Ranjbar (✉) • Z. Rahmani • R. Ghaderi
Intelligent System Research Group, Babol University of Technology,
P.O. Box 47135-484, Babol, Iran
e-mail: ehsan.razmju@gmail.com; a.ranjbar@nit.ac.ir; rahmaniz@gmail.com; r_ghaderi@nit.ac.ir

S. Momani
Department of Mathematics, The University of Jordan, Amman 11942, Jordan, Jordan

D. Baleanu et al. (eds.), *Fractional Dynamics and Control*,
DOI 10.1007/978-1-4614-0457-6_7, © Springer Science+Business Media, LLC 2012

generalized Lorenz chaotic system; the unified chaotic system is altered to the Lü chaotic system when $\alpha = 0.8$. For $0.8 < \alpha \leq 1$, the unified chaotic system is changed to the generalized Chen chaotic system.

Chen [3] considered that the parameter of the two unified chaotic systems is unknown. Hence, an adaptive controller was used to achieve synchronization based on Lyapunov stability theory. Chen [4] investigated the stabilization and synchronization of the unified chaotic system via an impulsive control method. Lu [10] used linear feedback and adaptive control to synchronize identical unified chaotic systems with only one controller. Ucar [18] used a nonlinear active controller to synchronize two coupled unified chaotic systems with three control inputs. Wang [20] proved that the unified chaotic system is equivalent to a passive system and asymptotically stabilized it at equilibrium points. Wang [21] studied the synchronization problem of two identical unified chaotic systems using three different methods. They used a linear feedback controller, a nonlinear feedback method, and an impulsive controller to synchronize the systems. In [24] based on the sliding mode theory, synchronization of two identical unified chaotic is discussed.

However, in this chapter, a linear state feedback controller stabilizes a fractional order unified chaotic system. An advantage of the proposed controller can be seen when it is used to stabilize a fractional order unified chaotic system, by means of increasing the stability region. In contrast, the application on the integer order system is shown to be failed.

The chapter is organized as follows: Sect. 2 includes the basic definition and preliminaries. A state feedback controller is proposed to stabilize the fractional order unified chaotic systems in Sect. 3. Results of numerical simulation are given in Sect. 4, to illustrate the effectiveness of the proposed controller. The chapter will be closed by a conclusion in Sect. 5.

2 Preliminary Definitions

2.1 Fractional Algebra

Among several definitions of fractional derivatives, the following Caputo-type definition [1] is more popular with respect the rest [17].

$$
{}_0D_t^q f(t) = \begin{cases} \frac{1}{\Gamma(m-q)} \int_0^t \frac{f^m(\tau)}{(t-\tau)^{q+1-m}} d\tau, & m-1 < q \leq m \\ \frac{d^m}{dt^m} f(t) & q = m \end{cases}
\tag{7.1}
$$

where m is the first integer number larger than q.

Definition 1. [6] A saddle point of index 2 is a saddle point with one stable eigenvalue and two unstable ones.

Definition 2. [2] Assume that a 3D fractional order chaotic system of $\dot{x} = f(x)$ displays a chaotic attractor. For every scroll existing in the chaotic attractor, this system has a saddle point of index 2 encircled by its respective scroll.

Theorem 1. *[11] Assume that a 3D chaotic system $\dot{x} = f(x)$ displays a chaotic attractor with n scrolls. Suppose Λ is a set of unstable eigenvalues of these n saddle points. A necessary condition for fractional system $D^q x = f(x)$ to exhibit an n-scroll chaotic attractor, similar to the chaotic attractor of system $\dot{x} = f(x)$, to keep the eigenvalues $\lambda \in \Lambda$ in the unstable region, satisfies:*

$$q > \frac{2}{\pi} \tan^{-1} \left(\frac{|\text{Im}(\lambda)|}{\text{Re}(\lambda)} \right), \qquad \forall \lambda \in \Lambda \tag{7.2}$$

Otherwise, at least one of these equilibriums becomes asymptotically stable and then attracts the nearby trajectories.

2.2 The Unified Chaotic System

[9] considered a kind of chaotic system which describes a class of unified form by:

$$\begin{cases} \frac{dx}{dt} = (25\alpha + 10)(y - x) \\ \frac{dy}{dt} = (28 - 35\alpha)x - xz + (29\alpha - 1)y \\ \frac{dz}{dt} = xy - \frac{8+\alpha}{3}z \end{cases} \tag{7.3}$$

where x, y, z are the state variables and $\alpha \in [0, 1]$ is a "homogeneity" parameter of the system. [9] calls (7.3) as unified chaotic system due to chaotic behavior for any $\alpha \in [0, 1]$. When $0 \leq \alpha < 0.8$, system (7.3) is called as the generalized Lorenz chaotic system. For $\alpha = 0.8$, it is called Lü chaotic system. Similarly, it is called generalized Chen chaotic system when $0.8 < \alpha \leq 1$. However, let us introduce a fractional version of dynamic (7.3) as in (7.4). Standard derivatives of (7.3) are accordingly replaced by the following fractional derivatives:

$$\begin{cases} \frac{d^q x}{dt^q} = (25\alpha + 10)(y - x) \\ \frac{d^q y}{dt^q} = (28 - 35\alpha)x - xz + (29\alpha - 1)y \\ \frac{d^q z}{dt^q} = xy - \frac{8+\alpha}{3}z \end{cases} \tag{7.4}$$

where q with $0 < q \leq 1$ is the fractional order. Chaos in the fractional order unified system of Chen, Lü, and Lorenz-Like for $q = 0.9, 0.95, 0.99$ are shown in [12].

From (7.4), a generalized scheme of the fractional order unified chaotic system can be given as follows:

$$\begin{cases} \frac{d^q x}{dt^q} = a(y - x) \\ \frac{d^q y}{dt^q} = bx - xz + cy \\ \frac{d^q z}{dt^q} = xy - dz \end{cases} \tag{7.5}$$

3 State Feedback Control

3.1 Design of the Controller for Fractional Order Chen System

The fractional order Chen system is given as follows [12]:

$$\begin{cases} \frac{d^q x}{dt^q} = a_1(y - x) \\ \frac{d^q y}{dt^q} = (c_1 - a_1)x - xz + c_1 y \\ \frac{d^q z}{dt^q} = xy - b_1 z \end{cases} \tag{7.6}$$

To obtain the Chen chaotic behavior, parameters in (7.6) are set to [12]:

$$a_1 = 40, b_1 = 3, c_1 = 28 \tag{7.7}$$

The equilibrium points of the Chen system are as follows:

$$O_1 = (0,0,0)$$
$$O_2 = (6.9282, 6.9282, 16)$$
$$O_3 = (-6.9282, -6.9282, 16) \tag{7.8}$$

From (7.6) the Jacobian matrix of the Chen system is achieved by:

$$J = \begin{bmatrix} -a_1 & a_1 & 0 \\ c_1 - a_1 - z & c_1 & -x \\ y & x & -b_1 \end{bmatrix} \tag{7.9}$$

Accordingly, the corresponding eigenvalues of the equilibrium (7.8) are obtained as:

$$O_1 \rightarrow \lambda_1 = -3, \lambda_2 = 20, \lambda_3 - 32$$
$$O_{2,3} \rightarrow \lambda_1 = -20.2304, \lambda_{2,3} = 2.6152 \pm 13.5268 j \tag{7.10}$$

From definition 1, $O_{2,3}$ are of saddle point of index 2. Therefore, from theorem 1 the fractional order Chen system becomes chaotic when:

$$q > \frac{2}{\pi} \tan^{-1} \left(\frac{|\text{Im}(\lambda_{2,3})|}{\text{Re}(\lambda_{2,3})} \right) = 0.8784 \qquad (7.11)$$

Otherwise the system is asymptotically stable. In order to stabilize the fractional order Chen system, a control input is added into the second state of the system, by the following:

$$\begin{cases} \frac{d^q x}{dt^q} = a_1(y - x) \\ \frac{d^q y}{dt^q} = (c_1 - a_1)x - xz + c_1 y + u \\ \frac{d^q z}{dt^q} = xy - b_1 z \end{cases} \qquad (7.12)$$

A linear state feedback controller is proposed to construct the input signal u as in the following form:

$$u = -(c_1 - a_1)x - k_1 y \qquad (7.13)$$

Where k_1 is a constant gain by $k_1 = 12.7$.

Theorem 2. *The proposed state feedback controller in (7.13) increases the stability region of the fractional order Chen system and stabilizes the system at their stable equilibrium points.*

Proof. Using the state feedback controller changes the equilibrium points and the Jacobian matrix J to:

$$O_1' = (0, 0, 0)$$
$$O_2' = (6.7749, 6.7749, 15.3)$$
$$O_3' = (-6.7749, -6.7749, 15.3) \qquad (7.14)$$

$$J = \begin{bmatrix} -a_1 & a_1 & 0 \\ -z & c_1 - k_1 & -x \\ y & x & -b_1 \end{bmatrix} \qquad (7.15)$$

The corresponding eigenvalues of the equilibrium points in (7.14) are:

$$O_1' \rightarrow \lambda_1 = -3, \lambda_2 = 15.3, \lambda_3 = -40$$
$$O_{2,3}' \rightarrow \lambda_1 = -28.0829, \lambda_{2,3} = 0.1915 \pm 11.4331 j \qquad (7.16)$$

Similarly, from the definition 1, $O_{2,3}'$ are of the saddle point of index 2. Hence, the fractional order Chen system becomes chaotic when:

$$q > \frac{2}{\pi} \tan^{-1} \left(\frac{|\text{Im}(\lambda_{2,3})|}{\text{Re}(\lambda_{2,3})} \right) = 0.9893 \qquad (7.17)$$

Otherwise the system is asymptotically stable. This means that for $q < 0.9893$ the fractional order Chen system is asymptotically stable. ∎

3.2   Design of the Controller for the Fractional Order Lü System

The fractional order Lü system is also given by [12]:

$$
\begin{cases}
\frac{d^q x}{dt^q} = a_1(y - x) \\
\frac{d^q y}{dt^q} = -xz + c_1 y \\
\frac{d^q z}{dt^q} = xy - b_1 z
\end{cases}
\tag{7.18}
$$

The chaos in the Lü dynamic occurs when parameters in (7.18) are set to [12]:

$$
a_1 = 35,\, b_1 = 3,\, c_1 = 30
\tag{7.19}
$$

From (7.18) and (7.19), the equilibrium points and the Jacobian matrix of the Lü system are, respectively, as follows:

$$
O_1 = (0,0,0)
$$
$$
O_2 = (9.4868, 9.4868, 30)
$$
$$
O_3 = (-9.4868, -9.4868, 30)
\tag{7.20}
$$

$$
J = \begin{bmatrix} -a_1 & a_1 & 0 \\ -z & c_1 & -x \\ y & x & -b_1 \end{bmatrix}
\tag{7.21}
$$

Then the corresponding eigenvalues of the equilibrium points in (7.20) are:

$$
O_1 \rightarrow \lambda_1 = -3,\, \lambda_2 = 30,\, \lambda_3 = -35
$$
$$
O_{2,3} \rightarrow \lambda_1 = -19.3701,\, \lambda_{2,3} = 5.6851 \pm 17.1149j
\tag{7.22}
$$

From definition 1, $O_{2,3}$ are of the saddle point of index 2. Thus, the fractional order Lü system becomes chaotic when:

$$
q > \frac{2}{\pi} \tan^{-1}\left(\frac{|\mathrm{Im}(\lambda_{2,3})|}{\mathrm{Re}(\lambda_{2,3})} \right) = 0.7958
\tag{7.23}
$$

Otherwise the system is asymptotically stable.

Similar to the previous section, to stabilize the fractional order Lü system, a controller is applied to the 2ns state, according to:

$$
\begin{cases}
\frac{d^q x}{dt^q} = a_1(y - x) \\
\frac{d^q y}{dt^q} = -xz + c_1 y + u \\
\frac{d^q z}{dt^q} = xy - b_1 z
\end{cases}
\tag{7.24}
$$

The linear state feedback controller u in the following form stabilizes the chaotic dynamic:

$$u = -k_1 y \tag{7.25}$$

where k_1 as a constant gain is set to $k_1 = 16.5$.

Theorem 3. *The proposed state feedback controller in (7.25) stabilizes the system at their stable equilibrium points while increasing the stability region of the fractional order Lü system.*

Proof. The state feedback controller in the fractional order Lü system similar to (7.24) the equilibrium points and Jacobian matrix are, respectively, achieved by:

$$O_1' = (0,0,0)$$
$$O_2' = (6.3639, 6.3639, 13.5)$$
$$O_3' = (-6.3639, -6.3639, 13.5) \tag{7.26}$$

$$J = \begin{bmatrix} -a_1 & a_1 & 0 \\ -z & c_1 - k_1 & -x \\ y & x & -b_1 \end{bmatrix} \tag{7.27}$$

Thus, the corresponding eigenvalues of the equilibrium points in (7.26) are:

$$O_1' \rightarrow \lambda_1 = -3, \lambda_2 = 13.5, \lambda_3 = -35$$
$$O_{2,3}' \rightarrow \lambda_1 = -24.863, \lambda_{2,3} = 0.1815 \pm 10.6767 j \tag{7.28}$$

Again from the definition 1, $O_{2,3}'$ are the saddle points of index 2. Therefore, the fractional order Lü system becomes chaotic when:

$$q > \frac{2}{\pi} \tan^{-1} \left(\frac{|\text{Im}(\lambda_{2,3})|}{\text{Re}(\lambda_{2,3})} \right) = 0.9892 \tag{7.29}$$

Otherwise the system is asymptotically stable. This means, for $q < 0.9892$ the fractional order Lü system is asymptotically stable. ∎

4 Simulation

A simulation approach has been carried out using SIMULINK™. Dormand–Prince solver is used to solve the system of differential equations during the simulation. Results of the unified chaotic Chen and Lü dynamics are shown for $q = 0.96, q = 0.98, q = 1$. Initial conditions of the states are selected as $(10, 15, 25)$. Simulation results show that the proposed state feedback controller stabilizes the fractional order unified chaotic systems while the behavior of the equivalent integer one still

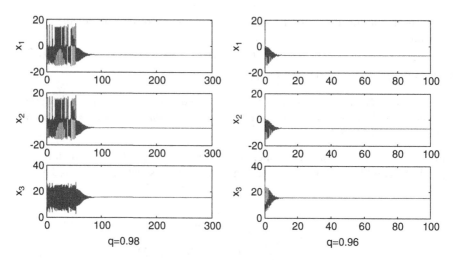

q=0.98   q=0.96

Fig. 7.1 Stabilization of the fractional order Chen system at their stable equilibrium points (O_2' and O_3'), via linear state feedback controller for $q = 0.96$ and $q = 0.98$

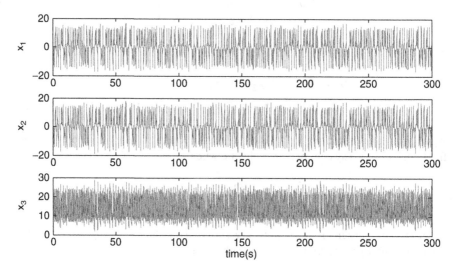

Fig. 7.2 Chaos behaviour in the integer order Chen system despite of using the state feedback controller

kept chaotic. Figure 7.1 shows that the fractional order Chen system is stabilized for $q = 0.96$ and $q = 0.98$ with state feedback controller in (7.13). Figure 7.2 shows the chaotic behavior of an integer order Chen system, despite of using the same state feedback controller in the system. Similar result is achieved in Fig. 7.3 when the fractional order Lü system is stabilized by the controller for $q = 0.96$ and $q = 0.98$. In the same way, Fig. 7.4 shows the chaotic behavior of integer order of the Lü system using the same state feedback controller.

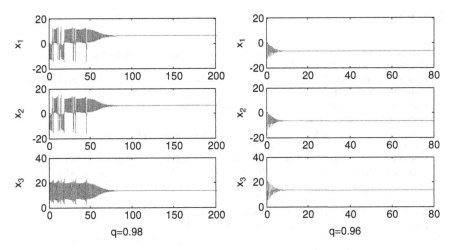

Fig. 7.3 Stabilization of the fractional order Lü system at their stable equilibrium points (O_2' and O_3'), via linear state feedback controller for $q = 0.96$ and $q = 0.98$.

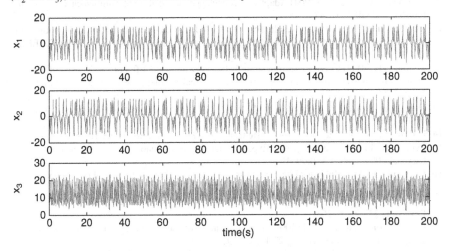

Fig. 7.4 Chaos behaviour in the integer order Lü system despite of using the state feedback controller

5   Conclusion

Three chaotic Lorenz, Chen, and Lü systems as unified systems will be separately shown unified by a same dynamic. These systems will separately be excited when a relevant parameter α is accordingly adjusted. A linear state feedback controller is gained to stabilize the unified chaotic systems at their stable equilibrium points. The controller also increases the stability region with respect to their integer order counterpart. Simulation approach is given to verify the outcome. The approach signifies the performance as well as the reliability of the proposed state feedback controller.

References

1. Caputo M (1967) Linear models of dissipation whose Q is almost frequency independent. Geophys J R Astron Soc 13:529–530
2. Cafagna D, Grassi G (2003) New 3-D-scroll attractors in hyperchaotic Chua's circuit forming a ring. Int J Bifurc Chaos 13:2889–2903
3. Chen SH, Lu JH (2002) Synchronization of an uncertain unified chaotic system via adaptive control. Chaos Solitons Fractals 14:643–647
4. Chen S, Yang Q, Wang C. (2004) Impulsive control and synchronization of unified chaotic systems. Chaos Solitons Fractals 20:751–758
5. Hosseinnia SH, Ghaderi R, Ranjbar A, Abdous F, Momani S (2010) Control of chaos via fractional-order state feedback controller. In: Baleanu D, Guvenc ZB, Machado JAT (eds) New trends in nanotechnology and fractional calculus applications. ISBN: 978-90-481-3292-8, DOI 10.1007/978-90-481-3293-5 46, pp 507–514
6. Khalil H (1992) Nonlinear systems. Macmillan, New York
7. Lurie BJ (1994) Tunable TID controller. US patent 5, 371, 670, December 6
8. Luce R, Kernevez JP (1991) Controllability of Lorenz equation. Int Ser Numer Math 97:257
9. Lü JH, Chen GR, Cheng DZ, Celikovsky S (2002) Bridge the gap between the Lorenz system and the Chen system. Int J Bifurcat Chaos 12:2917–2926
10. Lu J, Wu X, Han X, Lu J (2004) Adaptive feedback synchronization of a unified chaotic system. Phys Lett A 329:327–333
11. Matignon D (1996) Stability results for fractional differential equations with applications to control processing, in Computational Engineering in Systems Applications. IMACS, IEEE-SMC 2:963–968
12. Matouk AE (2009) Chaos synchronization between two different fractional systems of Lorenz family. Hindawi Publishing Corporation Mathematical Problems in Engineering, Article ID 572724, 11 p
13. Oustaloup A, Moreau X, Nouillant M (1996) The CRONE suspension. Control Eng Pract 4:1101–1108
14. Podlubny I (1999) Fractional-order systems and $PI^\lambda D^\mu$ controllers. IEEE Trans 44:208–214
15. Pyragas K, Tamasevieius A (1993) Experimental control of chaos by delayed self-controlling feedback. Phys Lett A 180:99–102
16. Raynaud HF, Zerga Inoh A (2000), State-space representation for fractional order controllers. Automatica 36:1017–1021
17. Tavazoei MS, Haeri M (2008) Chaotic attractors in incommensurate fractional order systems. Physica D 237:2628–2637
18. Ucar A, Lonngren K, Bai E (2006) Synchronization of the unified chaotic systems via active control. Chaos Solitons Fractals 27:1292–1297
19. Vincent TL, Yu J (1991) Control of a chaotic system. Dyn Contr 1:35–52
20. Wang F, Liu C (2007) Synchronization of unified chaotic system based on passive control. Phys D 225:55–60
21. Wang X, Song J (2008) Synchronization of the unified chaotic system, Nonlinear Anal 69:3409–3416
22. Yeap TH, Ahmed NU (1994) Feedback control of chaotic systems. Dyn Contr 4:97–114
23. Zeng Y, Singh SN (1997) Adaptive control of chaos in Lorenz system. Dyn Contr 7:143–154
24. Zribi M, Smaoui N, Salim H (2009) Synchronization of the unified chaotic systems using a sliding mode controller. Chaos Solitons Fractals 42:3197–3209

Part II
Fractional Variational Principles and Fractional Differential Equations

Chapter 8
Fractional Variational Calculus for Non-differentiable Functions

Agnieszka B. Malinowska

1 Introduction

The fractional calculus (FC) is one of the most interdisciplinary fields of mathematics, with many applications in physics and engineering. The history of FC goes back more than three centuries, when in 1695 the derivative of order $\alpha = 1/2$ was described by Leibniz. Since then, many different forms of fractional operators were introduced: the Grunwald–Letnikov, Riemann–Liouville, Riesz, and Caputo fractional derivatives [23, 34, 38]) and the more recent notions see [11, 19, 24, 25]. Fractional calculus is nowadays the realm of physicists and mathematicians, who investigate the usefulness of such non-integer order derivatives and integrals in different areas of physics and mathematics (see, e.g., [10, 18, 23]). It is a successful tool for describing complex quantum field dynamical systems, dissipation, and long-range phenomena that cannot be well illustrated using ordinary differential and integral operators (see, e.g., [15, 18, 24, 36]). Applications of FC are found in classical and quantum mechanics, field theories, variational calculus, and optimal control (see, e.g., [14, 17, 20]).

The calculus of variations is an old branch of optimization theory that has many applications both in physics and geometry. Apart from a few examples known since ancient times such as Queen Dido's problem (reported in *The Aeneid* by Virgil), the problem of finding optimal curves and surfaces has been posed first by physicists such as Newton, Huygens, and Galileo. Their contemporary mathematicians, starting with the Bernoulli brothers and Leibniz, followed by Euler and Lagrange, invented the calculus of variations of a functional in order to solve those problems. Fractional calculus of variations (FCV) unifies the calculus of variations and the fractional calculus, by inserting fractional derivatives into the variational integrals.

A.B. Malinowska (✉)
Faculty of Computer Science, Białystok University of Technology, 15-351 Białystok, Poland
e-mail: abmalinowska@ua.pt

D. Baleanu et al. (eds.), *Fractional Dynamics and Control*,
DOI 10.1007/978-1-4614-0457-6_8, © Springer Science+Business Media, LLC 2012

This occurs naturally in many problems of physics or mechanics, in order to provide more accurate models of physical phenomena. The FCV started in 1996 with the work of Riewe [36]. Riewe formulated the problem of the calculus of variations with fractional derivatives and obtained the respective Euler–Lagrange equations, combining both conservative and nonconservative cases. Nowadays the FCV is a subject under strong research. Different definitions for fractional derivatives and integrals are used, depending on the purpose under study. Investigations cover problems depending on Riemann–Liouville fractional derivatives (see, e.g., [6, 15, 16, 32]), the Caputo fractional derivative (see, e.g., [1, 7, 28, 30]), the symmetric fractional derivative (see, e.g., [24]), the Jumarie fractional derivative (see, e.g., [4, 5, 19–22, 27]), and others [2, 11, 14]. For applications of the FCV we refer the readers to [15, 20, 24, 35]. Although the literature of FCV is already vast, much remains to be done.

In this paper we study problems of FCV which are defined in terms of the Jumarie fractional derivatives and integrals. The Euler–Lagrange equations for such problems with and without constraints were recently shown in [4]. Here we develop further the theory by proving necessary optimality conditions for more general problems of FCV with a Lagrangian that may also depend on the unspecified endpoints $y(a)$, $y(b)$. More precisely, the problem under our study: to extremize a functional which is defined in terms of the Jumarie fractional operators and having no constraint on $y(a)$ and/or $y(b)$. The novelty is the dependence of the integrand L on the a priori unknown final values $y(a)$, $y(b)$. The new natural boundary conditions (8.5)–(8.6) have important implications in economics (see [12] and the references therein).

The paper is organized as follows. Section 2 presents the necessary definitions and concepts of Jumarie's fractional calculus. Our results are formulated, proved, and illustrated through examples in Sect. 3. Main results of the paper include necessary optimality conditions with the generalized natural boundary conditions (Theorem 8.1) that become sufficient under appropriate convexity assumptions (Theorem 8.2). We finish with Sect. 4 by providing conclusions.

2   Fractional Calculus

For an introduction to the classical fractional calculus we refer the readers to [11, 23, 34, 38]. In this section we briefly review the main notions and results from the recent fractional calculus proposed by Jumarie [19–21].

Definition 8.1. Let $f : [a, b] \to \mathbb{R}$ be a continuous function. The Jumarie fractional derivative of f is defined by

$$f^{(\alpha)}(t) := \frac{1}{\Gamma(-\alpha)} \int_0^t (t - \tau)^{-\alpha-1}(f(\tau) - f(a)) \, d\tau, \quad \alpha < 0,$$

where $\Gamma(z) = \int_0^\infty t^{z-1}e^{-t}\,dt$. For positive α, one will set

$$f^{(\alpha)}(t) = (f^{(\alpha-1)}(t))' = \frac{1}{\Gamma(1-\alpha)}\frac{d}{dt}\int_0^t (t-\tau)^{-\alpha}(f(\tau) - f(a))\,d\tau,$$

for $0 < \alpha < 1$, and

$$f^{(\alpha)}(t) := (f^{(\alpha-n)}(t))^{(n)}, \quad n \le \alpha < n+1, \quad n \ge 1.$$

The Jumarie fractional derivative has the following property:

- The αth derivative of a constant is zero.
- Assume that $0 < \alpha \le 1$, then the Laplace transform of $f^{(\alpha)}$ is

$$\mathcal{L}\{f^{(\alpha)}(t)\} = s^\alpha \mathcal{L}\{f(t)\} - s^{\alpha-1}f(0).$$

- $(g(t)f(t))^{(\alpha)} = g^{(\alpha)}(t)f(t) + g(t)f^{(\alpha)}(t), \quad 0 < \alpha < 1.$

Example 8.1. Let $f(t) = t^\gamma$. Then $f^{(\alpha)}(x) = \Gamma(\gamma+1)\Gamma^{-1}(\gamma+1-\alpha)t^{\gamma-\alpha}$, where $0 < \alpha < 1$ and $\gamma > 0$.

Example 8.2. The solution of the fractional differential equation

$$x^{(\alpha)}(t) = c, \quad x(0) = x_0, \quad c = constant,$$

is

$$x(t) = \frac{c}{\alpha!}t^\alpha + x_0,$$

with the notation $\alpha! := \Gamma(1+\alpha)$.

The integral with respect to $(dt)^\alpha$ is defined as the solution of the fractional differential equation

$$dy = f(x)(dx)^\alpha, \quad x \ge 0, \quad y(0) = y_0, \quad 0 < \alpha \le 1, \tag{8.1}$$

which is provided by the following result.

Lemma 8.1. *Let $f(t)$ denote a continuous function. The solution of the (8.1) is defined by the equality*

$$\int_0^t f(\tau)(d\tau)^\alpha = \alpha \int_0^t (t-\tau)^{\alpha-1}f(\tau)d\tau, \quad 0 < \alpha \le 1.$$

Example 8.3. Let $f(t) = 1$. Then $\int_0^t (d\tau)^\alpha = t^\alpha, 0 < \alpha \le 1$.

Example 8.4. The solution of the fractional differential equation

$$x^{(\alpha)}(t) = f(t), \quad x(0) = x_0$$

is

$$x(t) = x_0 + \Gamma^{-1}(\alpha) \int_0^t (t-\tau)^{\alpha-1} f(\tau) d\tau.$$

One can easily generalize the previous definitions and results for functions with a domain $[a,b]$:

$$f^{(\alpha)}(t) = \frac{1}{\Gamma(1-\alpha)} \frac{d}{dt} \int_a^t (t-\tau)^{-\alpha} (f(\tau) - f(a)) d\tau$$

and

$$\int_a^t f(\tau)(d\tau)^{\alpha} = \alpha \int_a^t (t-\tau)^{\alpha-1} f(\tau) d\tau.$$

For the discussion to follow, we will need the following formula of integration by parts:

$$\int_a^b u^{(\alpha)}(t) v(t) \, (dt)^{\alpha} = \alpha! [u(t)v(t)]_a^b - \int_a^b u(t) v^{(\alpha)}(t) \, (dt)^{\alpha}, \qquad (8.2)$$

where $\alpha! := \Gamma(1+\alpha)$.

3 Main Results

Let us consider the functional defined by

$$\mathcal{J}(y) = \int_a^b L(x, y(x), y^{(\alpha)}(x), y(a), y(b)) \, (dx)^{\alpha},$$

where $L(\cdot, \cdot, \cdot, \cdot, \cdot) \in C^1([a,b] \times \mathbb{R}^4; \mathbb{R})$ and $x \to \partial_3 L(t)$ has continuous α-derivative. The fractional problem of the calculus of variations under consideration has the form

$$\mathcal{J}(y) \longrightarrow \text{extr}$$
$$(y(a) = y_a), \quad (y(b) = y_b),$$
$$y(\cdot) \in C^0. \qquad (8.3)$$

Using parentheses around the end-point conditions means that the conditions may or may not be present.

Along the work we denote by $\partial_i L$, $i = 1, \ldots, 5$, the partial derivative of function $L(\cdot, \cdot, \cdot, \cdot, \cdot)$ with respect to its ith argument.

The following lemma will be needed in the next subsection.

Lemma 8.2. *Let g be a continuous function and assume that*

$$\int_a^b g(x)h(x)\,(dx)^\alpha = 0$$

for every continuous function h satisfying $h(a) = h(b) = 0$. Then $g \equiv 0$.

Proof. Can be done in a similar way as the proof of the standard fundamental lemma of the calculus of variations (see, e.g., [40]).

3.1   Necessary Conditions

Next theorem gives necessary optimality conditions for the problem (8.3).

Theorem 8.1. *Let y be an extremizer to problem (8.3). Then, y satisfies the fractional Euler–Lagrange equation*

$$\partial_2 L(x, y(x), y^{(\alpha)}(x), y(a), y(b)) = \frac{d^\alpha}{dx^\alpha} \partial_3 L(x, y(x), y^{(\alpha)}(x), y(a), y(b)), \qquad (8.4)$$

for all $x \in [a, b]$. Moreover, if $y(a)$ is not specified, then;

$$\int_a^b \partial_4 L(x, y(x), y^{(\alpha)}(x), y(a), y(b))\,(dx)^\alpha = \alpha! \partial_3 L(a, y(a), y^{(\alpha)}(a), y(a), y(b)),$$
$$(8.5)$$

if $y(b)$ is not specified, then;

$$\int_a^b \partial_5 L(x, y(x), y^{(\alpha)}(x), y(a), y(b))\,(dx)^\alpha = -\alpha! \partial_3 L(b, y(b), y^{(\alpha)}(b), y(a), y(b)).$$
$$(8.6)$$

Proof. Suppose that y is an extremizer of $\mathcal{J}$ and consider the value of $\mathcal{J}$ at a nearby function $\tilde{y} = y + \varepsilon h$, where $\varepsilon \in \mathbb{R}$ is a small parameter and h is an arbitrary continuous function. We do not require $h(a) = 0$ or $h(b) = 0$ in case $y(a)$ or $y(b)$, respectively, is free (it is possible that both are free). Let $j(\varepsilon) = \mathcal{J}(y + \varepsilon h)$. Then a necessary condition for y to be an extremizer is given by $j'(0) = 0$. Hence,

$$\int_a^b \left[\partial_2 L(\cdot)h(x) + \partial_3 L(\cdot)h^{(\alpha)}(x) + \partial_4 L(\cdot)h(a) + \partial_5 L(\cdot)h(b) \right](dx)^\alpha = 0, \qquad (8.7)$$

where $(\cdot) = \left(x, y(x), y^{(\alpha)}(x), y(a), y(b)\right)$. Using integration by parts (8.2) to the second term we get

$$\int_a^b \left[\partial_2 L(\cdot) - \frac{\mathrm{d}^\alpha}{\mathrm{d}x^\alpha} \partial_3 L(\cdot) \right] (\mathrm{d}x)^\alpha + \alpha! \partial_3 L(\cdot)|_{x=b} h(b) - \alpha! \partial_3 L(\cdot)|_{x=a} h(a)$$

$$+ \int_a^b \left[\partial_4 L(\cdot) h(a) + \partial_5 L(\cdot) h(b) \right] (\mathrm{d}x)^\alpha = 0. \tag{8.8}$$

We first consider functions h such that $h(a) = h(b) = 0$. Then, by the Lemma 8.2 we deduce that

$$\partial_2 L(\cdot) = \frac{\mathrm{d}^\alpha}{\mathrm{d}x^\alpha} \partial_3 L(\cdot),$$

for all $x \in [a,b]$. Therefore, in order for y to be an extremizer to the problem (8.3), y must be a solution of the fractional Euler–Lagrange equation (8.4). But if y is a solution of (8.4), the first integral in expression (8.8) vanishes, and then condition (8.7) takes the form

$$h(b) \left[\int_a^b \partial_5 L(\cdot)(\mathrm{d}x)^\alpha + \alpha! \partial_3 L(\cdot)|_{x=b} \right] + h(a) \left[\int_a^b \partial_4 L(\cdot)\mathrm{d}x - \alpha! \partial_3 L(\cdot)|_{x=a} \right] = 0.$$

If $y(a) = y_a$ and $y(b) = y_b$ are given in the formulation of problem (8.3), then the latter equation is trivially satisfied since $h(a) = h(b) = 0$. When $y(b)$ is free, then (8.6) holds, when $y(a)$ is free, then (8.5) holds, since $h(a)$ or $h(b)$ is, respectively, arbitrary.

In the case L does not depend on $y(a)$ and $y(b)$, by Theorem 8.1 we obtain the following result.

Corollary 8.1. *[4, Theorem 1] Let y be an extremizer to problem*

$$\mathcal{J}(y) = \int_a^b L(x, y(x), y^{(\alpha)}(x))(\mathrm{d}x)^\alpha \longrightarrow extr.$$

Then, y satisfies the fractional Euler–Lagrange equation

$$\partial_2 L(x, y(x), y^{(\alpha)}(x)) = \frac{\mathrm{d}^\alpha}{\mathrm{d}x^\alpha} \partial_3 L(x, y(x), y^{(\alpha)}(x)),$$

for all $x \in [a,b]$. Moreover, if $y(a)$ is not specified, then;

$$\partial_3 L(a, y(a), y^{(\alpha)}(a)) = 0,$$

if $y(b)$ is not specified, then;

$$\partial_3 L(b, y(b), y^{(\alpha)}(b)) = 0.$$

Observe that if α goes to 1, then the operators $\frac{d^\alpha}{dx^\alpha}$, $(dx)^\alpha$ could be replaced with $\frac{d}{dx}$ and dx. Thus, in this case we obtain the corresponding result in the classical context of the calculus of variations (see [29, Corollary 1], [12, Theorem 2.1]).

Corollary 8.2. *[29, Corollary 1] If y is a local extremizer for*

$$\mathcal{J}(y) = \int_a^b L(x,y(x),y'(x),y(a),y(b))\,dx \longrightarrow extr$$

$$(y(a) = y_a), \quad (y(b) = y_b),$$

then

$$\frac{d}{dx}\partial_3 L(x,y(x),y'(x),y(a),y(b)) = \partial_2 L(x,y(x),y'(x),y(a),y(b)),$$

for all $x \in [a,b]$. Moreover, if $y(a)$ is free, then;

$$\partial_3 L(a,y(a),y'(a),y(a),y(b)) = \int_a^b \partial_5 L(x,y(x),y'(x),y(a),y(b))dx;$$

and if $y(b)$ is free, then,

$$\partial_3 L(b,y(b),y'(b),y(a),y(b)) = -\int_a^b \partial_6 L(x,y(x),y'(x),y(a),y(b))dx.$$

3.2   Sufficient Conditions

In this section we prove sufficient conditions for optimality. Similarly to what happens in the classical calculus of variations, some conditions of convexity (concavity) are in order.

Definition 8.2. Given a function L, we say that $L(\underline{x},y,z,t,u)$ is jointly convex (concave) in (y,z,t,u), if $\partial_i L$, $i = 2,\ldots,5$, exist and are continuous and verify the following condition:

$$L(x,y+y_1,z+z_1,t+t_1,u+u_1) - L(x,y,z,t,u)$$
$$\geq (\leq)\partial_2 L(\cdot)y_1 + \partial_3 L(\cdot)z_1 + \partial_4 L(\cdot)t_1 + \partial_5 L(\cdot)u_1,$$

where $(\cdot) = (x,y,z,t,u)$, for all (x,y,z,t,u), $(x,y+y_1,z+z_1,t+t_1,u+u_1) \in [a,b] \times \mathbb{R}^4$.

Theorem 8.2. *Let $L(\underline{x},y,z,t,u)$ be a jointly convex (concave) in (y,z,t,u). If y_0 satisfies conditions (8.4)–(8.6), then y_0 is a global minimizer (maximizer) to problem (8.3).*

Proof. We shall give the proof for the convex case. Since L is jointly convex in (y,z,t,u) for any continuous function $y_0 + h$, we have

$$\mathcal{J}(y_0+h) - \mathcal{J}(y_0) = \int_a^b \Big[L(x, y_0(x) + h(x), y_0^{(\alpha)}(x) + h^{(\alpha)}(x), y_0(a) + h(a), y_0(b)$$

$$+ h(b)) - L(x, y_0(x), y_0^{(\alpha)}(x), y_0(a), y_0(b)) \Big] (dx)^\alpha$$

$$\geq \int_a^b \Big[\partial_2 L(\cdot) h(x) + \partial_3 L(\cdot) h^{(\alpha)}(x) + \partial_4 L(\cdot) h(a)$$

$$+ \partial_5 L(\cdot) h(b) \Big] (dx)^\alpha,$$

where $(\cdot) = \Big(x, y_0(x), y_0^{(\alpha)}(x), y_0(a), y_0(b) \Big)$. We can now proceed analogously to the proof of Theorem 8.1. As the result we get

$$\mathcal{J}(y_0+h) - \mathcal{J}(y_0) \geq \int_a^b \Big[\partial_2 L(\cdot) - \frac{d^\alpha}{dx^\alpha} \partial_3 L(\cdot) \Big] (dx)^\alpha$$

$$h(b) \Big[\int_a^b \partial_5 L(\cdot)(dx)^\alpha + \alpha! \partial_3 L(\cdot)|_{x=b} \Big] + h(a) \Big[\int_a^b \partial_4 L(\cdot) dx - \alpha! \partial_3 L(\cdot)|_{x=a} \Big] = 0,$$

since y_0 satisfies conditions (8.4)–(8.6). Therefore, we obtain $\mathcal{J}(y_0+h) \geq \mathcal{J}(y_0)$.

3.3   Examples

We shall provide examples in order to illustrate our main results.

Example 8.5. Consider the following problem

$$\mathcal{J}(y) = \int_0^1 \left\{ \left[\frac{x^\alpha}{\Gamma(\alpha+1)} (y^{(\alpha)})^2 - 2x^\alpha y^{(\alpha)} \right]^2 \right.$$

$$\left. + (y(0)-1)^2 + (y(1)-2)^2 \right\} (dx)^\alpha \longrightarrow \text{extr}.$$

The Euler–Lagrange equation associated to this problem is

$$\frac{d^\alpha}{dx^\alpha} \left(2 \left[\frac{x^\alpha}{\Gamma(\alpha+1)} (y^{(\alpha)})^2 - 2x^\alpha y^{(\alpha)} \right] \cdot \left[\frac{2x^\alpha}{\Gamma(\alpha+1)} y^{(\alpha)} - 2x^\alpha \right] \right) = 0. \qquad (8.9)$$

Let $y = x^\alpha + b$, where $b \in \mathbb{R}$. Since $y^{(\alpha)} = \Gamma(\alpha+1)$, it follows that y is a solution of (8.9). In order to determine b we use the generalized natural boundary conditions (8.5)–(8.6), which can be written for this problem as,

$$\int_0^1 (y(0)-1)(dx)^\alpha = 0,$$

$$\int_0^1 (y(1)-2)(dx)^\alpha = 0.$$

Hence, $\tilde{y} = x^\alpha + 1$ is a candidate solution. We remark that the $\tilde{y}$ is not differentiable in $[0, 1]$.

Example 8.6. Consider the following problem:

$$\mathcal{J}(y) = \int_0^1 \left[(y^{(\alpha)}(x))^2 + \gamma y^2(0) + \lambda (y(1) - 1)^2 \right] (dx)^\alpha \longrightarrow \min, \qquad (8.10)$$

where $\gamma, \lambda \in \mathbb{R}^+$. For this problem, the fractional Euler–Lagrange equation and the generalized natural boundary conditions (see Theorem 8.1) are given, respectively, as

$$2 \frac{d^\alpha}{dx^\alpha} y^{(\alpha)}(x) = 0, \qquad (8.11)$$

$$\int_0^1 \gamma y(0)(dx)^\alpha = \alpha! y^{(\alpha)}(0), \qquad (8.12)$$

$$\int_0^1 \lambda (y(1) - 1)(dx)^\alpha = -\alpha! y^{(\alpha)}(1). \qquad (8.13)$$

Solving (8.11)–(8.13) we obtain that

$$\bar{y}(x) = \frac{\gamma \lambda \alpha!}{\gamma \lambda + (\alpha!)^2 (\lambda + \gamma)} x^\alpha + \frac{(\alpha!)^2 \lambda}{\gamma \lambda + (\alpha!)^2 (\lambda + \gamma)}$$

is a candidate for minimizer. Observe that problem (8.10) satisfies assumptions of Theorem 8.2. Therefore $\bar{y}$ is a global minimizer to this problem. We note that when α goes to 1 problem (8.10) tends to

$$\mathcal{K}(y) = \int_0^1 \left[(y'(x))^2 + \gamma y^2(0) + \lambda (y(1) - 1)^2 \right] dx \longrightarrow \min.$$

with the solution

$$y(x) = \frac{\gamma \lambda}{\gamma \lambda + \lambda + \gamma} x + \frac{\lambda}{\gamma \lambda + \lambda + \gamma}.$$

4 Conclusion

In recent years fractional calculus has played an important role in various fields such as physics, chemistry, biology, economics, modeling, identification, control theory and signal processing (see, e.g., [3, 9, 13, 25, 26, 33, 37]). The fractional operators are non-local, therefore they are suitable for constructing models possessing memory. This gives several possible applications of the FCV, e.g., in describing non-local properties of physical systems in mechanics (see, e.g., [10, 24, 35]) or electrodynamics (see, e.g., [8, 39]). The Jumarie fractional derivative is quite suitable to describe dynamics evolving in a space which exhibit coarse-grained phenomenon. When the point in this space is not infinitely thin but rather has a thickness, then it would be better to replace dx by $(dx)^\alpha$, $0 < \alpha < 1$, where α characterizes the grade of the phenomenon. The fractal feature of the space is transported on time, and so both space and time are fractal. Thus, the increment of time of the dynamics of the system is not dx but $(dx)^\alpha$. In this note we generalize some previous results of the FCV (which are defined in terms of the Jumarie fractional derivatives and integrals) by proving optimality conditions for problems of FCV with a Lagrangian density depending on the free end-points. The advantage of using the Jumarie fractional derivative lies in the fact that this derivative is defined for continuous functions, non-differentiable (see, Example 8.5). Note that the integrand in problem (8.3) depends upon the a priori unknown final values $y(a)$ and $y(b)$. The present paper indicates how such problems may be solved.

Acknowledgements This work is partially supported by the *Portuguese Foundation for Science and Technology* (FCT) through the *Systems and Control Group* of the R&D Unit CIDMA, and partially by BUT Grant S/WI/2/2011. The author is grateful to Delfim F. M. Torres for inspiring discussions and useful comments.

References

1. Agrawal OP (2006) Fractional variational calculus and the transversality conditions. J Phys A 39:10375–10384
2. Almeida R, Torres DFM (2009) Calculus of variations with fractional derivatives and fractional integrals. Appl Math Lett 22:1816–1820
3. Almeida R, Torres DFM (2009) Hölderian variational problems subject to integral constraints. J Math Anal Appl 359:674–681
4. Almeida R, Torres DFM (2011) Fractional variational calculus for nondifferentiable functions. Comput Math Appl 61:3097–3104
5. Almeida R, Malinowska AB, Torres DFM (2010) A fractional calculus of variations for multiple integrals. Application to vibrating string. J Math Phys 51:033503
6. Atanacković TM, Konjik S, Pilipović S (2008) Variational problems with fractional derivatives: Euler–Lagrange equations. J Phys A 41:095201
7. Baleanu D (2008) Fractional constrained systems and caputo derivatives. J Comput Nonlinear Dynam 3:199–206
8. Baleanu D, Golmankhaneh AK, Golmankhaneh AK, Baleanu, M.C. (2009) Fractional electromagnetic equations using fractional forms. Int J Theor Phy. 48:3114–3123
9. Baleanu D, Guvenc ZB, Machado JAT (2010) New Trends in Nanotechnology and Fractional Calculus Applications, Springer Science Business Media
10. Carpinteri A, Mainardi F (1997) Fractals and fractional calculus in continuum mechanics. Springer, Vienna
11. Cresson J (2007) Fractional embedding of differential operators and Lagrangian systems. J Math Phys 48:033504

12. Cruz PAF, Torres DFM, Zinober ASI (2010) A non-classical class of variational problems. Int J Math Model Numerical Optimisation 1:227–236
13. Debnath L (2003) Recent applications of fractional calculus to science and engineering. Int J Math Math Sci 54:3413–3442
14. El-Nabulsi RA, Torres DFM (2007) Necessary optimality conditions for fractional action-like integrals of variational calculus with Riemann–Liouville derivatives of order (α, β). Math Meth Appl Sci 30:1931–1939
15. El-Nabulsi RA, Torres DFM (2008) Fractional actionlike variational problems. J Math Phys 49:053521
16. Frederico GSF, Torres DFM (2007) A formulation of Noether's theorem for fractional problems of the calculus of variations. J Math Anal Appl 334:834–846
17. Frederico GSF, Torres, DFM (2008) Fractional conservation laws in optimal control theory. Nonlinear Dynam 53:215–222
18. Hilfer R (2000) Applications of fractional calculus in physics. World Scientific Publishing, River Edge
19. Jumarie G (2005) On the representation of fractional Brownian motion as an integral with respect to $(dt)^a$. Appl Math Lett 18:739–748
20. Jumarie G (2007) Fractional Hamilton-Jacobi equation for the optimal control of nonrandom fractional dynamics with fractional cost function. J Appl Math Comput 23:215–228
21. Jumarie G (2009) Table of some basic fractional calculus formulae derived from a modified Riemann–Liouville derivative for non-differentiable functions. Appl Math Lett 22: 378–385
22. Jumarie G (2010) Analysis of the equilibrium positions of nonlinear dynamical systems in the presence of coarse-graining disturbance in space, J Appl Math Comput 32:329–351
23. Kilbas AA, Srivastava HM, Trujillo JJ (2006) Theory and applications of fractional differential equations. Elsevier, Amsterdam
24. Klimek M (2002) Lagrangean and Hamiltonian fractional sequential mechanics. Czechoslovak J Phys 52:1247–1253
25. Kolwankar KM, Gangal AD (1997) Holder exponents of irregular signals and local fractional derivatives. Pramana J Phys 48:49–68
26. Machado JAT, Silva MF, Barbosa RS, Jesus IS, Reis CM, Marcos MG, Galhano AF (2010) Some applications of fractional calculus in engineering. Mathematical Problems in Engineering, doi:10.1155/2010/639801
27. Malinowska AB, Sidi Ammi MR, Torres DFM (2010) Composition functionals in fractional calculus of variations, Commun Frac Calc 1:32–40
28. Malinowska AB, Torres DFM (2010) Generalized natural boundary conditions for fractional variational problems in terms of the Caputo derivative. Comput Math Appl 59:3110–3116
29. Malinowska AB, Torres DFM (2010) Natural Boundary Conditions in the Calculus of Variations. Math Meth Appl Sc. 33:1712–1722
30. Malinowska AB, Torres DFM (2011) Fractional calculus of variations for a combined Caputo derivative. Fract Calc Appl Anal 14(4), in press
31. Miller KS, Ross B (1993) An introduction to the fractional calculus and fractional differential equations. Wiley, New York
32. Odzijewicz T, Torres DFM 2010 Calculus of variations with fractional and classical derivatives. In: Podlubny I, Vinagre Jara BM, Chen YQ, Feliu Batlle V, Tejado Balsera I (ed) Proceedings of FDA'10, The 4th IFAC Workshop on Fractional Differentiation and its Applications, Badajoz, Spain, p. 5, 18–20, Article no. FDA10-076
33. Ortigueira MD, Machado JAT (2006) Fractional calculus applications in signals and systems. Signal Process 86:2503–2504
34. Podlubny I (1999) Fractional differential equations. Academic Press, San Diego, CA, USA
35. Rabei EM, Ababneh BS (2008) Hamilton-Jacobi fractional mechanics. J Math Anal Appl 344:799–805
36. Riewe F (1996) Nonconservative Lagrangian and Hamiltonian mechanics. Phys Rev E 53:1890–1899

37. Ross B (1975) Fractional calculus and its applications, Springer, Berlin
38. Samko SG, Kilbas AA, Marichev OI (1993) Fractional integrals and derivatives. Translated from the 1987 Russian original, Gordon and Breach, Yverdon
39. Tarasov VE (2008) Fractional vector calculus and fractional Maxwell's equations. Ann Phy 323:2756–2778
40. van Brunt B (2004) The calculus of variations. Springer, New York

Chapter 9
Fractional Euler–Lagrange Differential Equations via Caputo Derivatives

Ricardo Almeida, Agnieszka B. Malinowska, and Delfim F.M. Torres

1 Introduction

Fractional calculus plays an important role in many different areas, and has proven to be a truly multidisciplinary subject [20, 26]. It is a mathematical field as old as the calculus itself. In a letter dated 30th September 1695, Leibniz posed the following question to L'Hopital: "Can the meaning of derivative be generalized to derivatives of non-integer order?" Since then, several mathematicians had investigated Leibniz's challenge, prominent among them were Liouville, Riemann, Weyl, and Letnikov. There are many applications of fractional calculus, for example, in viscoelasticity, electrochemistry, diffusion processes, control theory, heat conduction, electricity, mechanics, chaos and fractals, and signals and systems [12, 22].

Several methods to solve fractional differential equations are available, using Laplace and Fourier transforms, truncated Taylor series, and numerical approximations. In Almeida and Torres [7] a new direct method to find exact solutions of fractional variational problems is proposed, based on a simple but powerful idea introduced by Leitmann, that does not involve solving (fractional) differential equations [32]. By an appropriate coordinate transformation, we rewrite the initial problem to an equivalent simpler one; knowing the solution for the new equivalent problem, and since there exists an one-to-one correspondence between the minimizers (or maximizers) of the new problem with the ones of the original, we determine the desired solution. For a modern account on Leitmann's direct method see [25,26].

R. Almeida (✉) • D.F.M. Torres
Center for Research and Development in Mathematics and Applications, Department of Mathematics, University of Aveiro, 3810-193 Aveiro, Portugal
e-mail: ricardo.almeida@ua.pt; delfim@ua.pt

A.B. Malinowska
Faculty of Computer Science, Białystok University of Technology, 15-351 Białystok, Poland
e-mail: abmalinowska@ua.pt

D. Baleanu et al. (eds.), *Fractional Dynamics and Control*,
DOI 10.1007/978-1-4614-0457-6_9, © Springer Science+Business Media, LLC 2012

The calculus of variations is a field of mathematics that deals with extremizing functionals [33]. The variational functionals are often formed as definite integrals involving unknown functions and their derivatives. The fundamental problem consists to find functions $y(x)$, $x \in [a,b]$, that extremize a given functional when subject to boundary conditions $y(a) = y_a$ and $y(b) = y_b$. Since this can be a hard task, one wishes to study necessary and sufficient optimality conditions. The simplest example is the following one: what is the shape of the curve $y(x)$, $x \in [a,b]$, joining two fixed points y_a and y_b, that has the minimum possible length? The answer is obviously the straight line joining y_a and y_b. One can obtain it solving the corresponding Euler–Lagrange necessary optimality condition. If the boundary condition $y(b) = y_b$ is not fixed, that is, if we are only interested in the minimum length, the answer is the horizontal straight line $y(x) = y_a$, $x \in [a,b]$ (free endpoint problem). In this case we need to complement the Euler–Lagrange equation with an appropriate natural boundary condition. For a general account on Euler–Lagrange equations and natural boundary conditions, we refer the readers to [23, 24] and references therein. Another important family of variational problems is the isoperimetric one [5]. The classical isoperimetric problem consists to find a continuously differentiable function $y = y(x)$, $x \in [a,b]$, satisfying given boundary conditions $y(a) = y_a$ and $y(b) = y_b$, which minimizes (or maximizes) a functional

$$I(y) = \int_a^b L(x, y(x), y'(x)) \, dx$$

subject to the constraint

$$\int_a^b g(x, y(x), y'(x)) \, dx = l.$$

The most famous isoperimetric problem can be posed as follows. Amongst all closed curves with a given length, which one encloses the largest area? The answer, as we know, is the circle. The general method to solve such problems involves an Euler–Lagrange equation obtained via the concept of Lagrange multiplier (see, e.g., [4]).

The fractional calculus of variations is a recent field, initiated in 1997, where classical variational problems are considered but in presence of some fractional derivative or fractional integral [31]. In the past few years an increasing of interest has been put on finding necessary conditions of optimality for variational problems with Lagrangians involving fractional derivatives [1, 9–11, 16–19, 27, 28], fractional derivatives and fractional integrals [3, 6, 15], classical and fractional derivatives [29], as well as fractional difference operators [13, 14]. A good introduction to the subject is given in the monograph of Klimek [21]. Here we consider unconstrained and constrained fractional variational problems via Caputo operators.

2 Preliminaries and Notations

There exist several definitions of fractional derivatives and fractional integrals, for example, Riemann–Liouville, Caputo, Riesz, Riesz–Caputo, Weyl, Grunwald–Letnikov, Hadamard, and Chen. Here we review only some basic features of Caputo's fractional derivative. For proofs and more on the subject, we refer the readers to [20, 26].

Let $f : [a,b] \to \mathbb{R}$ be an integrable function, $\alpha > 0$, and Γ be the Euler gamma function. The left and right Riemann–Liouville fractional integral operators of order α are defined by[1]

$$_aI_x^\alpha[f] := x \mapsto \frac{1}{\Gamma(\alpha)} \int_a^x (x-t)^{\alpha-1} f(t)\,\mathrm{d}t$$

and

$$_xI_b^\alpha[f] := x \mapsto \frac{1}{\Gamma(\alpha)} \int_x^b (t-x)^{\alpha-1} f(t)\,\mathrm{d}t,$$

respectively. The left and right Riemann–Liouville fractional derivative operators of order α are, respectively, defined by

$$_aD_x^\alpha := \frac{\mathrm{d}^n}{\mathrm{d}x^n} \circ {}_aI_x^{n-\alpha}$$

and

$$_xD_b^\alpha := (-1)^n \frac{\mathrm{d}^n}{\mathrm{d}x^n} \circ {}_xI_b^{n-\alpha},$$

where $n = [\alpha] + 1$. Interchanging the composition of operators in the definition of Riemann–Liouville fractional derivatives, we obtain the left and right Caputo fractional derivatives of order α:

$$_a^C D_x^\alpha := {}_aI_x^{n-\alpha} \circ \frac{\mathrm{d}^n}{\mathrm{d}x^n}$$

and

$$_x^C D_b^\alpha := {}_xI_b^{n-\alpha} \circ (-1)^n \frac{\mathrm{d}^n}{\mathrm{d}x^n}.$$

---

[1] Along the work we use round brackets for the arguments of functions, and square brackets for the arguments of operators. By definition, an operator receives a function and returns another function.

Theorem 9.1. *Assume that f is of class C^n on $[a,b]$. Then its left and right Caputo derivatives are continuous on the closed interval $[a,b]$.*

One of the most important results for the proof of necessary optimality conditions, is the integration by parts formula. For Caputo derivatives the following relations hold.

Theorem 9.2. *Let $\alpha > 0$, and $f,g : [a,b] \to \mathbb{R}$ be C^n functions. Then,*

$$\int_a^b g(x) \cdot {}_a^C D_x^\alpha [f](x) \mathrm{d}x = \int_a^b f(x) \cdot {}_x D_b^\alpha [g](x) \mathrm{d}x$$

$$+ \sum_{j=0}^{n-1} \left[{}_x D_b^{\alpha+j-n} [g](x) \cdot {}_x D_b^{n-1-j} [f](x) \right]_a^b$$

and

$$\int_a^b g(x) \cdot {}_x^C D_b^\alpha [f](x) \mathrm{d}x = \int_a^b f(x) \cdot {}_a D_x^\alpha [g](x) \mathrm{d}x$$

$$+ \sum_{j=0}^{n-1} \left[(-1)^{n+j} {}_a D_x^{\alpha+j-n} [g](x) \cdot {}_a D_x^{n-1-j} [f](x) \right]_a^b,$$

where ${}_a D_x^k = {}_a I_x^{-k}$ and ${}_x D_b^k = {}_x I_b^{-k}$ whenever $k < 0$.

In the particular case when $0 < \alpha < 1$, we get from Theorem 9.2 that

$$\int_a^b g(x) \cdot {}_a^C D_x^\alpha [f](x) \mathrm{d}x = \int_a^b f(x) \cdot {}_x D_b^\alpha [g](x) \mathrm{d}x + \left[{}_x I_b^{1-\alpha} [g](x) \cdot f(x) \right]_a^b$$

and

$$\int_a^b g(x) \cdot {}_x^C D_b^\alpha [f](x) \mathrm{d}x = \int_a^b f(x) \cdot {}_a D_x^\alpha [g](x) \mathrm{d}x - \left[{}_a I_x^{1-\alpha} [g](x) \cdot f(x) \right]_a^b.$$

In addition, if f is such that $f(a) = f(b) = 0$, then

$$\int_a^b g(x) \cdot {}_a^C D_x^\alpha [f](x) \mathrm{d}x = \int_a^b f(x) \cdot {}_x D_b^\alpha [g](x) \mathrm{d}x$$

and

$$\int_a^b g(x) \cdot {}_x^C D_b^\alpha [f](x) \mathrm{d}x = \int_a^b f(x) \cdot {}_a D_x^\alpha [g](x) \mathrm{d}x.$$

Along the work, we denote by $\partial_i L$, $i = 1, \ldots, m$ ($m \in \mathbb{N}$), the partial derivative of function $L : \mathbb{R}^m \to \mathbb{R}$ with respect to its ith argument. For convenience of notation, we introduce the operator ${}_\alpha^C [\cdot]_\beta$ defined by

$$\substack{C\\\alpha}[y]_\beta := x \mapsto \left(x, y(x), \substack{C\\a}D^\alpha_x[y](x), \substack{C\\x}D^\beta_b[y](x)\right),$$

where $\alpha, \beta \in (0,1)$.

3   Euler–Lagrange Equations

The fundamental problem of the fractional calculus of variations is addressed in the following way: find functions $y \in \mathcal{E}$,

$$\mathcal{E} := \left\{y \in C^1([a,b]) \,|\, y(a) = y_a \text{ and } y(b) = y_b\right\},$$

that maximize or minimize the functional

$$J(y) = \int_a^b \left(L \circ \substack{C\\\alpha}[y]_\beta\right)(x)\mathrm{d}x. \tag{9.1}$$

As usual, the Lagrange function L is assumed to be of class C^1 on all its arguments. We also assume that $\partial_3 L \circ \substack{C\\\alpha}[y]_\beta$ has continuous right Riemann–Liouville fractional derivative of order α and $\partial_4 L \circ \substack{C\\\alpha}[y]_\beta$ has continuous left Riemann–Liouville fractional derivative of order β for $y \in \mathcal{E}$.

In [1] a necessary condition of optimality for such functionals is proved. We remark that although functional (9.1) contains only Caputo fractional derivatives, the fractional Euler–Lagrange equation also contains Riemann–Liouville fractional derivatives.

Theorem 9.3 (Euler–Lagrange equation for (9.1)). *If y is a minimizer or a maximizer of J on $\mathcal{E}$, then y is a solution of the fractional differential equation*

$$\left(\partial_2 L \circ \substack{C\\\alpha}[y]_\beta\right)(x) + {}_xD^\alpha_b \left[\partial_3 L \circ \substack{C\\\alpha}[y]_\beta\right](x) + {}_aD^\beta_x \left[\partial_4 L \circ \substack{C\\\alpha}[y]_\beta\right](x) = 0 \tag{9.2}$$

for all $x \in [a,b]$.

Proof. Given $|\varepsilon| \ll 1$, consider $h \in V$ where

$$V := \left\{h \in C^1([a,b]) \,|\, h(a) = 0 \text{ and } h(b) = 0\right\},$$

and a variation of function y of type $y + \varepsilon h$. Define the real valued function $j(\varepsilon)$ by

$$j(\varepsilon) = J(y + \varepsilon h) = \int_a^b \left(L \circ \substack{C\\\alpha}[y + \varepsilon h]_\beta\right)(x)\mathrm{d}x.$$

Since $\varepsilon = 0$ is a minimizer or a maximizer of j, we have $j'(0) = 0$. Thus,

$$\int_a^b \Big[\big(\partial_2 L \circ {}_\alpha^C[y]_\beta\big)(x) \cdot h(x) + \big(\partial_3 L \circ {}_\alpha^C[y]_\beta\big)(x) \cdot {}_a^C D_x^\alpha[h](x)$$
$$+ \big(\partial_4 L \circ {}_\alpha^C[y]_\beta\big)(x) \cdot {}_x^C D_b^\beta[h](x) \Big] dx = 0.$$

We obtain equality (9.2) integrating by parts and applying the classical fundamental lemma of the calculus of variations [33].   □

We remark that when $\alpha \to 1$, then (9.1) is reduced to a classical functional

$$J(y) = \int_a^b f(x, y(x), y'(x))\, dx,$$

and the fractional Euler–Lagrange equation (9.2) gives the standard one:

$$\partial_2 f(x, y(x), y'(x)) - \frac{d}{dx}\partial_3 f(x, y(x), y'(x)) = 0.$$

Solutions to equation (9.2) are said to be *extremals* of (9.1).

4   The Isoperimetric Problem

The fractional isoperimetric problem is stated in the following way: find the minimizers or maximizers of functional J as in (9.1), over all functions $y \in \mathcal{E}$ satisfying the fractional integral constraint

$$I(y) = \int_a^b \big(g \circ {}_\alpha^C[y]_\beta\big)(x) dx = l.$$

Similarly as L, g is assumed to be of class C^1 with respect to all its arguments, function $\partial_3 g \circ {}_\alpha^C[y]_\beta$ is assumed to have continuous right Riemann–Liouville fractional derivative of order α and $\partial_4 g \circ {}_\alpha^C[y]_\beta$ continuous left Riemann–Liouville fractional derivative of order β for $y \in \mathcal{E}$. A necessary optimality condition for the fractional isoperimetric problem is given in [8].

Theorem 9.4. *Let y be a minimizer or maximizer of J on $\mathcal{E}$, when restricted to the set of functions $z \in \mathcal{E}$ such that $I(z) = l$. In addition, assume that y is not an extremal of I. Then, there exists a constant λ such that y is a solution of*

$$\big(\partial_2 F \circ {}_\alpha^C[y]_\beta\big)(x) + {}_xD_b^\alpha[\partial_3 F \circ {}_\alpha^C[y]_\beta](x) + {}_aD_x^\beta[\partial_4 F \circ {}_\alpha^C[y]_\beta](x) = 0 \qquad (9.3)$$

for all $x \in [a, b]$, where $F = L + \lambda g$.

Proof. Given $h_1, h_2 \in V$, $|\varepsilon_1| \ll 1$ and $|\varepsilon_2| \ll 1$, consider

$$j(\varepsilon_1, \varepsilon_2) = \int_a^b \left(L \circ {}_\alpha^C [y + \varepsilon_1 h_1 + \varepsilon_2 h_2]_\beta \right)(x) dx$$

and

$$i(\varepsilon_1, \varepsilon_2) = \int_a^b \left(g \circ {}_\alpha^C [y + \varepsilon_1 h_1 + \varepsilon_2 h_2]_\beta \right)(x) dx - l.$$

Since y is not an extremal for I, there exists a function h_2 such that

$$\left. \frac{\partial i}{\partial \varepsilon_2} \right|_{(0,0)} \neq 0,$$

and by the implicit function theorem, there exists a C^1 function $\varepsilon_2(\cdot)$, defined in some neighborhood of zero, such that

$$i(\varepsilon_1, \varepsilon_2(\varepsilon_1)) = 0.$$

Applying the Lagrange multiplier rule (see, e.g., [33, Theorem 4.1.1]) there exists a constant λ such that

$$\nabla(j(0,0) + \lambda i(0,0)) = \mathbf{0}.$$

Differentiating j and i at $(0,0)$, and integrating by parts, we prove the theorem. □

Example 9.1. Let $\bar{y}(x) = E_\alpha(x^\alpha)$, $x \in [0,1]$, where E_α is the Mittag-Leffler function. Then ${}_0^C D_x^\alpha [\bar{y}] = \bar{y}$. Consider the following fractional variational problem:

$$J(y) = \int_0^1 \left({}_0^C D_x^\alpha [y](x) \right)^2 dx \longrightarrow \text{extr},$$

$$I(y) = \int_0^1 \bar{y}(x) {}_0^C D_x^\alpha [y](x) dx = l,$$

$$y(0) = 1, \quad y(1) = y_1,$$

with $l := \int_0^1 (\bar{y}(x))^2 dx$ and $y_1 := E_\alpha(1)$. In this case function F of Theorem 9.4 is

$$F(x, y, v, w) = v^2 + \lambda \bar{y}(x) v$$

and the fractional Euler–Lagrange equation (9.3) is

$$_x D_1^\alpha [2 {}_0^C D_x^\alpha [y] + \lambda \bar{y}](x) = 0.$$

A solution to this problem is $\lambda = -2$ and $y(x) = \bar{y}(x)$, $x \in [0,1]$.

The case when y is an extremal of I is also included in the results of [8].

Theorem 9.5. *If y is a minimizer or a maximizer of J on $\mathcal{E}$, subject to the isoperimetric constraint $I(y) = l$, then there exist two constants λ_0 and λ, not both zero, such that*

$$\left(\partial_2 K \circ {}^C_\alpha[y]_\beta\right)(x) + {}_x D^\alpha_b \left[\partial_3 K \circ {}^C_\alpha[y]_\beta\right](x) + {}_a D^\beta_x \left[\partial_4 K \circ {}^C_\alpha[y]_\beta\right](x) = 0$$

for all $x \in [a,b]$, where $K = \lambda_0 L + \lambda g$.

Proof. The same as the proof of Theorem 9.4, but now using the abnormal Lagrange multiplier rule (see, e.g., [33, Theorem 4.1.3]). □

5   Transversality Conditions

We now give the *natural boundary conditions* (also known as *transversality conditions*) for problems with the terminal point of integration free as well as y_b.

Let

$$\mathcal{F} := \left\{(y,x) \in C^1([a,b]) \times [a,b] \,|\, y(a) = y_a\right\}.$$

The type of functional we consider now is

$$J(y,T) = \int_a^T \left(L \circ {}^C_\alpha[y]\right)(x)\,dx, \tag{9.4}$$

where the operator ${}^C_\alpha[\cdot]$ is defined by

$${}^C_\alpha[y] := x \mapsto \left(x, y(x), {}^C_a D^\alpha_x[y](x)\right).$$

These problems are investigated in [1] and more general cases in [2].

Theorem 9.6. *Suppose that $(y,T) \in \mathcal{F}$ minimizes or maximizes J defined by (9.4) on $\mathcal{F}$. Then*

$$\left(\partial_2 L \circ {}^C_\alpha[y]\right)(x) + {}_x D^\alpha_T \left[\partial_3 L \circ {}^C_\alpha[y]\right](x) = 0 \tag{9.5}$$

for all $x \in [a,T]$. Moreover, the following transversality conditions hold:

$$\left(L \circ {}^C_\alpha[y]\right)(T) = 0, \quad {}_x I^{1-\alpha}_T \left[\partial_3 L \circ {}^C_\alpha[y]\right](T) = 0.$$

Proof. The result is obtained by considering variations $y + \varepsilon h$ of function y and variations $T + \varepsilon \triangle T$ of T as well, and then applying the Fermat theorem, integration by parts, Leibniz's rule, and using the arbitrariness of h and $\triangle T$. □

Transversality conditions for several other situations can be easily obtained. Some important examples are:

- If T is fixed but $y(T)$ is free, then besides the Euler–Lagrange equation (9.5) one obtains the transversality condition

$$_xI_T^{1-\alpha}\left[\partial_3 L\circ{}_\alpha^C[y]\right](T)=0.$$

- If $y(T)$ is given but T is free, then the transversality condition is

$$\left(L\circ{}_\alpha^C[y]\right)(T)-y'(T)\cdot{}_xI_T^{1-\alpha}\left[\partial_3 L\circ{}_\alpha^C[y]\right](T)=0.$$

- If $y(T)$ is not given but is restricted to take values on a certain given curve ψ, that is, $y(T)=\psi(T)$, then

$$\left(\psi'(T)-y'(T)\right)\cdot{}_xI_T^{1-\alpha}\left[\partial_3 L\circ{}_\alpha^C[y]\right](T)+\left(L\circ{}_\alpha^C[y]\right)(T)=0.$$

Acknowledgements Work supported by *FEDER* funds through *COMPETE* — Operational Programme Factors of Competitiveness ("Programa Operacional Factores de Competitividade") and by Portuguese funds through the *Center for Research and Development in Mathematics and Applications* (University of Aveir) and the Portuguese Foundation for Science and Technology ("FCT — Fundação para a Ciência e a Tecnologia"), within project PEst-C/MAT/UI4106/2011 with COMPETE number FCOMP-01-0124-FEDER-022690. Agnieszka Malinowska is also supported by Białystok University of Technology grant S/WI/2/2011.

References

1. Agrawal OP (2007) Generalized Euler-Lagrange equations and transversality conditions for fvps in terms of the Caputo derivative. J Vib Contr 13(9–10):1217–1237
2. Almeida R, Malinowska AB Generalized transversality conditions in fractional calculus of variations. (Submitted)
3. Almeida R, Malinowska AB, Torres DFM (2010) A fractional calculus of variations for multiple integrals with application to vibrating string. J Math Phys 51(3):033503, 12 pp
4. Almeida R, Torres DFM (2009) Hölderian variational problems subject to integral constraints. J Math Anal Appl 359(2):674–681
5. Almeida R, Torres DFM (2009) Isoperimetric problems on time scales with nabla derivatives. J Vib Contr 15(6):951–958
6. Almeida R, Torres DFM (2009) Calculus of variations with fractional derivatives and fractional integrals. Appl Math Lett 22(12):1816–1820
7. Almeida R, Torres DFM (2010) Leitmann's direct method for fractional optimization problems. Appl Math Comput 217(3):956–962
8. Almeida R, Torres DFM (2011) Necessary and sufficient conditions for the fractional calculus of variations with Caputo derivatives. Comm Nonlinear Sci Numer Simult 16(3):1490–1500
9. Atanacković TM, Konjik S, Pilipović S (2007) Variational problems with fractional derivatives: Euler-Lagrange equations. J Phys A 41(9):095201, 12 pp
10. Baleanu D (2008) Fractional constrained systems and caputo derivatives. J Comput Nonlinear Dynam 3(2):021102

11. Baleanu D (2008) New applications of fractional variational principles. Rep Math Phys 61(2):199–206
12. Baleanu D, Güvenç ZB, Tenreiro Machado JA (2010) New trends in nanotechnology and fractional calculus applications. Springer, New York
13. Bastos NRO, Ferreira RAC, Torres DFM (2011) Necessary optimality conditions for fractional difference problems of the calculus of variations. Discrete Contin Dyn Syst 29(2):417–437
14. Bastos NRO, Ferreira RAC, Torres DFM (2011) Discrete-time fractional variational problems. Signal Process 91(3):513–524
15. El-Nabulsi RA, Torres DFM (2007) Necessary optimality conditions for fractional action-like integrals of variational calculus with Riemann-Liouville derivatives of order (α, β). Math Methods Appl Sci 30(15):1931–1939
16. El-Nabulsi RA, Torres DFM (2008) Fractional actionlike variational problems. J Math Phys 49(5):053521, 7 pp
17. Frederico GSF, Torres DFM (2007) A formulation of Noether's theorem for fractional problems of the calculus of variations. J Math Anal Appl 334(2):834–846
18. Frederico GSF, Torres DFM (2008) Fractional conservation laws in optimal control theory. Nonlinear Dynam 53(3):215–222
19. Frederico GSF, Torres DFM (2010) Fractional Noether's theorem in the Riesz-Caputo sense. Appl Math Comput 217(3):1023–1033
20. Kilbas AA, Srivastava HM, Trujillo JJ (2006) Theory and applications of fractional differential equations. Elsevier, Amsterdam
21. Klimek M (2009) On solutions of linear fractional differential equations of a variational type. Czestochowa Series Monographs 172, Czestochowa University of Technology, Czestochowa
22. Magin R, Ortigueira MD, Podlubny I, Trujillo J (2011) On the fractional signals and systems. Signal Process 91(3):350–371
23. Malinowska AB, Torres DFM (2010) Natural boundary conditions in the calculus of variations. Math Methods Appl Sci 33(14):1712–1722
24. Malinowska AB, Torres DFM (2010) Generalized natural boundary conditions for fractional variational problems in terms of the Caputo derivative. Comput Math Appl 59(9):3110–3116
25. Malinowska AB, Torres DFM (2010) Leitmann's direct method of optimization for absolute extrema of certain problems of the calculus of variations on time scales. Appl Math Comput 217(3):1158–1162
26. Malinowska AB, Torres DFM (2010) The Hahn quantum variational calculus. J Optim Theory Appl 147(3):419–442
27. Mozyrska D, Torres DFM (2010) Minimal modified energy control for fractional linear control systems with the Caputo derivative. Carpathian J Math 26(2):210–221
28. Mozyrska D, Torres DFM (2011) Modified optimal energy and initial memory of fractional continuous-time linear systems. Signal Process 91(3):379–385
29. Odzijewicz T, Malinowska AB, Torres DFM (2011) Fractional variational calculus with classical and combined Caputo derivatives. Nonlinear Anal (in press) DOI: 10.1016/j.na.2011.01.010
30. Podlubny I (1999) Fractional differential equations. Academic Press, San Diego, CA
31. Riewe F (1997) Mechanics with fractional derivatives. Phys Rev E 55(3):3581–3592
32. Torres DFM, Leitmann G (2008) Contrasting two transformation-based methods for obtaining absolute extrema. J Optim Theory Appl 137(1):53–59
33. van Brunt B (2004) The calculus of variations. Universitext, Springer, New York

Chapter 10
Strict Stability of Fractional Perturbed Systems in Terms of Two Measures

Coşkun Yakar, Mustafa Bayram Gücen, and Muhammed Çiçek

1 Introduction

The history of fractional order derivative was started by L' Hospital and Leibnitz [2, 5, 10, 11]. The question of L' Hospital to Leibnitz gave a new theory to mathematicians. Then, the mathematicians has developed the fractional calculus and the theory of fractional differential equations. The improvement of these studies are continued still and gained new results in mathematics and this results are also applied for different disciplines such as physics, chemistry and engineering [5, 7, 16].

The application of Lyapunov's second method in stability of differential equations see [4–6, 13, 16–18] that has the advantage of not requiring behavior of solutions of the system which investigates. Recently, the stability with initial time difference in terms of two measures [15, 19, 20] and the properties of fractional differential equations [5, 16], has been investigated. The strict stability criteria of differential equations in [14] and initial time difference strict stability worked and obtained comparison results and the appropriate definitions are presant in the work of Yakar [17].

In this paper, we investigate strict stability criteria with initial time difference in terms of two measures on fractional order differential equations and we have used the definition of Caputo's fractional order derivative because of some advantages which we express. We look into the strict stability criteria in terms of two measures with initial time difference of a perturbed fractional order differential system with respect to an unperturbed fractional order differential system which have different

C. Yakar (✉) • M. Çiçek
Gebze Institute of Technology, Gebze, Kocaeli, Turkey 141-41400
e-mail: cyakar@gyte.edu.tr; mcicek@gyte.edu.tr

M.B. Gücen
Yıldız Technical University, Davutpaşa, Esenler İstanbul, Turkey 34210
e-mail: mgucen@yildiz.edu.tr

D. Baleanu et al. (eds.), *Fractional Dynamics and Control*,
DOI 10.1007/978-1-4614-0457-6_10, © Springer Science+Business Media, LLC 2012

initial time and initial position. Using Lyapunov functions and comparison principle have been given sufficient conditions for the strict stability of dynamic systems on fractional order differential equations.

2   Preliminaries

In this study, we have used Caputo's fractional order derivative. But we have three definition of fractional order derivative: Caputo, Reimann–Liouville and Grünwald–Letnikov. The definition of Caputo's and Reimann–Liouville's fractional derivatives

$$^cD^qx = \frac{1}{\Gamma(1-q)} \int_{t_0}^{t} (t-s)^{-q} x'(s) ds, \ t_0 \leq t \leq T, \tag{10.1}$$

$$D^qx = \frac{1}{\Gamma(p)} \left(\frac{d}{dt} \int_{\tau_0}^{t} (t-s)^{p-1} x(s) ds \right), \ t_0 \leq t \leq T, \tag{10.2}$$

order of $0 < q < 1$,and $p+q = 1$ where Γ denotes the Gamma function.

It can't be denied that the fractional derivative of Riemann-Liouville is important for the development of fractional calculus and fractional order differential equations. But in mathematical modeling of some applications of various areas, there is a difficulty to interpret the initial condition required for the initial value problems of fractional order differential equations. The main advantage of Caputo's definition of fractional order derivative is that the initial conditions for fractional order differential equations with Caputo derivative take on the same form as that of ordinary differential equations with integer derivatives and another difference is that the Caputo derivative for a constant C is zero, while the Riemann–Liouville fractional derivative for a constant C is not zero but equals to $D^qC = \frac{C(t-t_0)^{-q}}{\Gamma(1-q)}$. By using (1) and therefore,

$$^cD^qx(t) = D^q[x(t) - x(t_0)], \tag{10.3}$$

$$^cD^qx(t) = D^qx(t) - \frac{x(t_0)}{\Gamma(1-q)} (t-t_0)^{-q}. \tag{10.4}$$

In particular, if $x(t_0) = 0$, the equality holds

$$^cD^qx(t) = D^qx(t) \tag{10.5}$$

and Caputo's derivative is defined for functions for which Riemann–Liouville fractional order derivative exists.

Let us write that Grünwald–Letnikov's notion of fractional order derivative in a convenient form

$$D_0^q x(t) = \lim_{\substack{h \to 0 \\ nh=t-t_0}} \frac{1}{h^q} [x(t) - S(x,h,r,q)],$$

where $S(x,h,r,q) = \sum_{r=1}^{n} (-1)^{r+1} \binom{q}{r} x(t-rh)$. If $x(t)$ is continuous and $\frac{dx(t)}{dt}$ exists and integrable, then Riemann–Liouville and Grünwald–Letnikov fractional order derivatives are connected by the relation

$$D^q x(t) = D_0^q x(t) = \frac{x(t_0)(t-t_0)^{-q}}{\Gamma(1-q)} + \int_{\tau_0}^{t} \frac{(t-s)^{-q}}{\Gamma(1-q)} \frac{d}{ds} x(s) ds. \qquad (10.6)$$

By using (3) implies that we have the relations among the Caputo, Riemann–Liouville and Grünwald–Letnikov fractional derivatives

$$^c D^q x(t) = D^q [x(t) - x(t_0)] \qquad (10.7)$$

$$^c D^q x(t) = \frac{1}{\Gamma(1-q)} \int_{\tau_0}^{t} (t-s)^{-q} \frac{dx(s)}{ds} ds$$

This relations of the definitions of the fractional order derivative are important to understand the properties of the solutions of fractional order differential equations.

3   Definition and Notation

Consider the differential systems

$$^c D^q x(t) = f(t,x), x(t_0) = x_0 \text{ for } t \geq t_0, t_0 \in \mathbb{R}_+, \qquad (10.8)$$

$$^c D^q y(t) = f(t,y), y(\tau_0) = y_0 \text{ for } t \geq \tau_0, \tau_0 \in \mathbb{R}_+, \qquad (10.9)$$

where $x_0 = \lim_{t \to t_0} D^{q-1} x(t)$ and $y_0 = \lim_{t \to \tau_0} D^{q-1} y(t)$ exist and the perturbed fractional order differential system with Caputo's derivative of (8)

$$^c D^q y(t) = F(t,y), y(\tau_0) = y_0 \text{ for } t \geq \tau_0, \qquad (10.10)$$

where $y_0 = \lim_{t \to \tau_0} D^{q-1} y(t)$ exist and $f, F \in C[[t_0, \tau_0 + T] \times \mathbb{R}^n, \mathbb{R}^n]$; satisfy a local Lipschitz condition on the set $\mathbb{R}_+ \times S\rho$, $S\rho = [x \in \mathbb{R}^n : \|x\| \leq \rho < \infty]$ and $f(t,0) = 0$ for $t \geq 0$. A special case of (10) is where $F(t,y) = f(t,y) + R(t,y)$ and $R(t,y)$ is the perturbation term. Assume that the existence and uniqueness of the solutions $x(t) = x(t,t_0,x_0)$ of (8) for $t \geq t_0$ and $y(t) = y(t,\tau_0,y_0)$ of (10) for $t \geq t_0$.

The basic existence and uniqueness result with the Lipschitz condition by using contraction mapping theorem and a weighted norm with Mittag-Leffler function in [6, 10–12]. We introduce definitions for a variety of classes of functions that we use in Sect. 4 and for generalized Dini-like derivatives and initial time difference strict stability in terms of two measures. All inequalities between vectors are componentwise.

Let us give the definition of the fractional strict stability in terms of two measures with initial time difference.

Definition 10.1. The solution $y(t,\tau_0,y_0)$ of the perturbed system (10) through (τ_0,y_0) is said to be initial time difference $(h_0 - h)$−strict stable in fractional case with respect to the solution $x(t - \eta, t_0, x_0)$, where $x(t,t_0,x_0)$ is any solution of the unperturbed system (8) for $t \geq \tau_0 \geq 0, t_0 \in \mathbb{R}_+$ and $\eta = \tau_0 - t_0$. If given any $\varepsilon_1 > 0$ and $\tau_0 \in \mathbb{R}_+$ there exist $\delta_1 = \delta_1(\varepsilon_1,\tau_0) > 0$ and $\delta_2 = \delta_2(\varepsilon_1,\tau_0) > 0$ such that

$$h(t, y(t,\tau_0,y_0) - x(t - \eta, t_0, x_0)) < \varepsilon_1 \text{ for } t \geq \tau_0$$

whenever $h(\tau_0, y_0 - x_0) < \delta_1$ and $h_0(\tau_0, \tau_0 - t_0) < \delta_2$ and, for $\delta_1^* < \delta_1$ and $\delta_2^* < \delta_2$ there exist $0 < \varepsilon_2 < \min\{\delta_1^*, \delta_2^*\}$ such that

$$h(t, y(t,\tau_0,y_0) - x(t - \eta, t_0, x_0)) > \varepsilon_2 \text{ for } t \geq \tau_0$$

whenever $h(\tau_0, y_0 - x_0) > \delta_1^*$ and $h_0(\tau_0, \tau_0 - t_0) > \delta_2^*$.

Definition 10.2. If δ_1, δ_2 and ε_2 in Definition 1 are independent of τ_0, then the solution $y(t,\tau_0,y_0)$ of the perturbed system (10) through (τ_0,y_0) is initial time difference $(h_0 - h)$−uniformly strict stable in fractional case with respect to the solution $x(t - \eta, t_0, x_0)$ for $t \geq \tau_0$.

Definition 10.3. The solution $y(t,\tau_0,y_0)$ of the system (10) through (τ_0,y_0) is said to be initial time difference $(h_0 - h)$−strictly attractive in fractional case with respect to the solution $x(t - \eta, t_0, x_0)$, where $x(t,t_0,x_0)$ is any solution of the system (8) for $t \geq \tau_0 \geq 0, t_0 \in \mathbb{R}_+$ and $\eta = \tau_0 - t_0$. If given any $\alpha_1 > 0, \gamma_1 > 0, \varepsilon_1 > 0$ and $\tau_0 \in \mathbb{R}_+$, for every $\alpha_2 < \alpha_1$ and $\gamma_2 < \gamma_1$, there exist $\varepsilon_2 < \varepsilon_1, T_1 = T_1(\varepsilon_1,\tau_0)$ and $T_2 = T_2(\varepsilon_1,\tau_0)$ such that

$$h(t, y(t,\tau_0,y_0) - x(t - \eta, t_0, x_0)) < \varepsilon_1, T_1 + \tau_0 \leq t \leq T_2 + \tau_0,$$

whenever $h(\tau_0, y_0 - x_0) < \alpha_1$ and $h_0(\tau_0, \tau_0 - t_0) < \gamma_1$ and

$$h(t, y(t,\tau_0,y_0) - x(t - \eta, t_0, x_0)) > \varepsilon_2, T_2 + \tau_0 \geq t \geq T_1 + \tau_0,$$

whenever $h(\tau_0, y_0 - x_0) > \alpha_2$ and $h_0(\tau_0, \tau_0 - t_0) > \gamma_2$.

If T_1 and T_2 in Definition 3 are independent of τ_0, then the solution $y(t,\tau_0,y_0)$ of the system (10) is initial time difference $(h_0 - h)$−strictly uniformly attractive in fractional case with respect to the solution $x(t - \eta, t_0, x_0)$ for $t \geq \tau_0$.

Definition 10.4. The solution $y(t,\tau_0,y_0)$ of the system (10) through (τ_0,y_0) is said to be initial time difference $(h_0 - h)$−strictly asymptotically stable in fractional case with respect to the solution $x(t - \eta, t_0, x_0)$ if Definition 3 satisfies and the solution $y(t,\tau_0,y_0)$ of the perturbed system (10) through (τ_0,y_0) is initial time difference $(h_0 - h)$−strictly stable in fractional case with respect to the solution $x(t - \eta, t_0, x_0)$ of the unperturbed system (8).

If T_1 and T_2 in Definition 3 are independent of τ_0, then the solution $y(t,\tau_0,y_0)$ of the system (10) is initial time difference $(h_0 - h)$−uniformly strictly asymptotically stable in fractional case with respect to the solution $x(t - \eta, t_0, x_0)$ for $t \geq \tau_0$.

Definition 10.5. For any real-valued function $V \in C[\mathbb{R}_+ \times \mathbb{R}^n, \mathbb{R}_+]$, we define the fractional order Dini derivatives in Caputo's sense

$$^{c}D_+^q V(t,x) = \lim_{h \to 0^+} \sup \frac{1}{h^q} [V(t,x) - V(t - h, x - h^q f(t,x))],$$

and

$$^{c}D_-^q V(t,x) = \lim_{h \to 0^-} \inf \frac{1}{h^q} [V(t,x) - V(t - h, x - h^q f(t,x))],$$

where $x(t) = x(t,t_0,x_0)$ for $(t,x) \in \mathbb{R}_+ \times \mathbb{R}^n$.

Definition 10.6. For a real-valued function $V(t,x) \in C[\mathbb{R}_+ \times \mathbb{R}^n, \mathbb{R}_+]$ we define the generalized fractional order derivatives (Dini-like derivatives) in Caputo's sense $^{c}_{*}D_+^q V(t,y - \tilde{x})$ and $^{c}_{*}D_-^q V(t,y - \tilde{x})$ as follows

$$^{c}_{*}D_+^q V(t,y - \tilde{x}) = \lim_{h \to 0^+} \sup \left[\frac{V(t,y - \tilde{x}) - V(t - h, y - \tilde{x} - h^q H(t,y,\tilde{x}))}{h^q} \right]$$

$$^{c}_{*}D_-^q V(t,y - \tilde{x}) = \lim_{h \to 0^-} \inf \left[\frac{V(t,y - \tilde{x}) - V(t - h, y - \tilde{x} - h^q H(t,y,\tilde{x}))}{h^q} \right],$$

where $H(t,y,\tilde{x}) = F(t,y) - \tilde{f}(t,\tilde{x}))$ for $(t,x) \in \mathbb{R}_+ \times \mathbb{R}^n$,

Definition 10.7. The class $\mathcal{K}$ is set of functions such that $\mathcal{K} := [a : a \in C[[0,\rho], \mathbb{R}_+],$ a is strictly increasing and $a(0) = 0$ and also $a(t) \to \infty$ as $t \to \infty]$.

Definition 10.8. A function $h(t,x)$ is said to belong to the class Γ if $h \in C[\mathbb{R}_+ \times \mathbb{R}^n, \mathbb{R}_+]$, $\inf_{(t,x)} h(t,x) = 0$ for all $(t,x) \in \mathbb{R}_+ \times \mathbb{R}^n$.

Definition 10.9. A function $h(t,x)$ is said to belong to the class Γ_0 if $h \in \Gamma$, $\sup_{t \in \mathbb{R}_+} h(t,x)$ exist for $x \in \mathbb{R}^n$.

4 Main Results

In this section we obtain the strict stability concepts in fractional case with initial time difference parallel to the Lyapunov's results.

Theorem 10.1. *Assume that*

(A_1) for each $\mu, 0 < \mu < \rho, V_\mu \in C[\mathbb{R}_+ \times S_\rho, \mathbb{R}_+]$ and V_μ is locally Lipschitzian in z and for $(t,z) \in \mathbb{R}_+ \times S_\rho$ and $h(t,z) \geq \mu$,

$$b_1(h(t,z)) \leq V_\mu(t,z) \leq a_1(h(t,z)), a_1, b_1 \in \mathcal{K}$$

$$_*^c D_+^q V_\mu(t,z) \leq 0;$$

(A_2) for each $\theta, 0 < \theta < \rho, V_\theta \in C[\mathbb{R}_+ \times S_\rho, \mathbb{R}_+]$ and V_θ is locally Lipschitzian in z and for $(t,z) \in \mathbb{R}_+ \times S_\rho$ and $h(t,z) \leq \theta$,

$$b_2(h(t,z)) \leq V_\theta(t,z) \leq a_2(h(t,z)), a_2, b_2 \in \mathcal{K}$$

$$_*^c D_+^q V_\theta(t,z) \geq 0; \tag{10.11}$$

where $z(t) = y(t,\tau_0,y_0) - x(t-\eta,t_0,x_0)$ for $t \geq \tau_0$, $y(t,\tau_0,y_0)$ the solution of the system (10) through (τ_0,y_0), $x(t,t_0,x_0)$ is any solution of the system (8) for $t \geq \tau_0 \geq t_0 > 0$, and $\eta = \tau_0 - t_0$.

Then the solution $y(t,\tau_0,y_0)$ of the perturbed system (10) is the initial time difference (h_0,h)−strictly stable in fractional case with respect to the solution $x(t-\eta,t_0,x_0)$ of the unperturbed system, where $x(t,t_0,x_0)$ is any solution of the system (8) for $t \geq \tau_0 \geq t_0 > 0$.

Proof. Let $0 < \varepsilon_1 < \rho$ and $\tau_0 \in \mathbb{R}_+$ and choose $\delta_1 = \delta_1(\varepsilon_1,\tau_0) > 0$ and $\delta_2 = \delta_2(\varepsilon_1,\tau_0) > 0$ such that

$$a_1(\delta_1) < b_1(\varepsilon_1), \tag{10.12}$$

since we have $b_1(\varepsilon_1) \leq a_1(\delta_1)$ in (A_1). Then we claim that

$$h(t,y(t,\tau_0,y_0) - x(t-\eta,t_0,x_0)) < \varepsilon_1 \text{ for } t \geq \tau_0, \tag{10.13}$$

whenever $h(\tau_0,y_0 - x_0) < \delta_1$ and $h_0(\tau_0,\tau_0 - t_0) < \delta_2$.

If (14) is not true, then there would exist $t_1 > t_2 > \tau_0$ and the solution of (8) and from (11) with $h(\tau_0,y_0 - x_0) < \delta_1, h_0(\tau_0,\tau_0 - t_0) < \delta_2$ satisfying

$$h(t_1,y(t_1) - \tilde{x}(t_1)) = \varepsilon_1, h(t_2,y(t_2) - \tilde{x}(t_2)) = \delta_1$$

and $\delta_1 \leq h(t,y(t) - \tilde{x}(t)) \leq \varepsilon_1$ for $t \in [t_2,t_1]$, where $\tilde{x}(t) = x(t-\eta,t_0,x_0)$.

Let us set $\mu = \delta_1$, and using (A_1) we get

$$
\begin{aligned}
b_1(\varepsilon_1) &= b_1(h(t_1, y(t_1) - \tilde{x}(t_1))) \\
&\leq V_\mu(t_1, y(t_1) - \tilde{x}(t_1)) \\
&\leq V_\mu(t_2, y(t_2) - \tilde{x}(t_2)) \\
&\leq a_1(h(t_2, y(t_2) - \tilde{x}(t_2))) \\
&= a_1(\delta_1)
\end{aligned}
$$

and we have the inequality

$$ b_1(\varepsilon_1) \leq a_1(\delta_1), $$

which contradicts with (13). Hence (14) is valid.

Now let $0 < \delta_1^* < \delta_1, 0 < \delta_2^* < \delta_2$ and choose $0 < h(\tau_0, y_0 - x_0) < \delta_1^* < \delta_1$ and $0 < h_0(\tau_0, \tau_0 - t_0) < \delta_2^* < \delta_2$ for $0 < \varepsilon_2 < \delta = \min\{\delta_1^*, \delta_2^*\}$ such that

$$ a_2(\varepsilon_2) < b_2(\delta). \tag{10.14} $$

Then we can prove that

$$ \varepsilon_2 < h(t, y(t, \tau_0, y_0) - x(t - \eta, t_0, x_0)) < \varepsilon_1 \quad \text{for } t \geq \tau_0, \tag{10.15} $$

whenever $\delta_1^* < h(\tau_0, y_0 - x_0) < \delta_1$ and $\delta_2^* < h_0(\tau_0, \tau_0 - t_0) < \delta_2$.

If (16) is not true, then there would exist $t_1 > t_2 > \tau_0$ and the solution of (8) and (22) with $\delta_1^* < h(\tau_0, y_0 - x_0) < \delta_1, \delta_2^* < h_0(\tau_0, \tau_0 - t_0) < \delta_2$ satisfying

$$ h(t_1, y(t_1) - \tilde{x}(t_1)) = \varepsilon_2, \tag{10.16} $$

$h(t_2, y(t_2) - \tilde{x}(t_2)) = \delta$ and $h(t, y(t) - \tilde{x}(t)) \leq \delta$, $t \in [t_2, t_1]$.

Let us set $\theta = \delta$ and using (A_2), we get

$$
\begin{aligned}
a_2(\varepsilon_2) &= a_2(h(t_1, y(t_1) - \tilde{x}(t_1))) \\
&\geq V_\theta(t_1, y(t_1) - \tilde{x}(t_1)) \\
&\geq V_\theta(t_2, y(t_2) - \tilde{x}(t_2)) \\
&\geq b_2(h(t_2, y(t_2) - \tilde{x}(t_2))) \\
&= b_2(\delta)
\end{aligned}
$$

and we have the inequality

$$ a_2(\varepsilon_2) \geq b_2(\delta) $$

which contradicts with (15). Thus (16) is valid.

Then the solution $y(t,\tau_0,y_0)$ of the perturbed system (10) through (τ_0,y_0) is initial time difference (h_0,h)−strictly stable in fractional case with respect to the solution of unperturbed system $x(t-\eta,t_0,x_0)$ for $t \geq \tau_0$.

This completes the proof of Theorem 1.   □

If δ_1,δ_2 and ε_2 in the proof of the Theorem 1 are chosen independent of τ_0, then the solution $y(t,\tau_0,y_0)$ of the perturbed system (10) is initial time difference (h_0-h)−strictly uniformly stable in fractional case with respect to the solution $x(t-\eta,t_0,x_0)$ for $t \geq \tau_0$.

Theorem 10.2. *Assume that*

(A_1) for each $\mu,0<\mu<\rho,V_\mu \in C[\mathbb{R}_+ \times S_\rho,\mathbb{R}_+]$ and V_μ is locally Lipschitzian in z and for $(t,z) \in \mathbb{R}_+ \times S_\rho$ and $h(t,z) \geq \mu$,

$$b_1(h(t,z)) \leq V_\mu(t,z) \leq a_1(h(t,z)), a_1,b_1 \in \mathcal{K},$$

$$^c_*D^q_+V_\mu(t,z) \leq -c_1(h(t,z)), c_1 \in \mathcal{K}; \tag{10.17}$$

(A_2) for each $\theta,0<\theta<\rho,V_\theta \in C[\mathbb{R}_+ \times S_\rho,\mathbb{R}_+]$ and V_θ is locally Lipschitzian in z and for $(t,z) \in \mathbb{R}_+ \times S_\rho$ and $h(t,z) \leq \theta$,

$$b_2(h(t,z)) \leq V_\theta(t,z) \leq a_2(h(t,z)), a_2,b_2 \in \mathcal{K},$$

$$^c_*D^q_+V_\theta(t,z) \geq -c_2(h(t,z)) c_2 \in \mathcal{K}; \tag{10.18}$$

where $z(t) = y(t,\tau_0,y_0) - x(t-\eta,t_0,x_0)$ for $t \geq \tau_0, y(t,\tau_0,y_0)$ of the perturbed system (10) through (τ_0,y_0) and $x(t-\eta,t_0,x_0)$, where $x(t,t_0,x_0)$ is any solution of the unperturbed system (8) for $t \geq \tau_0 \geq t_0 > 0$.

Then the solution $y(t,\tau_0,y_0)$ of the perturbed system (10) through (τ_0,y_0) is the initial time difference (h_0,h)−strictly uniformly asymptotically stable in fractional case with respect to the solution $x(t-\eta,t_0,x_0)$ of the unperturbed system, where $x(t,t_0,x_0)$ is any solution of the unperturbed system (8) for $t \geq \tau_0 \geq t_0 > 0$.

Proof. We note that (18) implies (11). However, (19) does not yield (12). Therefore, we get because of (18) only (h_0,h)−uniformly stability in fractional case of perturbed systems related to initial time difference with respect to unperturbed systems, that is, for given any $\varepsilon_1 \leq \rho$ and $\tau_0 \in \mathbb{R}_+$ there exist $\delta_{10} = \delta_{10}(\varepsilon_1) > 0$ and $\delta_{20} = \delta_{20}(\varepsilon_1) > 0$ such that

$$h(t,y(t,\tau_0,y_0) - x(t-\eta,t_0,x_0)) < \varepsilon_1 \text{ for } t \geq \tau_0$$

whenever $h(\tau_0,y_0-x_0) < \delta_{10}$ and $h_0(\tau_0,\tau_0-t_0) < \delta_{20}$.

To prove the conclusion of Theorem 2 we need to show that the solution $y(t,\tau_0,y_0)$ of the system (10) through (τ_0,y_0) is initial time difference (h_0,h)−strictly uniformly attractive in fractional case with respect to $x(t-\eta,t_0,x_0)$

for this purpose, let $\varepsilon_1 = \rho$ and set $\delta_{10} = \delta_1(\rho)$ and $\delta_{20} = \delta_2(\rho)$ so that (20) yields

$$h(t, y(t, \tau_0, y_0) - x(t - \eta, t_0, x_0)) < \rho \quad \text{for } t \geq \tau_0$$

whenever $h(\tau_0, y_0 - x_0) < \delta_{10}$ and $h_0(\tau_0, \tau_0 - t_0) < \delta_{20}$.

Let $h(\tau_0, y_0 - x_0) < \delta_{10}$ and $h_0(\tau_0, \tau_0 - t_0) < \delta_{20}$. We show, using standard argument, that there exists a $t^* \in [\tau_0, \tau_0 + T]$, we choose

$T = T(\varepsilon, \tau_0) \geq \left(\frac{a_1(\max\{\delta_{10}, \delta_{20}\})}{c_1(\min\{\delta_1, \delta_2\})} \Gamma(q+1) \right)^{\frac{1}{q}}$ where δ_{10} and δ_{20} are the numbers corresponding to ε_1 in (20) ,that is, in initial time difference (h_0, h)–uniformly stability in fractional case of perturbed system with respect to $x(t - \eta, t_0, x_0)$ such that $h(\tau_0, y(t^*, \tau_0, y_0) - x(t^* - \eta, t_0, x_0)) < \delta_1$ for any solutions of the system (10) with $h(\tau_0, y_0 - x_0) < \delta_{10}$ and $h_0(\tau_0, \tau_0 - t_0) < \delta_{20}$. If this is not true, we will have $h(t, y(t, \tau_0, y_0) - x(t - \eta, t_0, x_0)) \geq \delta_1$ for $t \in [\tau_0, \tau_0 + T]$. Then, $\mu = \delta_1$ and using (A_1) with (18), we have

$$0 < b_1(\delta_1) \leq b_1(h(\tau_0 + T, y(\tau_0 + T) - \tilde{x}(\tau_0 + T)))$$
$$\leq V_\mu(\tau_0 + T, y(\tau_0 + T) - \tilde{x}(\tau_0 + T))$$
$$\leq V_\mu(\tau_0, y_0 - x_0)$$
$$- \frac{1}{\Gamma(q)} \int_{\tau_0}^{\tau_0 + T} (t - s)^{q-1} c_1(h(s, y(s) - \tilde{x}(s))) ds$$
$$\leq a_1(\max\{\delta_{10}, \delta_{20}\})$$
$$- \frac{c_1(\min\{\delta_1, \delta_2\})}{\Gamma(q)} \int_{\tau_0}^{\tau_0 + T} (t - s)^{q-1} ds$$
$$\leq a_1(\max\{\delta_{10}, \delta_{20}\}) - \frac{c_1(\min\{\delta_1, \delta_2\})}{\Gamma(q+1)} T^q$$
$$\leq 0$$

in view of the choice of T. This contradiction implies that there exist a $t^* \in [\tau_0, \tau_0 + T]$ satisfying $h(\tau_0, y(t^*, \tau_0, y_0) - x(t^* - \eta, t_0, x_0)) < \delta_1$. Due to the (h_0, h)–uniform stability in fractional case $y(t, \tau_0, y_0)$ of the perturbed system with initial time difference with respect to $x(t - \eta, t_0, x_0)$, this yields that

$$h(t, y(t, \tau_0, y_0) - x(t - \eta, t_0, x_0)) < \varepsilon_1, t \geq \tau_0 + T \geq t^*$$

which implies that there exists a $\tau_0 < T_1 < T$ such that

$$h(\tau_0 + T, y(\tau_0 + T) - x(\tau_0 + T - \eta)) = \varepsilon_1$$

Now, for any $\delta_{12}, 0 < \delta_{12} < \delta_{10}$ and $0 < \delta_{12} < \delta_{20}$ we choose ε_2 such that $b_2(\delta_{12}) > a_2(\varepsilon_2)$ and $0 < \varepsilon_2 < \varepsilon_1 < \delta_{12}$.

Suppose that $\delta_{12} < h(\tau_0, y_0 - x_0) < \min\{\delta_{10}, \delta_{20}\}$ and $\delta_{12} < h_0(\tau_0, \tau_0 - t_0) < \min\{\delta_{10}, \delta_{20}\}$. Let us define $\tau = \left[\frac{\Gamma(q)(b_2(\varepsilon_1) - a_2(\varepsilon_2))}{c_2(\varepsilon_1)}\right]^{\frac{1}{q}}$, and $T_2 = T_1 + \tau$. Since $h(t, y(t) - \tilde{x}(t)) \leq \varepsilon_1$ for $t \geq \tau_0 + T_1$, choosing $\theta = \varepsilon_1$ and using (A_2) with (19) we have for $t \in [\tau_0 + T_1, \tau_0 + T_2]$,

$$a_2(\|y(t) - \tilde{x}(t)\|) \geq V_\theta(t, y(t) - \tilde{x}(t))$$

$$\geq V_\theta(\tau_0 + T_1, y(\tau_0 + T_1) - \tilde{x}(\tau_0 + T_1))$$

$$- \frac{1}{\Gamma(q)} \int_{\tau_0 + T_1}^{t} (t-s)^{q-1} c_2(h(s, y(s) - \tilde{x}(s))) ds$$

$$\geq b_2(\varepsilon_1)$$

$$- \frac{1}{\Gamma(q)} \int_{\tau_0 + T_1}^{t} (t-s)^{q-1} c_2(h(s, y(s) - \tilde{x}(s))) ds$$

$$\geq b_2(\varepsilon_1) - \frac{c_2(\varepsilon_1)}{\Gamma(q)}[t - (\tau_0 + T_1)]^q.$$

Since, $t - (\tau_0 + T_1) > \tau$ and a_2^{-1} exists, it follows that

$$a_2(\|y(t) - \tilde{x}(t)\|) > b_2(\varepsilon_1) - \frac{c_2(\varepsilon_1)}{\Gamma(q)}\left[\frac{\Gamma(q)(b_2(\varepsilon_1) - a_2(\varepsilon_2))}{c_2(\varepsilon_1)}\right]$$

$$= a_2(\varepsilon_2).$$

This yields that,

$$h(t, y(t, \tau_0, y_0) - x(t - \eta, t_0, x_0)) \geq \varepsilon_2 \text{ for } t \in [\tau_0 + T_1, \tau_0 + T_2]$$

and therefore,

$$\varepsilon_2 < h(t, y(t, \tau_0, y_0) - x(t - \eta, t_0, x_0)) < \varepsilon_1, t \in [\tau_0 + T_1, \tau_0 + T_2].$$

Then the solution $y(t, \tau_0, y_0)$ of the perturbed system (10) through (τ_0, y_0) is initial time difference (h_0, h)−strictly uniformly asymptotically stable in fractional case with respect to the solution $x(t - \eta, t_0, x_0)$, where $x(t, t_0, x_0)$ is any solution of the unperturbed system (8) for $t \geq \tau_0 \geq t_0 > 0$. This completes the proof.   □

Before we express the comparison result in fractional case, we need to give uncoupled comparison fractional order differential systems and to define (h_0, h)−strictly stability in fractional case of comparison fractional order differential systems. Consider the uncoupled comparison fractional order differential systems:

$$\begin{cases} (i)^c D^q u_1 = g_1(t, u_1), & u_1(\tau_0) = u_{10} \geq 0, \\ (ii)^c D^q u_2 = g_2(t, u_2), & u_2(\tau_0) = u_{20} \geq 0, \end{cases} \quad (10.19)$$

where $g_1, g_2 \in C\left[\mathbb{R}_+^2, \mathbb{R}\right]$. The fractional order comparison system (21) is said to be (h_0, h)−strictly stable in fractional case:

   If given any $\varepsilon_1 > 0$ and $t \geq \tau_0, \tau_0 \in \mathbb{R}_+$, there exist a $\delta_1 > 0$ such that

$$h_0(\tau_0, u_{10}) < \delta_1 \text{ implies } h(t, u_1(t)) < \varepsilon_1 \text{ for } t \geq \tau_0,$$

and for every $\delta_2 \leq \delta_1$ there exists an $\varepsilon_2, 0 < \varepsilon_2 < \delta_2$ such that

$$h_0(\tau_0, u_{20}) > \delta_2 \text{ implies } h(t, u_2(t)) > \varepsilon_2 \text{ for } t \geq \tau_0.$$

Here, $u_1(t)$ and $u_2(t)$ are any solutions of (i) in (21) and (ii) in (21); respectively. Following this theorem based on this definition and an another theorem is formulated in terms of comparison principle.

Theorem 10.3. *Assume that*
   (A_1) *for each* $\mu, 0 < \mu < \rho, V_\mu \in C[\mathbb{R}_+ \times S_\rho, \mathbb{R}_+]$ *and* V_μ *is locally Lipschitzian in z and for* $(t, z) \in \mathbb{R}_+ \times S_\rho$ *and* $h(t, z) \geq \eta$,

$$b_1(h(t, z)) \leq V_\mu(t, z) \leq a_1(h(t, z)), a_1, b_1 \in \mathcal{K},$$
$$^c_* D_+^q V_\mu(t, z) \leq g_1(t, V_\mu(t, z)); \tag{10.20}$$

   (A_2) *for each* $\theta, 0 < \theta < \rho, V_\theta \in C[\mathbb{R}_+ \times S_\rho, \mathbb{R}_+]$ *and* V_θ *is locally Lipschitzian in z and for* $(t, z) \in \mathbb{R}_+ \times S_\rho$ *and* $h(t, z) \leq \theta$,

$$b_2(h(t, z)) \leq V_\theta(t, z) \leq a_2(h(t, z)), a_2, b_2 \in \mathcal{K},$$
$$^c_* D_+^q V_\theta(t, z) \geq g_2(t, V_\theta(t, z)); \tag{10.21}$$

where $g_2(t, u) \leq g_1(t, u), g_1, g_2 \in C[\mathbb{R}_+^2, \mathbb{R}], g_1(t, 0) \equiv g_2(t, 0) \equiv 0$ *and* $z(t) = y(t, \tau_0, y_0) - x(t - \eta, t_0, x_0)$ *for* $t \geq \tau_0, y(t, \tau_0, y_0)$ *of the system (10) through* (τ_0, y_0) *and* $x(t - \eta, t_0, x_0)$, *where* $x(t, t_0, x_0)$ *is any solution of the system (8) for* $t \geq \tau_0 \geq t_0 > 0$.
   Then any (h_0, h)−*strict stability concept in fractional case of the comparison system implies the corresponding* (h_0, h)−*strict stability concept in fractional case of the solution* $y(t, \tau_0, y_0)$ *of the perturbed system (10) through* (τ_0, y_0) *with respect to the solution* $x(t - \eta, t_0, x_0)$ *of the unperturbed system (8) with initial time difference where* $x(t, t_0, x_0)$ *is any solution of the unperturbed system (8) for* $t \geq \tau_0 \geq t_0 > 0$.

Proof. First we will prove the case of initial time difference (h_0, h)−strictly uniformly stability in fractional case of the perturbed system with respect to the unperturbed system. Suppose that the comparison differential systems in (21) is (h_0, h)− strictly uniformly stable in fractional case, then for any given $\varepsilon_1, 0 < \varepsilon_1 < \delta$, there exist a $\delta^* > 0$ such that

$$0 < u_{10} < \delta^* \text{ implies that } u_1(t, \tau_0, u_{10}) < b_1(\varepsilon_1) \text{ for } t \geq \tau_0, \tag{10.22}$$

where $u_1(t) = u_1(t, \tau_0, u_{10})$ is the solution of (21).

For this $\varepsilon_1 > 0$, we choose $\delta_1 > 0$ and $\delta_{11} > 0$, such that $a_1(\delta_1) \leq \delta^*$ and $\delta_1^* < \varepsilon_1$ where $\delta_1^* = \max\{\delta_1, \delta_{11}\}$, then we claim that

$$h(\tau_0, y_0 - x_0) < \delta_1, h_0(\tau_0, \tau_0 - t_0) < \delta_{11}, \tag{10.23}$$

$$h(t, y(t, \tau_0, y_0) - x(t - \eta, t_0, x_0)) < \varepsilon_1, t \geq \tau_0.$$

If it is not true, then there exist t_1 and t_2, $t_2 > t_1 > \tau_0$ and a solution of

$$^cD^q z = \tilde{f}(t, z), z(\tau_0) = y_0 - x_0 \text{ for } t \geq \tau_0$$

with $h_0(\tau_0, \tau_0 - t_0) < \delta_{11}$ and $h(\tau_0, y_0 - x_0) < \delta_1$.

$$h(t_1, y(t_1, \tau_0, y_0) - x(t_1 - \eta, t_0, x_0)) = \delta_1^*,$$

$$h(t_2, y(t_2, \tau_0, y_0) - x(t_2 - \eta, t_0, x_0)) = \varepsilon_1 \text{ and}$$

$$\delta_1^* < h(t, y(t, \tau_0, y_0) - x(t - \eta, t_0, x_0)) < \varepsilon_1 \text{ for } [t_1, t_2].$$

Choosing $\mu = \delta_1^*$ and using the theory of differential inequalities, together with (A_1), we obtain (22) and (24)

$$b_1(\varepsilon_1) = b_1(h(t_2, y(t_2, \tau_0, y_0) - x(t_2 - \eta, t_0, x_0)))$$

$$\leq V_\mu(t_2, y(t_2, \tau_0, y_0) - x(t_2 - \eta, t_0, x_0))$$

$$\leq r(t_2, t_1, V_\mu(t_1, y(t_1, \tau_0, y_0) - x(t_1 - \eta, t_0, x_0)))$$

$$\leq r(t_2, t_1, a_1(\delta_1))$$

$$\leq r(t_2, t_1, \delta^*)$$

$$< b_1(\varepsilon_1).$$

which is a contradiction. Here $r(t, t_0, u_{10})$ is the maximal solution of (21). Hence, (25) is true and we have initial time difference (h_0, h)−strictly uniformly stability in fractional case.

Now, we shall prove initial time difference (h_0, h)−strictly uniformly attractive in fractional case.

For any given δ_2, $\varepsilon_2 > 0$, $\delta_2 < \delta^*$ we choose $\overline{\delta}_2$ and $\overline{\varepsilon}_2$ such that $a_1(\delta_2) < \overline{\delta}_2$ and $b_1(\varepsilon_2) \geq \overline{\varepsilon}$. For these $\overline{\delta}_2$ and $\overline{\varepsilon}_2$, since (21) is strictly uniformly attractive in fractional case, for any $\overline{\delta}_3 < \overline{\delta}_2$ there exist $\overline{\varepsilon}_3$ and T_1 and T_2 (we assume $T_2 < T_1$) such that $\overline{\delta}_3 < u_{10} = u_{20} < \overline{\delta}_2$ implies

$$r(t, \tau_0, u_{10}) \leq r(t, \tau_0, \overline{\delta}_2) < \overline{\varepsilon}_2,$$

$$\rho(t, \tau_0, u_{20}) \geq \rho(t, \tau_0, \overline{\delta}_3) > \overline{\varepsilon}_2,$$

where $r(t, \tau_0, u_{10})$ and $\rho(t, \tau_0, u_{20})$ is the maximal solution and minimal solution of (21) (i) and (ii); respectively.

Now, for any δ_3, let $b_2(\delta_3) \geq \overline{\delta_3}$. We choose ε_3 such that $a_2(\varepsilon_3) < \overline{\varepsilon_3}$. Then by using comparison principle in fractional case (21), (i) and (A_1), we have

$$b_1(h(t,y(t,\tau_0,y_0)) - x(t-\eta,t_0,x_0)))$$
$$\leq V_\mu(t,y(t,\tau_0,y_0)) - x(t-\eta,t_0,x_0)))$$
$$\leq r(t,\tau_0,V_\mu(\tau_0,y_0-x_0))$$
$$\leq r(t,\tau_0,a_1(h(\tau_0,y_0-x_0)))$$
$$\leq r(t,\tau_0,\overline{\delta_2})$$
$$< \overline{\varepsilon_2} \leq b_1(\varepsilon_2)$$

$$b_1(h(t,y(t,\tau_0,y_0)) - x(t-\eta,t_0,x_0))) < b_1(\varepsilon_2), \qquad (10.24)$$

which implies that $h(t,y(t,\tau_0,y_0)) - x(t-\eta,t_0,x_0)) < \varepsilon_2$ for $t \in [\tau_0 + T_2, \tau_0 + T_1]$.

Similarly, by using comparison principle in fractional case (21), (ii) and (A_2), we get

$$a_2(h(t,y(t,\tau_0,y_0)) - x(t-\eta,t_0,x_0)))$$
$$\geq V_\theta(t,y(t,\tau_0,y_0)) - x(t-\eta,t_0,x_0))$$
$$\geq \rho(t,\tau_0,V_\theta(\tau_0,y_0-x_0))$$
$$\geq \rho(t,\tau_0,b_2(h(\tau_0,y_0-x_0)))$$
$$\geq \rho(t,\tau_0,b_2(\delta_3))$$
$$\geq \rho(t,\tau_0,\overline{\delta_3})$$
$$> \overline{\varepsilon_3} \geq a_2(\varepsilon_3)$$

$$a_2(h(t,y(t,\tau_0,y_0)) - x(t-\eta,t_0,x_0))) > a_2(\varepsilon_3), \qquad (10.25)$$

which implies that for $h(t,y(t,\tau_0,y_0)) - x(t-\eta,t_0,x_0)) > \varepsilon_3$ for $\tau_0 + T_2 < t < \tau_0 + T_1$. Thus (26) and (27) yield that

$$\varepsilon_3 < h(t,y(t,\tau_0,y_0)) - x(t-\eta,t_0,x_0)) < \varepsilon_2, \ \tau_0 + T_2 < t < \tau_0 + T_1$$

provided that $\delta_2 < h(\tau_0,y_0-x_0) < \delta_1$ and $\delta^* < h_0(\tau_0,\tau_0-t_0) < \delta_{11}$. Hence, the solution $y(t,\tau_0,y_0)$ of the perturbed system of (10) through (τ_0,y_0) is initial time difference (h_0,h)–strictly uniformly attractive in fractional case with respect to the solution $x(t-\eta,t_0,x_0)$ of the unperturbed system where $x(t,t_0,x_0)$ is any solution of the unperturbed system of (8) for $t \geq \tau_0 \geq t_0 > 0$. Therefore the proof is completed. □

References

1. Caputo M (1967) Linear models of dissipation whose Q is almost independent. II Geophys J Roy Astron 13:529–539
2. Kilbas AA, Srivastava HM, Trujillo JJ (2006) Theory and applications of fractional differential equations. Elsevier, Amsterdam, North-Holland, The Netherlands
3. Lakshmikantham V, Leela S (1969) Differential and integral inequalities, vol 1. Academic, New York
4. Lakshmikantham V, Leela S, Martynyuk AA (1989) Stability analysis of nonlinear systems. Marcel Dekker, New York
5. Lakshmikantham V, Leela S, Vasundhara Devi J (2009) Theory of fractional dynamical systems. Cambridge Scientific Publishers, Cambridge
6. Lakshmikantham V, Mohapatra RN (2001) Strict stability of differential equations. Nonlinear Anal 46(7):915–921
7. Lakshmikantham V, Vatsala AS (1999) Differential inequalities with time difference and application. J Inequalities Appl 3:233–244
8. Lakshmikanthama V, Vatsala AS (2008) Basic theory of fractional differential equations. Nonlinear Anal 69:2677–2682
9. Oldham KB, Spanier J (1974) The fractional calculus. Academic, New York
10. Podlubny I (1999) Fractional differential equations. Academic, New York
11. Samko S, Kilbas A, Marichev O (1993) Fractional integrals and derivatives: Theory and applications. Gordon and Breach, 1006 p, ISBN 2881248640
12. Shaw MD, Yakar C (1999) Generalized variation of parameters with initial time difference and a comparison result in term Lyapunov-like functions. Int J Non-linear Diff Equations Theory Methods Appl 5:86–108
13. Shaw MD, Yakar C (2000) Stability criteria and slowly growing motions with initial time difference. Probl Nonlinear Anal Eng Syst 1:50–66
14. Sivasundaram S (2001) The strict stability of dynamic systems on time scales. J Appl Math Stoch Anal 14(2):195–204
15. Yakar C (2007) Boundedness Criteria in Terms of Two Measures with Initial Time Difference. Dynamics of Continuous, Discrete and Impulsive Systems. Series A: Mathematical Analysis. Watam Press. Waterloo, pp 270–275. DCDIS 14 (S2) 1–305
16. Yakar C (2010) Fractional Differential Equations in Terms of Comparison Results and Lyapunov Stability with Initial Time Difference. J Abstr Appl Anal AAA/762857. Vol 3. Volume 2010, Article ID 762857, 16 pages. doi:10.1155/2010/762857
17. Yakar C (2007) Strict stability criteria of perturbed systems with respect to unperturbed systems in term of initial time difference. Proceedings of the conference on complex analysis and potential theory. World Scientific Publishing, pp 239–248
18. Yakar C, Shaw MD (2005) A Comparison result and lyapunov stability criteria with initial time difference. Dyn Continuous Discrete Impulsive Syst A Math Anal 12(6):731–741
19. Yakar C, Shaw MD (2008) Initial time difference stability in terms of two measures and variational comparison result. Dyn Continuous Discrete Impulsive Syst A Math Anal 15:417–425
20. Yakar C, Shaw MD (2009) Practical stability in terms of two measures with initial time difference. Nonlinear Anal Theory Methods Appl 71:e781–e785

Chapter 11
Initial Time Difference Strict Stability of Fractional Dynamic Systems

Coşkun Yakar and Mustafa Bayram Gücen

1 Introduction

The theory of fractional differential equations has been realized that it has very useful models for studying and understanding the various disciplines and processes of engineering applications. The development and improvement of this theory are very important for mathematics and other areas of modern science. The history of fractional order derivative and fractional order differential equations goes back to the Seventeenth century. It had always attracted the interest of many famous mathematicians, including L' Hospital, Leibnitz, Liouville, Riemann, Grünwald and Letnikov [2, 5, 8–10]. In recent decades, fractional order differential equations have been found to be a powerful tool for some fields, such as, physics, mechanics, and engineering. It was realized that the derivatives of non-integer order provide an perfect framework for modeling of the real world applications in related disciplines from physics, chemistry and engineering [2, 10].

The strict stability criteria had been studied by Sivasundaram [13] and Yakar [16].

We have investigated the strict stability criteria between two unperturbed differential systems with different initial time and initial position of fractional order with initial time difference. The differential operators are taken in the Riemann–Liouville and Caputo's sense and the initial conditions are specified according to Caputo's suggestion [1], thus allowing for interpretation in a physically meaningful way [2, 5, 9, 10]. The initial time difference stability is very important for dynamic systems. It has been worked by Lakshmikantham and Vatsala [7] and Yakar

C. Yakar (✉)
Gebze Institute of Technology, Gebze, Kocaeli, Turkey 141-41400
e-mail: cyakar@gyte.edu.tr

M.B. Gücen
Yıldız Technical University, Davutpaşa, Esenler İstanbul, Turkey 34210
e-mail: mgucen@yildiz.edu.tr

D. Baleanu et al. (eds.), *Fractional Dynamics and Control*,
DOI 10.1007/978-1-4614-0457-6_11, © Springer Science+Business Media, LLC 2012

[11, 12, 14–19]. We develop initial time difference fractional strict stability criteria for unperturbed fractional order differential systems with Caputo's derivative. We establish a comparative results for unperturbed fractional order differential systems with respect to another unperturbed fractional order differential systems which have different initial position and initial time. The difference of these systems is that they have different initial conditions. The Lyapunov stability is with respect to null solution. The difference of these definitions and results from Lyapunov stability [4–6] is that these systems stability investigates with respect to another unperturbed fractional differential systems which have different initial position and initial time.

2 Definition and Notation

The definition of Caputo's and Reimann–Liouville's fractional derivatives are as follows:

$$^{c}D^{q}x = \frac{1}{\Gamma(1-q)} \int_{t_0}^{t} (t-s)^{-q} x'(s)ds, t_0 \leq t \leq \tau_0 + T, \qquad (11.1)$$

$$D^{q}x = \frac{1}{\Gamma(p)} \left(\frac{d}{dt} \int_{t_0}^{t} (t-s)^{p-1} x(s)ds \right), t_0 \leq t \leq \tau_0 + T, \qquad (11.2)$$

order of $0 < q < 1$, and $p + q = 1$ where Γ denotes the Gamma function.

The main advantage of Caputo's approach is that the initial conditions for fractional order differential equations with Caputo derivative take on the same form as that of ordinary differential equations with integer derivatives and another difference is that the Caputo derivative for a constant C is zero, while the Riemann–Liouville fractional derivative for a constant C is not zero but equals to $D^{q}C = \frac{C(t-t_0)^{-q}}{\Gamma(1-q)}$. By using (1) and therefore,

$$^{c}D^{q}x(t) = D^{q}[x(t) - x(t_0)], \qquad (11.3)$$

$$^{c}D^{q}x(t) = D^{q}x(t) - \frac{x(t_0)}{\Gamma(1-q)} (t-t_0)^{-q}. \qquad (11.4)$$

In particular, if $x(t_0) = 0$, we obtain

$$^{c}D^{q}x(t) = D^{q}x(t). \qquad (11.5)$$

Hence, we can see that Caputo's derivative is defined for the functions for which Riemann–Liouville fractional order derivative exists.

Consider the initial value problems of the fractional order differential equations with Caputo's fractional derivative

$$^{c}D^{q}x = f(t,x), \; x(t_0) = x_0 \quad \text{for } t \geq t_0, \; t_0 \in \mathbb{R}_{+}, \tag{11.6}$$

$$^{c}D^{q}y = f(t,y), \; y(\tau_0) = y_0 \quad \text{for } t \geq \tau_0 \geq t_0, \tag{11.7}$$

where $x_0 = \lim_{t \to t_0} D^{q-1}x(t)$ and $y_0 = \lim_{t \to \tau_0} D^{q-1}y(t)$ exist and $f \in C[[t_0, \tau_0 + T] \times \mathbb{R}^n, \mathbb{R}^n]$; satisfy a local Lipschitz condition on the set $\mathbb{R}_{+} \times S\rho$, $S\rho = [x \in \mathbb{R}^n : \|x\| \leq \rho < \infty]$ and $f(t,0) = 0$ for $t \geq 0$.

We assume that we have sufficient conditions to the existence and uniqueness of solutions through (t_0, x_0) and (τ_0, y_0). If $f \in C[[t_0, \tau_0 + T] \times \mathbb{R}^n, \mathbb{R}^n]$ and $x(t)$ is the solution of the system (6) where $^{c}D^{q}x$ is the Caputo fractional order derivative of x as in (1), then it also satisfies the Volterra fractional order integral equation

$$x(t) = x_0 + \frac{1}{\Gamma(q)} \int_{t_0}^{t} (t-s)^{q-1} f(s,x(s)) ds, \; t_0 \leq t \leq \tau_0 + T \tag{11.8}$$

and that is every solution of (6) is also a solution of (8), for details, see [3].

Let us give the definition of the strict stability criteria for unperturbed fractional differential systems with the initial time difference.

Definition 11.1. The solution $y(t, \tau_0, y_0)$ of the fractional order differential system (7) through (τ_0, y_0) is said to be the initial time difference strict stable in fractional case with respect to the solution $x(t - \eta, t_0, x_0)$, where $x(t, t_0, x_0)$ is any solution of the fractional order differential system (6) for $t \geq \tau_0 \geq 0, t_0 \in \mathbb{R}_{+}$ and $\eta = \tau_0 - t_0$. If given any $\varepsilon_1 > 0$ and $\tau_0 \in \mathbb{R}_{+}$ there exist $\delta_1 = \delta_1(\varepsilon_1, \tau_0) > 0$ and $\delta_2 = \delta_2(\varepsilon_1, \tau_0) > 0$ such that

$$\|y(t, \tau_0, y_0) - x(t - \eta, t_0, x_0)\| < \varepsilon_1 \; \text{for } t \geq \tau_0$$

whenever $\|y_0 - x_0\| < \delta_1$ and $|\tau_0 - t_0| < \delta_2$ and, for $\delta_1^* < \delta_1$ and $\delta_2^* < \delta_2$ there exist $\varepsilon_2 < \min\{\delta_1^*, \delta_2^*\}$ such that

$$\|y(t, \tau_0, y_0) - x(t - \eta, t_0, x_0)\| > \varepsilon_2 \; \text{for } t \geq \tau_0$$

whenever $\|y_0 - x_0\| > \delta_1^*$ and $|\tau_0 - t_0| > \delta_2^*$.

Definition 11.2. If δ_1, δ_2 and ε_2 in Definition 1 are independent of τ_0, then the solution $y(t, \tau_0, y_0)$ of the system (7) is the initial time difference uniformly strict stable in fractional case with respect to the solution $x(t - \eta, t_0, x_0)$ for $t \geq \tau_0$.

Definition 11.3. The solution $y(t, \tau_0, y_0)$ of the system (7) through (τ_0, y_0) is said to be the initial time difference strictly attractive in fractional case with respect to the solution $x(t - \eta, t_0, x_0)$, where $x(t, t_0, x_0)$ is any solution of the system (6) for

$t \geq \tau_0 \geq 0, t_0 \in \mathbb{R}_+$ and $\eta = \tau_0 - t_0$. If given any $\alpha_1 > 0, \gamma_1 > 0, \varepsilon_1 > 0$ and $\tau_0 \in \mathbb{R}_+$, for every $\alpha_2 < \alpha_1$ and $\gamma_2 < \gamma_1$, there exist $\varepsilon_2 < \varepsilon_1, T_1 = T_1(\varepsilon_1, \tau_0)$ and $T_2 = T_2(\varepsilon_1, \tau_0)$ such that

$$\|y(t, \tau_0, y_0) - x(t - \eta, t_0, x_0)\| < \varepsilon_1 \quad \text{for } T_1 + \tau_0 \leq t \leq T_2 + \tau_0$$

whenever $\|y_0 - x_0\| < \alpha_1$ and $|\tau_0 - t_0| < \gamma_1$ and

$$\|y(t, \tau_0, y_0) - x(t - \eta, t_0, x_0)\| > \varepsilon_2 \quad \text{for } T_2 + \tau_0 \geq t \geq T_1 + \tau_0$$

whenever $\|y_0 - x_0\| > \alpha_2$ and $|\tau_0 - t_0| > \gamma_2$.

If T_1 and T_2 in Definition 3 are independent of τ_0, then the solution $y(t, \tau_0, y_0)$ of the system (7) is the initial time difference uniformly strictly attractive stable with respect to the solution $x(t - \eta, t_0, x_0)$ for $t \geq \tau_0$.

Definition 11.4. The solution $y(t, \tau_0, y_0)$ of the system (7) through (τ_0, y_0) is said to be initial time difference strictly asymptotically stable in fractional case with respect to the solution $x(t - \eta, t_0, x_0)$ if Definition 3 satisfies and the solution $y(t, \tau_0, y_0)$ of the system (7) through (τ_0, y_0) is the initial time difference strictly stable with respect to the solution $x(t - \eta, t_0, x_0)$.

If T_1 and T_2 in Definition 3 are independent of τ_0, then the solution $y(t, \tau_0, y_0)$ of the system (7) is the initial time difference uniformly strictly asymptotically stable in fractional case with respect to the solution $x(t - \eta, t_0, x_0)$ for $t \geq \tau_0$.

Definition 11.5. For any real-valued function $V \in C[\mathbb{R}_+ \times \mathbb{R}^n, \mathbb{R}_+]$, we define the fractional order Dini derivatives in Caputo's sense

$$^cD_+^q V(t, x) = \lim_{h \to 0^+} \sup \frac{1}{h^q} [V(t, x) - V(t - h, x - h^q f(t, x))], \qquad (11.9)$$

where $x(t) = x(t, t_0, x_0)$ for $(t, x) \in \mathbb{R}_+ \times \mathbb{R}^n$.

Definition 11.6. For a real-valued function $V(t, x) \in C[\mathbb{R}_+ \times \mathbb{R}^n, \mathbb{R}_+]$, we define the generalized fractional order derivatives (Dini-like derivatives) in Caputo's sense $^c_*D_+^q V(t, y - \tilde{x})$ as follows:

$$^c_*D_+^q V(t, y - \tilde{x}) \qquad (11.10)$$

$$= \lim_{h \to 0^+} \sup \left[\frac{V(t, y - \tilde{x}) - V(t - h, y - \tilde{x} - h^q(f(t, y) - \tilde{f}(t, \tilde{x})))}{h^q} \right] \qquad (11.11)$$

for $(t, x) \in \mathbb{R}_+ \times \mathbb{R}^n$, where $\tilde{x} = x(t - \eta, t_0, x_0)$.

Definition 11.7. $\mathcal{K}$ is said to be the class $\mathcal{K}$ set of functions such that

$$\mathcal{K} := [a : a \in C([0, \rho], \mathbb{R}_+), a \text{ is strictly monotone increasing and } a(0) = 0].$$

3   Main Results

In this section we obtain the strict stability concepts with initial time difference for fractional differential equations parallel to the Lyapunov's results.

Theorem 11.1. *Assume that*
   (A_1) for each μ ,$0 < \mu < \rho, V_\mu \in C[\mathbb{R}_+ \times S_\rho, \mathbb{R}_+]$ and V_μ is locally Lipschitzian in z and for $(t,z) \in \mathbb{R}_+ \times S_\rho$ and $\|z\| \geq \mu$,

$$b_1(\|z\|) \leq V_\mu(t,z) \leq a_1(\|z\|), a_1, b_1 \in \mathcal{K}$$
$$^c_*D^q_+ V_\mu(t,z) \leq 0; \tag{11.12}$$

   (A_2) for each $\theta, 0 < \theta < \rho, V_\theta \in C[\mathbb{R}_+ \times S_\rho, \mathbb{R}_+]$ and V_θ is locally Lipschitzian in z and for $(t,z) \in \mathbb{R}_+ \times S_\rho$ and $\|z\| \leq \theta$,

$$b_2(\|z\|) \leq V_\theta(t,z) \leq a_2(\|z\|), a_2, b_2 \in \mathcal{K}$$
$$^c_*D^q_+ V_\theta(t,z) \geq 0; \tag{11.13}$$

where $z(t) = y(t, \tau_0, y_0) - x(t - \eta, t_0, x_0)$ for $t \geq \tau_0, y(t, \tau_0, y_0)$ of the system (7) through (τ_0, y_0) and $x(t - \eta, t_0, x_0)$, where $x(t, t_0, x_0)$ is any solution of the system (6) for $t \geq \tau_0 \geq 0, t_0 \in \mathbb{R}_+$ and $\eta = \tau_0 - t_0$.
   Then the solution $y(t, \tau_0, y_0)$ of the system (7) is the initial time difference strictly stable in fractional case with respect to $x(t - \eta, t_0, x_0)$ of the system (6) for $t \geq \tau_0 \geq 0, t_0 \in \mathbb{R}_+$ and $\eta = \tau_0 - t_0$.

Proof. Let us assume that $0 < \varepsilon_1 < \rho$ and $\tau_0 \in \mathbb{R}_+$. Let us choose that $\delta_1 = \delta_1(\varepsilon_1, \tau_0) > 0$ such that

$$a_1(\delta_1) < b_1(\varepsilon_1) \tag{11.14}$$

since we have $b_1(\varepsilon_1) \leq a_1(\delta_1)$ in (A_1). Then we claim that

$$\|y(t, \tau_0, y_0) - x(t - \eta, t_0, x_0)\| < \varepsilon_1 \text{ for } t \geq \tau_0 \tag{11.15}$$

whenever $\|y_0 - x_0\| < \delta_1$ and $|\tau_0 - t_0| < \delta_2$.
   If (14) is not true, then there exist $t_1 > t_2 > \tau_0$ and the solution of (6) and by using (11) with $\|y_0 - x_0\| < \delta_1, |\tau_0 - t_0| < \delta_2$ satisfying

$$\|y(t_1) - \tilde{x}(t_1)\| = \varepsilon_1, \|y(t_2) - \tilde{x}(t_2)\| = \delta_1$$

and $\delta_1 \leq \|y(t) - \tilde{x}(t)\| \leq \varepsilon_1, t \in [t_2, t_1]$ where $\tilde{x}(t) = x(t - \eta, t_0, x_0)$. Let us set $\mu = \delta_1$, we can obtain that

$$b_1(\varepsilon_1) = b_1(\|y(t_1) - \tilde{x}(t_1); \|) \leq V_\mu(t_1, y(t_1) - \tilde{x}(t_1));$$
$$\leq V_\mu(t_2, y(t_2) - \tilde{x}(t_2));$$

$$\leq a_1(\|y(t_2) - \tilde{x}(t_2)\|) = a_1(\delta_1),$$

$$b_1(\varepsilon_1) \leq a_1(\delta_1);$$

which contradicts with (13). Hence, (14) is valid.

Now let $0 < \delta_1^* < \delta_1, 0 < \delta_2^* < \delta_2$ and $\varepsilon_2 < \delta = \min\{\delta_1^*, \delta_2^*\}$ such that

$$a_2(\varepsilon_2) < b_2(\delta), \tag{11.16}$$

since we have $a_2(\varepsilon_2) \geq b_2(\delta)$ in (A_2). Then we can prove that

$$\varepsilon_2 < \|y(t, \tau_0, y_0) - x(t - \eta, t_0, x_0)\| < \varepsilon_1 \quad \text{for } t \geq \tau_0 \tag{11.17}$$

whenever $\delta_1^* < \|y_0 - x_0\| < \delta_1$ and $\delta_2^* < |\tau_0 - t_0| < \delta_2$.

In fact, if (16) is not true, then there would exist $t_1 > t_2 > \tau_0$ and the solution of (6) and by using (12) with $\delta_1^* < \|y_0 - x_0\| < \delta_1, \delta_2^* < |\tau_0 - t_0| < \delta_2$ satisfying

$$\|y(t_1) - \tilde{x}(t_1)\| = \varepsilon_2, \|y(t_2) - \tilde{x}(t_2)\| = \delta \quad \text{and} \quad \|y(t) - \tilde{x}(t)\| \leq \delta \quad \text{for } t \in [t_2, t_1]. \tag{11.18}$$

Let us set $\theta = \delta$ and by using (A_2), we get

$$a_2(\varepsilon_2) = a_2(\|y(t_1) - \tilde{x}(t_1)\|) \geq V_\theta(t_1, y(t_1) - \tilde{x}(t_1));$$

$$\geq V_\theta(t_2, y(t_2) - \tilde{x}(t_2));$$

$$\geq b_2(\|y(t_2) - \tilde{x}(t_2)\|) = b_2(\delta);$$

$$a_2(\varepsilon_2) \geq b_2(\delta),$$

which contradicts with (15). Thus (16) is valid. Then the solution $y(t, \tau_0, y_0)$ of the system (7) through (τ_0, y_0) is the initial time difference strictly stable in fractional case with respect to the solution $x(t - \eta, t_0 x_0)$ for $t \geq \tau_0$. This completes the proof of Theorem 1.   □

If δ_1, δ_2 and ε_2 is independent of τ_0, then the solution $y(t, \tau_0, y_0)$ of the system (7) is initial time difference uniformly strict stable in fractional case with respect to the solution $x(t - \eta, t_0, x_0)$ for $t \geq \tau_0$.

Theorem 11.2. *Assume that*

(A_1) *for each μ, $0 < \mu < \rho, V_\mu \in C[\mathbb{R}_+ \times S_\rho, \mathbb{R}_+]$ and V_μ is locally Lipschitzian in z and for $(t, z) \in \mathbb{R}_+ \times S_\rho$ and $\|z\| \geq \mu$,*

$$b_1(\|z\|) \leq V_\mu(t, z) \leq a_1(\|z\|), a_1, b_1 \in \mathcal{K},$$

$$_*^c D_+^q V_\mu(t, z) \leq -c_1(\|z\|), c_1 \in \mathcal{K}; \tag{11.19}$$

(A_2) *for each* $\theta, 0 < \theta < \rho, V_\theta \in C[\mathbb{R}_+ \times S_\rho, \mathbb{R}_+]$ *and* V_θ *is locally Lipschitzian in z and for* $(t,z) \in \mathbb{R}_+ \times S_\rho$ *and* $\|z\| \leq \theta,$

$$b_2(\|z\|) \leq V_\theta(t,z) \leq a_2(\|z\|), a_2, b_2 \in \mathcal{K},$$

$$^c_* D_+^q V_\theta(t,z) \geq -c_2(\|z\|), c_2 \in \mathcal{K}; \tag{11.20}$$

where $z(t) = y(t, \tau_0, y_0) - x(t - \eta, t_0, x_0)$ *for* $t \geq \tau_0, y(t, \tau_0, y_0)$ *of the system (7) through* (τ_0, y_0) *and* $x(t - \eta, t_0, x_0)$, *where* $x(t, t_0, x_0)$ *is any solution of the system (6) for* $t \geq \tau_0 \geq 0, t_0 \in \mathbb{R}_+$ *and* $\eta = \tau_0 - t_0$.

Then the solution $y(t, \tau_0, y_0)$ *of the system (7) through* (τ_0, y_0) *is the initial time difference uniformly strictly asymptotically stable in fractional case with respect to* $x(t - \eta, t_0, x_0)$ *of the solution of the system (6) for* $t \geq \tau_0 \geq 0, t_0 \in \mathbb{R}_+$ *and* $\eta = \tau_0 - t_0$.

Proof. We note that (18) implies (11). However, (19) does not yield (12). As a result of these, we obtain because of (18) only uniformly stability of unperturbed systems with initial time difference with respect to $x(t - \eta, t_0, x_0)$ that is for given any $\varepsilon_1 \leq \rho$ and $\tau_0 \in \mathbb{R}_+$ there exist $\delta_{10} = \delta_{10}(\varepsilon_1) > 0$ and $\delta_{20} = \delta_{20}(\varepsilon_1) > 0$ such that

$$\|y(t, \tau_0, y_0) - x(t - \eta, t_0, x_0)\| < \varepsilon_1 \text{ for } t \geq \tau_0 .$$

whenever $\|y_0 - x_0\| < \delta_{10}$ and $|\tau_0 - t_0| < \delta_{20}$.

To prove the conclusion of Theorem 2 we need to show that the solution $y(t, \tau_0, y_0)$ of the system (7) through (τ_0, y_0) for $t \geq \tau_0$ is strictly uniformly attractive in fractional case with respect to $x(t - \eta, t_0, x_0)$ for this purpose and for $t \geq \tau_0$, let $\varepsilon_1 = \rho$ and set $\delta_{10} = \delta_1(\rho)$ and $\delta_{20} = \delta_2(\rho)$ so that (20) yields $\|y(t, \tau_0, y_0) - x(t - \eta, t_0, x_0)\| < \rho$ for $t \geq \tau_0$ whenever $\|y_0 - x_0\| < \delta_1$ and $|\tau_0 - t_0| < \delta_2$.

Let $\|y_0 - x_0\| < \delta_{10}$ and $|\tau_0 - t_0| < \delta_{20}$. We show, using standard argument, that there exists a $t^* \in [\tau_0, \tau_0 + T]$, we choose $T = T(\varepsilon, \tau_0) \geq \left(\frac{a_1(\max\{\delta_{10}, \delta_{20}\})}{c_1(\min\{\delta_1, \delta_2\})} \Gamma(q+1) \right)^{\frac{1}{q}}$ where δ_{10} and δ_{20} are the numbers corresponding to ε_1 in (20), that is, in stability of unperturbed systems with initial time difference with respect to $x(t - \eta, t_0, x_0)$ such that $\|y(t^*, \tau_0, y_0) - x(t^* - \eta, t_0, x_0)\| < \delta_1$, $t^* \geq \tau_0$ for any solutions of the systems (6) and (7) with $\|y_0 - x_0\| < \delta_{10}$ and $|\tau_0 - t_0| < \delta_{20}$. If this is not true, we will have $\|y(t^*, \tau_0, y_0) - x(t^* - \eta, t_0, x_0)\| \geq \delta_1$ for $t^* \in [\tau_0, \tau_0 + T]$. Then, $\mu = \delta_1$ and using (A_1) with (18), we have in view of the choice of T,

$$0 < b_1(\delta_1) \leq b_1(\|y(\tau_0 + T) - \tilde{x}(\tau_0 + T)\|)$$

$$\leq V_\mu(\tau_0 + T, y(\tau_0 + T) - \tilde{x}(\tau_0 + T))$$

$$\leq V_\mu(\tau_0, y_0 - x_0) - \frac{1}{\Gamma(q)} \int_{\tau_0}^{\tau_0 + T} (t - s)^{q-1} c_1(\|y(s) - \tilde{x}(s)\|) ds$$

$$\leq a_1(\max\{\delta_{10}, \delta_{20}\}) - \frac{c_1(\min\{\delta_1, \delta_2\})}{\Gamma(q)} \int_{\tau_0}^{\tau_0 + T} (t - s)^{q-1} ds$$

$$\leq a_1(\max\{\delta_{10},\delta_{20}\}) - \frac{c_1(\min\{\delta_1,\delta_2\})}{\Gamma(q+1)}T^q$$

$$\leq 0.$$

This contradiction implies that there exist a $t^* \in [\tau_0, \tau_0 + T]$ satisfying $\|y(t^*,\tau_0,y_0) - x(t^* - \eta,t_0,x_0)\| < \delta_1$ for $t^* \geq \tau_0$. Because of the uniform stability $y(t,\tau_0,y_0)$ of (7) with initial time difference with respect to $x(t - \eta,t_0,x_0)$ related to the solution of (6), this yields

$$\|y(t,\tau_0,y_0) - x(t - \eta,t_0,x_0)\| < \varepsilon_1 \text{ for } t \geq \tau_0 + T \geq t^*$$

which implies that there exists a $\tau_0 < T_1 < T$ such that

$$\|y(\tau_0 + T,\tau_0,y_0) - x(\tau_0 + T - \eta,t_0,x_0)\| = \varepsilon_1.$$

Now, for any $\delta_{12}, 0 < \delta_{12} < \delta_{10}$ and $0 < \delta_{12} < \delta_{20}$ we can choose ε_2 such that $b_2(\varepsilon_1) > a_2(\varepsilon_2)$ and $0 < \varepsilon_2 < \varepsilon_1 < \delta_{12}$.

Suppose $\delta_{12} < \|y_0 - x_0\| < \min\{\delta_{10},\delta_{20}\}$ and $\delta_{12} < |\tau_0 - t_0| < \min\{\delta_{10},\delta_{20}\}$.

Let us define $\tau = \left[\frac{\Gamma(q)(b_2(\varepsilon_1)-a_2(\varepsilon_2))}{c_2(\varepsilon_1)}\right]^{\frac{1}{q}}$, and $T_2 = T_1 + \tau$.

Since, $\|y(t,\tau_0,y_0) - x(t - \eta,t_0,x_0)\| \leq \varepsilon_1$ for $t \geq \tau_0 + T_1$, choosing $\theta = \varepsilon_1$ and using (A_2) with (19) we have for $t \in [\tau_0 + T_1, \tau_0 + T_2]$,

$$a_2(\|y(t) - \tilde{x}(t)\|) \geq V_\theta(t,y(t) - \tilde{x}(t))$$

$$\geq V_\theta(\tau_0 + T_1, y(\tau_0 + T_1) - \tilde{x}(\tau_0 + T_1))$$

$$- \frac{1}{\Gamma(q)}\int_{\tau_0+T_1}^{t}(t - s)^{q-1}c_2(\|y(s) - \tilde{x}(s)\|)ds$$

$$\geq b_2(\varepsilon_1)$$

$$- \frac{1}{\Gamma(q)}\int_{\tau_0+T_1}^{t}(t - s)^{q-1}c_2(\|y(s) - \tilde{x}(s)\|)ds$$

$$\geq b_2(\varepsilon_1) - \frac{c_2(\varepsilon_1)}{\Gamma(q)}[t - (\tau_0 + T_1)]^q.$$

Since, $t - (\tau_0 + T_1) > \tau$ and a_2^{-1} exists, it follows that

$$a_2(\|y(t) - \tilde{x}(t)\|) > b_2(\varepsilon_1) - \frac{c_2(\varepsilon_1)}{\Gamma(q)}\left[\frac{\Gamma(q)(b_2(\varepsilon_1) - a_2(\varepsilon_2))}{c_2(\varepsilon_1)}\right] = a_2(\varepsilon_2).$$

This yields

$$\|y(t,\tau_0,y_0) - x(t - \eta,t_0,x_0)\| > \varepsilon_2 \text{ for } t \in [\tau_0 + T_1, \tau_0 + T_2]$$

and therefore,

$$\varepsilon_2 < \|y(t,\tau_0,y_0) - x(t - \eta,t_0,x_0)\| < \varepsilon_1 \text{ for } t \in [\tau_0 + T_1, \tau_0 + T_2].$$

This completes the proof. Then the solution $y(t, \tau_0, y_0)$ of the system (7) through (τ_0, y_0) is the initial time difference uniformly strictly asymptotically stable in fractional case with respect to the solution $x(t - \eta, t_0, x_0)$, where $x(t, t_0, x_0)$ is any solution of the system (6) for $t \geq \tau_0 \geq 0, t_0 \in \mathbb{R}_+$ and $\eta = \tau_0 - t_0$.   □

Before we prove the general result in terms of the comparison principle. Let us consider the uncoupled comparison fractional differential systems in Caputo's sense:

$$\begin{cases} (i)^c D^q u_1 = g_1(t, u_1), & u_1(\tau_0) = u_{10} \geq 0, \\ (ii)\ ^c D^q u_2 = g_2(t, u_2), & u_2(\tau_0) = u_{20} \geq 0, \end{cases} \tag{11.21}$$

where $g_1, g_2 \in C\left[\mathbb{R}_+^2, \mathbb{R}\right]$. The comparison system (21) is said to be strictly stable in fractional case:

If given any $\varepsilon_1 > 0$ and $t \geq \tau_0, \tau_0 \in \mathbb{R}_+$, there exist a $\delta_1 > 0$ such that $u_{10} \leq \delta_1$ implies $u_1(t) < \varepsilon_1$ for $t \geq \tau_0$ and for every $\delta_2 < \delta_1$ there exists an $\varepsilon_2 > 0$, $0 < \varepsilon_2 < \delta_2$ such that $u_{20} \geq \delta_2$ implies $u_2(t) > \varepsilon_2$ for $t \geq \tau_0$. Here, $u_1(t)$ and $u_2(t)$ are any solutions of (i) and (ii) in (21); respectively.

The comparison system (21) is said to be strictly attractive in fractional case:

If given any $\alpha_1 > 0, \gamma_1 > 0, \varepsilon_1 > 0$ and $\tau_0 \in \mathbb{R}_+$, for every $\alpha_2 < \alpha_1$, there exist $\varepsilon_2 < \varepsilon_1, T_1 = T_1(\varepsilon_1, \tau_0) > 0$ and $T_2 = T_2(\varepsilon_1, \tau_0) > 0$ such that $u_1(t, \tau_0, u_0) < \varepsilon_1$ for $T_1 + \tau_0 \leq t \leq T_2 + \tau_0$ when $u_{10} \leq \alpha_1$ and $u_2(t, \tau_0, u_0) > \varepsilon_2$ for $T_2 + \tau_0 \geq t \geq T_1 + \tau_0$ when $u_{20} \geq \alpha_2$. If T_1 and T_2 are independent of τ_0, then the comparison system (21) is the initial time difference uniformly strictly attractive in fractional case for $t \geq \tau_0$.

Following the main result based on this definition another result is formulated in terms of comparison principle.

Theorem 11.3. *Assume that*
(A_1) for each $\mu, 0 < \mu < \rho, V_\mu \in C[\mathbb{R}_+ \times S_\rho, \mathbb{R}_+]$ and V_μ is locally Lipschitzian in z and for $(t, z) \in \mathbb{R}_+ \times S_\rho$ and $\|z\| \geq \mu$,

$$b_1(\|z\|) \leq V_\mu(t, z) \leq a_1(\|z\|), a_1, b_1 \in \mathscr{K},$$

$$^c_* D^q_+ V_\mu(t, z) \leq g_1(t, V_\mu(t, z)); \tag{11.22}$$

(A_2) for each $\theta, 0 < \theta < \rho, V_\theta \in C[\mathbb{R}_+ \times S_\rho, \mathbb{R}_+]$ and V_θ is locally Lipschitzian in z and for $(t, z) \in \mathbb{R}_+ \times S_\rho$ and $\|z\| \leq \theta$,

$$b_2(\|z\|) \leq V_\theta(t, z) \leq a_2(\|z\|), a_2, b_2 \in \mathscr{K},$$

$$^c_* D^q_+ V_\theta(t, z) \geq g_2(t, V_\theta(t, z)); \tag{11.23}$$

where $g_2(t, u) \leq g_1(t, u), g_1, g_2 \in C[\mathbb{R}_+^2, \mathbb{R}], g_1(t, 0) = g_2(t, 0) = 0$ and $z(t) = y(t, \tau_0, y_0) - x(t - \eta, t_0, x_0)$ for $t \geq \tau_0, y(t, \tau_0, y_0)$ of the system (7) through (τ_0, y_0) and $x(t - \eta, t_0, x_0)$, where $x(t, t_0, x_0)$ is any solution of the system (6) for $t \geq \tau_0 \geq 0, t_0 \in \mathbb{R}_+$ and $\eta = \tau_0 - t_0$. Then any strict stability concept in fractional *case of the comparison system implies the corresponding strict stability concept in fractional case of the solution* $y(t,\tau_0,y_0)$ *of the system (7) through* (τ_0,y_0) *with respect to the solution* $x(t-\eta,t_0,x_0)$ *of the system (6) with initial time difference where* $x(t,t_0,x_0)$ *is any solution of the system (6) for* $t \geq \tau_0 \geq 0$, $t_0 \in \mathbb{R}_+$.

Proof. We will only prove the case of strict uniformly asymptotically stability in fractional case. Suppose the comparison fractional differential systems in (21) is strictly uniformly asymptotically stable in fractional case, then for any given ε_1, $0 < \varepsilon_1 < \delta$, there exist a $\delta^* > 0$ such that $u_{10} \leq \delta^*$ implies that $u_1(t,\tau_0,u_{10}) < b_1(\varepsilon_1)$ for $t \geq \tau_0$. For this ε_1 we choose δ_1 and δ_{11}, such that $a_1(\delta_1^*) \leq \delta^*$ and $\delta_1^* < \varepsilon_1$ where $\delta_1^* = \max\{\delta_1,\delta_{11}\}$, then we claim that $\|y_0 - x_0\| < \delta_1, |\tau_0 - t_0| < \delta_{11}$ imply that

$$\|y(t,\tau_0,y_0) - x(t-\eta,t_0,x_0)\| < \varepsilon_1, t \geq \tau_0. \tag{11.24}$$

If it is not true, then there exist t_1 and t_2, $t_2 > t_1 > \tau_0$ and a solution $z(t)$ of

$$^cD^q z = \tilde{f}(t,z),\ z(\tau_0) = y_0 - x_0, t \geq \tau_0,$$

with $|\tau_0 - t_0| < \delta_{11}$ and $\|y_0 - x_0\| < \delta_1, \|y(t,\tau_0,y_0) - x(t-\eta,t_0,x_0)\| < \delta_1^*$, $\|y(t,\tau_0,y_0) - x(t-\eta,t_0,x_0)\| = \varepsilon_1$ and $\delta_1^* \leq \|y(t,\tau_0,y_0) - x(t-\eta,t_0,x_0)\| < \varepsilon_1$ for $[t_1,t_2]$.

Choosing $\mu = \delta_1^*$ and using the theory of differential inequalities we get

$$b_1(\varepsilon_1) = b_1(\|y(t_2,\tau_0,y_0) - x(t_2 - \eta,t_0,x_0)\|)$$
$$\leq V_\mu(t_2,y(t_2,\tau_0,y_0) - x(t_2 - \eta,t_0,x_0))$$
$$\leq r(t_2,t_1,V_\mu(t_1,y(t_1,\tau_0,y_0) - x(t_1 - \eta,t_0,x_0)))$$
$$\leq r(t_2,t_1,a_1(\delta_1^*))$$
$$\leq r(t_2,t_1,\delta^*)$$
$$< b_1(\varepsilon_1)$$
$$b_1(\varepsilon_1) < b_1(\varepsilon_1)$$

which is a contradiction. Here $r(t,\tau_0,u_{10})$ is the maximal solution of (21). Hence, (24) is true and we have uniformly stability in fractional case with initial time difference. Now, we shall prove strictly uniformly attractive in fractional case with initial time difference.

For any given δ_2, $\varepsilon_2 > 0$, $\delta_2 < \delta^*$, we choose $\overline{\delta}_2$ and $\overline{\varepsilon}_2$ such that $a_1(\delta_2) < \overline{\delta}_2$ and $b_1(\varepsilon_2) \geq \overline{\varepsilon}_2$. For these $\overline{\delta}_2$ and $\overline{\varepsilon}_2$, since (21) is strictly uniformly attractive in fractional case, for any $\overline{\delta}_3 < \overline{\delta}_2$ there exist $\overline{\varepsilon}_3$ and T_1 and T_2 (we assume $T_2 < T_1$) such that $\overline{\delta}_3 < u_{10} = u_{20} < \overline{\delta}_2$ implies

$$r(t,\tau_0,u_{10}) \leq r(t,\tau_0,\overline{\delta}_2) < \overline{\varepsilon}_2,$$
$$\rho(t,\tau_0,u_{20}) \geq \rho(t,\tau_0,\overline{\delta}_3) > \overline{\varepsilon}_2,$$

where $r(t,\tau_0,u_{10})$ and $\rho(t,\tau_0,u_{20})$ is the maximal solution and minimal solution of (i) and (ii) (21); respectively.

Now, for any δ_3, let $b_2(\delta_3) \geq \overline{\delta_3}$. We choose ε_3 such that $a_2(\varepsilon_3) < \overline{\varepsilon_3}$. Then by using comparison principle (21) (i) and (A_1), we have

$$b_1\left(\|y(t,\tau_0,y_0) - x(t-\eta,t_0,x_0)\|\right) \leq V_\mu(t,y(t,\tau_0,y_0)$$
$$-x(t-\eta,t_0,x_0))$$
$$\leq r(t,\tau_0,V_\mu(\tau_0,y_0-x_0))$$
$$\leq r(t,\tau_0,a_1(\|y_0-x_0\|))$$
$$\leq r(t,\tau_0,\overline{\delta_2})$$
$$< \overline{\varepsilon_2} \leq b_1(\varepsilon_2)$$
$$b_1\left(\|y(t,\tau_0,y_0) - x(t-\eta,t_0,x_0)\|\right) < b_1(\varepsilon_2),$$

since b_1^{-1} exists which implies that $\|y(t,\tau_0,y_0) - x(t-\eta,t_0,x_0)\| < \varepsilon_2$ for $t \in [\tau_0 + T_2, \tau_0 + T_1]$. Similarly, by using comparison principle in (21) (ii) and (A_2), we get

$$a_2\left(\|y(t,\tau_0,y_0) - x(t-\eta,t_0,x_0)\|\right) \geq V_\theta(t,y(t,\tau_0,y_0)$$
$$-x(t-\eta,t_0,x_0))$$
$$\geq \rho(t,\tau_0,V_\theta(\tau_0,y_0-x_0))$$
$$\geq \rho(t,\tau_0,b_2(\delta_3))$$
$$\geq \rho(t,\tau_0,\overline{\delta_3})$$
$$> \overline{\varepsilon_3} \geq a_2(\varepsilon_3)$$
$$a_2\left(\|y(t,\tau_0,y_0) - x(t-\eta,t_0,x_0)\|\right) > a_2(\varepsilon_3),$$

since a_2^{-1} exists which implies that for $\|y(t,\tau_0,y_0) - x(t-\eta,t_0,x_0)\| > \varepsilon_3$ for $t \in [\tau_0 + T_2, \tau_0 + T_1]$. Hence, the solution $y(t,\tau_0,y_0)$ of the system (7) through (τ_0,y_0) is strictly uniformly attractive in fractional case with respect to the solution $x(t-\eta,t_0,x_0)$ is any solution of the system (6) for $t \geq \tau_0 \geq 0$, $t_0 \in \mathbb{R}_+$. The proof is completed. $\square$

References

1. Caputo M (1967) Linear models of dissipation whose Q is almost independent. II Geophys J Roy Astron 13:529–539
2. Kilbas AA, Srivastava HM, Trujillo JJ (2006) Theory and applications of fractional differential equations. Elsevier, Amsterdam, North-Holland, The Netherlands
3. Lakshmikantham V, Leela S (1969) Differential and integral inequalities, vol 1. Academic, New York

4. Lakshmikantham V, Leela S, Martynyuk AA (1989) Stability analysis of nonlinear systems. Marcel Dekker, New York
5. Lakshmikantham V, Leela S, Vasundhara Devi J (2009) Theory of fractional dynamical systems. Cambridge Scientific Publishers, Cambridge
6. Lakshmikantham V, Mohapatra RN (2001) Strict stability of differential equations. Nonlinear Anal 46(7):915–921
7. Lakshmikanthama V, Vatsala AS (2008) Basic theory of fractional differential equations. Nonlinear Anal 69:2677–2682
8. Oldham KB, Spanier J (1974) The fractional calculus. Academic, New York
9. Podlubny I (1999) Fractional differential equations. Academic, New York
10. Samko S, Kilbas A, Marichev O (1993) Fractional integrals and derivatives: theory and applications. Gordon and Breach, 1006 pages, ISBN 2881248640
11. Shaw MD, Yakar C (1999) Generalized variation of parameters with initial time difference and a comparison result in term Lyapunov-like functions. Int J Nonlinear Diff Equations Theory Methods Applications 5:86–108
12. Shaw MD, Yakar C (2000) Stability criteria and slowly growing motions with initial time difference. Probl Nonlinear Anal Eng Syst 1:50–66
13. Sivasundaram S (2001) The strict stability of dynamic systems on time scales. J Appl Math Stoch Anal 14(2):195–204
14. Yakar C (2007) Boundedness criteria in terms of two measures with initial time difference. Dyn Continuous Discrete Impulsive Syst A Math Anal. Watam Press. Waterloo. Page: 270–275. DCDIS 14 (S2) 1–305
15. Yakar C (2010) Fractional Differential equations in terms of comparison results and lyapunov stability with initial time difference. J Abstr Appl Anal AAA/762857. Vol 3. Volume 2010, Article ID 762857, 16 pages. doi:10.1155/2010/762857
16. Yakar C (2007) Strict stability criteria of perturbed systems with respect to unperturbed systems in term of initial time difference. Proceedings of the conference on complex analysis and potential theory. World Scientific Publishing. Page: 239–248
17. Yakar C, Shaw MD (2005) A comparison result and lyapunov stability criteria with initial time difference. Dyn Continuous Discrete Impulsive Syst A Math Anal 12(6)731–741
18. Yakar C, Shaw MD (2008) Initial time difference stability in terms of two measures and variational comparison result. Dyn Continuous Discrete Impulsive Syst A Math Anal 15:417–425
19. Yakar C, Shaw MD (2009) Practical stability in terms of two measures with initial time difference. Nonlinear Anal Theory Methods Appl 71:e781–e785

Chapter 12
A Fractional Order Dynamical Trajectory Approach for Optimization Problem with HPM

Fırat Evirgen and Necati Özdemir

1 Introduction

Optimization theory is aimed to find out the optimal solution of problems which are defined mathematically from a model that arise in wide range of scientific and engineering disciplines. Many methods and algorithms have been developed for this purpose since the late 1940s. The penalty function methods are classical methods for solving nonlinear programming (NLP) problem by transforming it to the unconstrained problem, see Luenberger [1] and Sun [2] for details. Furthermore, dynamical trajectory approaches based on differential equations system are alternative methods for NLP problems. In this type of methods an optimization problem is formulated as a system of ordinary differential equations (ODEs) so that the equilibrium point of this system converges to the local minimum of the optimization problem. The methods based on ODEs for solving optimization problems have been first proposed by Arrow and Hurwicz [3] and then improved by Rosen [4], Fiacco and Mccormick [5], and Yamashita [6]. Recently, Wang et al. [7], Jin et al. [8] and Özdemir and Evirgen [9, 10] have made studies in differential equation approach for solving optimization problems.

In last decade, fractional calculus has drawn a wide attention from many physicists and mathematicians, because of its interdisciplinary application and physical meaning, e.g. [11–13]. Fractional calculus deals with the generalization of differentiation and integration of noninteger order. Several analytical and numerical methods have been proposed for solving fractional differential equations (FDEs). Some commonly used techniques are summarized as follows. The variational iteration method (VIM) was first introduced by He [14], and applied to FDEs [15]. The Adomian decomposition method (ADM) [16, 17] is applied to various problems. Also, the homotopy perturbation method (HPM) is an another successful analytical

F. Evirgen (✉) • N. Özdemir
Department of Mathematics, Balıkesir University, Çağış Campus, 10145 Balıkesir, Turkey
e-mail: fevirgen@balikesir.edu.tr; nozdemir@balikesir.edu.tr

D. Baleanu et al. (eds.), *Fractional Dynamics and Control*,
DOI 10.1007/978-1-4614-0457-6_12, © Springer Science+Business Media, LLC 2012

approximate technique, which provides a solution to linear and nonlinear problems, see [18, 19]. The HPM yields a very rapid convergent series solution, and usually a few iterations lead to very accurate approximation of the exact solution [18–23]. The reason of this success is mainly based on combination of the traditional perturbation method and homotopy techniques. The HPM is used to solve a wide range of differential equations in the literature. Abdulaziz et al. [24] used HPM for solving system of FDEs. Momani and Odibat presented HPM for fractional order partial differential equation [25] and fractional quadratic Riccati differential equation was described in Odibat and Momani [26]. Baleanu et al. have solved linear and nonlinear Schrodinger equations by HPM [27]. Chowdhury and Hashim [28] have employed HPM for solving Klein Gordon equation. Furthermore, some techniques are adapted to the HPM for getting the essential behavior of the differential equation system for large time t, such as multistage and Padé approximants. The adaptation of HPM with multistage strategy for numerical and analytical solution of a system of ODEs was introduced by Hashim and Chowdhury [29]. Applications of multistage HPM for solving chaotic systems and biochemical reaction model were illustrated in [30–32], respectively.

This paper constructs a system of FDEs which is proposed to solve NLP problem with equality constraints. In order to see the coincidence between the steady state solution of the system of FDEs and the optimal solution of the NLP problem in a long time t period, we used the multistage strategy.

The paper is organized as follows. In Sect. 2, the fundamentals of optimization problem, fractional calculus and HPM are briefly reviewed. In Sect. 3, the multistage HPM is adapted to the nonlinear system of FDEs for solving NLP problem. In Sect. 4, the applicability and efficiency of multistage HPM is illustrated by comparison among traditional HPM and fourth order Runge–Kutta (RK4) method on some numerical examples. And finally some concluding remarks are given in Sect. 5.

2   Preliminaries

2.1   Optimization Problem

Consider the NLP problem with equality constraints defined by

$$\begin{aligned} &\text{minimize } f(x),\\ &\text{subject to } x \in M \end{aligned} \tag{12.1}$$

with

$$M = \{x \in \mathbb{R}^n \,|\, h(x) = 0\},$$

where $f : \mathbb{R}^n \longrightarrow \mathbb{R}$ and $h = (h_1, h_2, \ldots, h_p)^T : \mathbb{R}^n \longrightarrow \mathbb{R}^p$ $(p \leq n)$. It is assumed that the functions in problem are at least twice continuously differentiable, that a solution exists, and which $\nabla h(x)$ has full rank. To obtain a solution of (12.1), the penalty function method solves a sequence of unconstrained optimization problems. A well-known penalty function for this problem is given by

$$F(x, \mu) = f(x) + \mu \frac{1}{\gamma} \sum_{l=1}^{p} (h_l(x))^{\gamma}, \quad l = 1, 2, \ldots, p, \tag{12.2}$$

where $\gamma > 0$ is constant and $\mu > 0$ is an auxiliary penalty variable. The corresponding unconstrained optimization problem of (12.1) is defined as follows:

$$\min F(x, \mu) \quad subject\ to\ \ x \in \mathbb{R}^n. \tag{12.3}$$

Further information about NLP problem can be found in Luenberger [1] and Sun [2].

2.2   Fractional Calculus

Now we will give some definitions and properties of the fractional calculus [11–13]. We begin with the Riemann–Liouville definition of the fractional integral of order $\alpha > 0$, which is given as

$$I^{\alpha} f(x) = \frac{1}{\Gamma(\alpha)} \int_0^x (x - t)^{\alpha - 1} f(t) \mathrm{d}t, \quad x > 0,$$

where $\Gamma(.)$ is the Gamma function.

Most commonly encountered fractional derivatives are Riemann–Liouville and Caputo fractional derivative. The definitions of these two derivatives are given as:
Riemann–Liouville fractional derivative (RLFD)

$$D^{\alpha} f(x) = D^m \left(I^{m-\alpha} f(x) \right) = \frac{1}{\Gamma(m - \alpha)} \left(\frac{\mathrm{d}}{\mathrm{d}t} \right)^m \int_0^x (x - t)^{m - \alpha - 1} f(t) \mathrm{d}t,$$

Caputo fractional derivative (CFD)

$${}^{C}D^{\alpha} f(x) = I^{m-\alpha} \left(D^m f(x) \right) = \frac{1}{\Gamma(m - \alpha)} \int_0^x (x - t)^{m - \alpha - 1} \left(\frac{\mathrm{d}}{\mathrm{d}t} \right)^m f(t) \mathrm{d}t,$$

where $m - 1 < \alpha \leqslant m$ and $m \in \mathbb{N}$. Note that D^m is the usual integer differential operator of order m. Furthermore,

$$I^\alpha D^\alpha f(x) = f(x) - \sum_{s=0}^{m-1} f^{(s)}(0^+) \frac{x^s}{s!}, \quad m - 1 < \alpha \leqslant m, \tag{12.4}$$

is satisfied.

2.3 Homotopy Perturbation Method

The brief outline of HPM is given in general by He in [18, 20]. For convenience, consider the following nonlinear differential equation

$$L(u) + N(u) = f(r), \quad r \in \Omega, \tag{12.5}$$

with boundary condition

$$B\left(u, \frac{\partial u}{\partial n}\right) = 0, \quad r \in \Gamma,$$

where L is a linear operator, while N is nonlinear operator, B is a boundary operator, Γ is the boundary of the domain Ω and $f(r)$ is a known analytic function. The He's homotopy perturbation technique defines the homotopy $v(r, p) : \Omega \times [0, 1] \to \mathbb{R}$ which satisfies

$$H(v, p) = (1 - p)[L(v) - L(u_0)] + p[L(v) - N(v) - f(r)] = 0, \tag{12.6}$$

where $p \in [0, 1]$ is an embedding parameter, u_0 is an initial approximation which satisfies the boundary conditions. The changing process of p from zero to unity is just that of $v(r, p)$ from u_0 to $u(r)$. The basic assumption is that the solution of (12.6) can be expressed as a power series in p:

$$v = v_0 + p v_1 + p^2 v_2 + \cdots$$

The approximate solution of nonlinear equation (12.5), therefore can be readily obtained:

$$u = \lim_{p \to 1} v = v_0 + v_1 + v_2 + \cdots \tag{12.7}$$

The convergence of the series (12.7) has been proved in [19, 21] and the asymptotic behavior of the series is given in [22, 23].

2.4   The Runge–Kutta Method

The Runge–Kutta method is one of the well known numerical methods for differential equations. The fourth order Runge–Kutta method computes the approximate solutions of the problem $\dot{x} = f(t, x)$ by the following iterative equations:

$$x_{n+1} = x_n + \frac{1}{6} h (k_1 + 2k_2 + 2k_3 + k_4),$$

$$k_1 = f(t_n, x_n),$$

$$k_2 = f\left(t_n + \frac{1}{2} h, x_n + \frac{1}{2} h k_1\right),$$

$$k_3 = f\left(t_n + \frac{1}{2} h, x_n + \frac{1}{2} h k_2\right),$$

$$k_4 = f(t_n + h, x_n + h k_3),$$

where h is the fixed step size $t_i - t_{i-1}$ and x_n is the estimated value of the solution at the time t_n.

3   Multistage HPM for System of FDEs

In this section we solve NLP problems which are governed by system of fractional differential equation. Consider the unconstrained optimization problem (12.3), an approach based on fractional dynamic system can be described by the following FDEs

$$^C D^\alpha x(t) = -\nabla_x F(x, \mu), \tag{12.8}$$

subjected to the initial conditions

$$x(t_0) = x(0),$$

where $^C D^\alpha$ is the fractional derivative in Caputo sense of x of order α $(0 < \alpha \leqslant 1)$.

Note that, a point x_e is called an equilibrium point of (12.8) if it satisfies the right hand side of (12.8). For convenience of reader, we reformulate fractional dynamic system (12.8) as follows:

$$^C D^\alpha x_i(t) = g_i(t, \mu, x_1, x_2, \ldots, x_n), \quad i = 1, 2, \ldots, n. \tag{12.9}$$

The steady state solution of the nonlinear system of FDEs (12.9) must be coincided with local optimal solution of the NLP problem (12.1).

In order to find the solution of system (12.9), we use multistage HPM. Because the multistage strategy is provided to reach steady state solution in whole time horizon rather than traditional HPM. According to (12.6), we have constructed the following homotopy:

$$^CD^\alpha\, x_i(t) = pg_i(t,\mu,x_1,x_2,\ldots,x_n),\qquad(12.10)$$

where $i = 1,2,\ldots,n$ and $p \in [0,1]$. If $p = 0$, (12.10) becomes the linear equation

$$^CD^\alpha\, x_i(t) = 0,$$

and when $p = 1$, the homotopy (12.10) turns out to be the original system given in (12.9).

We assume that the system (12.9) is defined on the time interval $t \in [0,T]$. We divide the time interval into N equal length subintervals $\Delta T = T_j - T_{j-1}$, $j = 1,2,\ldots,N$ with $T_0 = 0$ and $T_N = T$. Using the parameter p, we expand the solution x_i in the following form:

$$x_i(t) = x_{i,0}(t) + px_{i,1}(t) + p^2 x_{i,2}(t) + \cdots,\quad i = 1,2,3,\ldots,n.\qquad(12.11)$$

Also, we take the initial approximations as below

$$x_{1,0}(t) = x_1(t^*),\ x_{2,0}(t) = x_2(t^*),\ldots,x_{n,0}(t) = x_n(t^*),\qquad(12.12)$$

where t^* is the left end point of each subinterval and initial conditions as

$$x_{1,1}(t^*) = 0,\ x_{2,1}(t^*) = 0,\ldots,x_{n,1}(t^*) = 0$$

$$\vdots$$

$$x_{1,K}(t^*) = 0,\ x_{2,K}(t^*) = 0,\ldots,x_{n,K}(t^*) = 0$$

$$\vdots$$

Substituting (12.11) into (12.10), and equating the coefficient of the terms with identical power of p, we get

$$p^0 : {}^CD^\alpha\, x_{i,0}(t) = 0$$
$$p^1 : {}^CD^\alpha\, x_{i,1}(t) = g_{i,1}(t,\mu,x_{1,0},\ldots,x_{n,0})$$

$$\vdots$$

$$p^K : {}^CD^\alpha\, x_{i,K}(t) = g_{i,K}(t,\mu,x_{1,0},\ldots,x_{n,0};x_{1,1},\ldots,x_{n,1};\ldots;x_{1,K-1},\ldots,x_{n,K-1})$$

$$\vdots$$

$$(12.13)$$

where $i = 1,2,3,\ldots,n$ and the function $g_{i,1},g_{i,2},\ldots$ satisfy the following equation:

$$g_i(t,\mu,x_{1,0} + px_{1,1} + \cdots, x_{2,0} + px_{2,1} + \cdots, x_{n,0} + px_{n,1} + \cdots)$$
$$= g_{i,1}(t,\mu,x_{1,0},\ldots,x_{n,0}) + pg_{i,2}(t,\mu,x_{1,0},\ldots,x_{n,0};x_{1,1},\ldots,x_{n,1})$$
$$+ p^2 g_{i,3}(t,\mu,x_{1,0},\ldots,x_{n,0};x_{1,1},\ldots,x_{n,1};x_{1,2},\ldots,x_{n,2}) + \cdots.$$

For solving the linear system (12.13), we apply the inverse operator I^α both side of equations. Therefore the components $x_{i,k}$ $(i = 1, 2, \ldots, n \,; k = 0, 1, 2, \ldots)$ of the multistage HPM can be determined. In order to carry out the iterations for every subinterval, we have to clarify initial approximations (12.12). For this purpose we set $t^* = t_0$. In multistage HPM, the iterations provide appropriate value of solutions by means of the previous K-term approximations $\Phi_{i,K}$ of the preceding subinterval. Consequently, the approximation solution of (12.9) can be denoted as follows:

$$x_i(t) = \Phi_{i,K} = \sum_{k=0}^{K-1} x_{i,k}, \quad 1 \leqslant i \leqslant n. \tag{12.14}$$

Here the effectiveness and the applicability of the approach especially depend on choosing ΔT and the number of term in approximate solution (12.14).

4   Numerical Implementation

To illustrate the effectiveness of the multistage HPM according to the HPM and fourth order Runge–Kutta method, some test problems are taken from Hock and Schittkowski [33, 34]. Methods are coded in Maple and digits of the variables are set to 15 in all the calculations done in this paper.

Example 12.1.   Consider the following NLP problem [34, Problem No: 216]

$$\begin{aligned} &\text{minimize } f(x) = 100\left(x_1^2 - x_2\right)^2 + (x_1 - 1)^2, \\ &\text{subject to } h(x) = x_1(x_1 - 4) - 2x_2 + 12 = 0. \end{aligned} \tag{12.15}$$

The optimal solution is $x^* = (2, 4)^T$. For solving the above problem, we convert it to an unconstrained optimization problem with quadratic penalty function (12.2) for $\gamma = 2$, then we have

$$F(x, \mu) = 100\left(x_1^2 - x_2\right)^2 + (x_1 - 1)^2 + \frac{1}{2}\mu\left(x_1(x_1 - 4) - 2x_2 + 12\right)^2,$$

where $\mu \in \mathbb{R}^+$ is an auxiliary penalty variable. The corresponding nonlinear system of FDEs from (12.8) is defined as

$$\left. \begin{aligned} {}^C D^\alpha x_1(t) &= -400(x_1^2 - x_2)x_1 - 2(x_1 - 1) - \mu(2x_1 - 4)(x_1^2 - 4x_1 - 2x_2 + 12), \\ {}^C D^\alpha x_2(t) &= 200(x_1^2 - x_2) + 2\mu(x_1^2 - 4x_1 - 2x_2 + 12), \end{aligned} \right\}$$

$$\tag{12.16}$$

where $0 < \alpha \leqslant 1$. The initial conditions are $x_1(0) = 0$ and $x_2(0) = 0$. Utilizing the homotopy (12.10) with auxiliary penalty variable $\mu = 800$ and step size $\Delta T = 0.00001$, the terms of the multistage HPM solutions (12.14) are acquired. In Fig. 12.1(a)–(b) we show the approximate–exact solution x_1 and x_2 of the

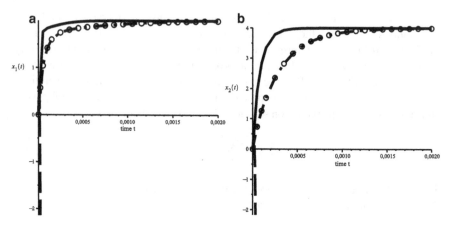

Fig. 12.1 Comparison of x_1 (**a**) and x_2 (**b**); *Dash*: HPM for $\alpha = 0.9$, *Dashdot*: MHPM ($\Delta T = 0.00001$) for $\alpha = 1$, *Solidline*: MHPM($\Delta T = 0.00001$) for $\alpha = 0.9$, *Open circle*: RK4($\Delta T = 0.00001$) for $\alpha = 1$

Table 12.1 Comparison of $x(t)$ between HPM and MHPM with RK4 solutions for different value of α

	HPM ($\alpha = 0.9$)		MHPM ($\alpha = 0.9$)		MHPM ($\alpha = 1$)		RK4 ($\alpha = 1$)	
t	$x_1(t)$	$x_2(t)$	$x_1(t)$	$x_2(t)$	$x_1(t)$	$x_2(t)$	$x_1(t)$	$x_2(t)$
0.000	0.0000	0.0000	0.0000	0.0000	0.0000	0.0000	0.0000	0.0000
0.001	$-0.69E+07$	$-0.11E+07$	1.9991	3.9996	1.9338	3.8549	1.9338	3.8549
0.002	$-0.84E+08$	$-0.14E+08$	1.9993	3.9998	1.9916	3.9915	1.9916	3.9915
0.003	$-0.36E+09$	$-0.62E+08$	1.9993	3.9998	1.9986	3.9992	1.9986	3.9992
0.004	$-0.10E+10$	$-0.17E+09$	1.9993	3.9998	1.9993	3.9997	1.9992	3.9997
0.005	$-0.23E+10$	$-0.39E+09$	1.9993	3.9998	1.9994	3.9998	1.9993	3.9998

problem (12.15) for the derivative order $\alpha = 1$ and $\alpha = 0.9$. We see that for $\alpha = 1$ and $\alpha = 0.9$ our solutions obtained using the multistage HPM are in good agreement with the RK4 method solution on $x^* = (2,4)^T$. Furthermore, the numerical results in Table 12.1 show that the multistage HPM for $\alpha = 0.9$ has better performance than for $\alpha = 1$. Clearly, the MHPM($\alpha = 0.9$) iterations converge faster than MHPM($\alpha = 1$) and RK4. So, it seems fractional order dynamical systems more realistic than integer order one for finding optimal solution of NLP problem.

Example 12.2. Consider the equality constrained optimization problem [33, Problem No: 79]

$$\text{minimize } f(x) = (x_1 - 1)^2 + (x_1 - x_2)^2 + (x_2 - x_3)^2 + (x_3 - x_4)^4 + (x_4 - x_5)^4$$
$$\text{subject to } h_1(x) = x_1 + x_2^2 + x_3^3 - 2 - 3\sqrt{2} = 0,$$
$$h_2(x) = x_2 - x_3^2 + x_4 + 2 - 2\sqrt{2} = 0,$$
$$h_3(x) = x_1 x_5 - 2 = 0.$$

$$(12.17)$$

Table 12.2 Comparison of $x(t)$ between HPM and MHPM with RK4 solutions for different value of α

	HPM ($\alpha = 0.9$)		MHPM ($\alpha = 0.9$)		MHPM ($\alpha = 1$)		RK4 ($\alpha = 1$)	
t	$x_1(t)$	$x_2(t)$	$x_1(t)$	$x_2(t)$	$x_1(t)$	$x_2(t)$	$x_1(t)$	$x_2(t)$
0	2	2	2	2	2	2	2	2
2	$0.160E+8$	$0.620E+8$	1.198931	1.369223	1.182161	1.352495	1.191010	1.359541
10	$0.288E+9$	$0.112E+10$	1.191090	1.362530	1.191050	1.362499	1.191082	1.362524
15	$0.594E+9$	$0.230E+10$	1.191090	1.362530	1.191084	1.362498	1.191090	1.362530
20	$0.100E+10$	$0.388E+10$	1.191090	1.362530	1.191082	1.362472	1.191090	1.362530
30	$0.209E+10$	$0.811E+10$	1.191090	1.362530	1.191113	1.362541	1.191090	1.362530

Table 12.3 Comparison of $x(t)$ between HPM and MHPM with RK4 solutions for different value of α

	HPM ($\alpha = 0.9$)		MHPM ($\alpha = 0.9$)		MHPM ($\alpha = 1$)		RK4 ($\alpha = 1$)	
t	$x_3(t)$	$x_4(t)$	$x_3(t)$	$x_4(t)$	$x_3(t)$	$x_4(t)$	$x_3(t)$	$x_4(t)$
0	2	2	2	2	2	2	2	2
2	$0.301E+9$	$-0.420E+7$	1.468744	1.616076	1.478320	1.661326	1.474039	1.641529
10	$0.546E+10$	$-0.756E+8$	1.472774	1.634738	1.472792	1.634827	1.472778	1.634755
15	$0.113E+11$	$-0.156E+9$	1.472774	1.634738	1.472786	1.634792	1.472774	1.634738
20	$0.191E+11$	$-0.263E+9$	1.472774	1.634738	1.472798	1.634853	1.472774	1.634738
30	$0.395E+11$	$-0.550E+9$	1.472774	1.634738	1.472765	1.634750	1.472774	1.634738

The solution of (12.17) is $x^* \approx (1.191127, 1.362603, 1.472818, 1.635017, 1.679081)^T$, and this is not an exact solution. The equality constrained optimization problem (12.17) is transformed to an unconstrained optimization problem by using quadratic penalty function (12.2) for $\gamma = 2$ as follows:

$$F(x, \mu) = f(x) + \frac{1}{2}\mu \sum_{l=1}^{3} (h_l(x))^2,$$

where $\mu \in \mathbb{R}^+$ is an auxiliary penalty variable.

The corresponding nonlinear system of FDEs from (12.8) is defined as

$$^C D^\alpha x(t) = -\nabla f(x) - \mu \nabla h(x) h(x), \tag{12.18}$$

where $0 < \alpha \leqslant 1$. The initial condition is $x(0) = (2,2,2,2,2)^T$ that is not feasible. Using the homotopy (12.10) with auxiliary penalty variable $\mu = 75$ and step size $\Delta T = 0.0001$, the multistage HPM approximate–exact solutions (12.14) are obtained. In Tables 12.2–12.4, the comparison of the x_i, $i = 1, 2, 3, 4, 5$ solutions between the HPM for $\alpha = 0.9$, the multistage HPM for $\alpha = 0.9$ and $\alpha = 1$ with the classical RK4 method are given, respectively. Here, the solutions continuously depends on the order of fractional derivative. Furthermore, our approximate solutions using the multistage HPM are in good agreement with the RK4 method solution and the optimal solution of the optimization problem (12.17).

Table 12.4 Comparison of $x(t)$ between HPM and MHPM with RK4 solutions for different value of α

	HPM ($\alpha = 0.9$)	MHPM ($\alpha = 0.9$)	MHPM ($\alpha = 1$)	RK4 ($\alpha = 1$)
t	$x_5(t)$	$x_5(t)$	$x_5(t)$	$x_5(t)$
0	2	2	2	2
2	$0.102E+7$	1.668076	1.691867	1.679209
10	$0.183E+8$	1.679130	1.679187	1.679140
15	$0.378E+8$	1.679130	1.679136	1.679130
20	$0.637E+8$	1.679130	1.679142	1.679130
30	$0.133E+9$	1.679130	1.679093	1.679130

5 Conclusions

In the present work, the HPM has been successfully used to obtain approximate analytical solutions of NLP problems. Initially, the NLP problem is reformulated by a system of FDEs. In order to see the essential behavior of the system of FDEs, the multistage strategy is adapted to the HPM. The numerical comparison among the fourth order Runge–Kutta (RK4), the multistage HPM ($\alpha = 0.9$ and $\alpha = 1$) and HPM ($\alpha = 0.9$) shows that the multistage HPM ($\alpha = 0.9$) performs rapid covergency to the optimal solutions of the optimization problems. Consequently, these results verify the efficiency of the multistage HPM as a practical tool for solving NLP problem.

References

1. Luenberger DG (1973) Introduction to linear and nonlinear programming. Addison-Wesley, California
2. Sun W, Yuan YX (2006) Optimization theory and methods: Nonlinear programming. Springer, New York
3. Arrow KJ, Hurwicz L, Uzawa H (1958) Studies in linear and non-linear programming. Stanford University Press, California
4. Rosen JB (1961) The gradient projection method for nonlinear programming: Part II nonlinear constraints. SIAM J Appl Math 9:514–532
5. Fiacco AV, Mccormick GP (1968) Nonlinear programming: Sequential unconstrained minimization techniques. Wiley, New York
6. Yamashita H (1976) Differential equation approach to nonlinear programming. Math Program 18:155–168
7. Wang S, Yang XQ, Teo KL (2003) A unified gradient flow approach to constrained nonlinear optimization problems. Comput Optim Appl 25:251–268
8. Jin L, Zhang L-W, Xiao X (2007) Two differential equation systems for equality-constrained optimization. Appl Math Comput 190:1030–1039
9. Özdemir N, Evirgen F (2009) Solving NLP problems with dynamic system approach based on smoothed penalty function. Selçuk J Appl Math 10:63–73
10. Özdemir N, Evirgen F (2010) A dynamic system approach to quadratic programming problems with penalty method. Bull Malays Math Sci Soc 33:79–91

11. Miller KS, Ross B (1993) An introduction to the fractional calculus and fractional differential equations. Wiley, New York
12. Oldham KB, Spanier J (1974) The fractional calculus. Academic, New York
13. Podlubny I (1999) Fractional differential equations. Academic, New York
14. He JH (1997) Variational iteration method for delay differential equations. Commun Nonlinear Sci Numer Simul 2:235–236
15. He JH (1998) Approximate analytical solution for seepage flow with fractional derivative in prous media. Comput Methods Appl Mech Eng 167: 57–68
16. Adomian G (1988) A review of the decomposition method in applied mathematics. J Math Anal Appl 135:501–544
17. Adomian G (1994) Solving frontier problems of physics: the decomposition method. Kluwer, Boston
18. He JH (1999) Homotopy perturbation technique. Comput Meth Appl Mech Eng 178:257–262
19. He JH (2000) A coupling method of homotopy technique and perturbation technique for nonlinear problems. Int J Nonlinear Mech 35:37–43
20. He JH (2003) Homotopy perturbation method: A new nonlinear analytical technique. Appl Math Comput 135:73–79
21. He JH (2004) Comparison of homotopy perturbation method and homotopy analysis method. Appl Math Comput 156:527–539
22. He JH (2006) New interpretation of homotopy perturbation method. Int J Mod Phys B 20:2561–2568
23. He JH (2006) Some asymptotic methods for strongly nonlinear equations. Int J Mod Phys B 20:1141–1199
24. Abdulaziz O, Hashim I, Momani S (2008) Solving systems of fractional differential equations by homotopy-perturbation method. Phys Lett A 372:451–459
25. Momani S, Odibat Z (2007) Homotopy perturbation method for nonlinear partial differential equations of fractional order. Phys Lett A 365:345–350
26. Odibat Z, Momani S (2008) Modified homotopy perturbation method: Application to quadratic riccati differential equation of fractional order. Chaos Solitons Fractals 36:167–174
27. Baleanu D, Golmankhaneh Alireza K, Golmankhaneh Ali K (2009) Solving of the fractional non-linear and linear schroedinger equations by homotopy perturbation method. Romanian J Phys 52:823–832
28. Chowdhury MSH, Hashim I (2009) Application of homotopy-perturbation method to Klein-Gordon and sine-Gordon equations. Chaos Solitons Fractals 39:1928–1935
29. Hashim I, Chowdhury MSH (2008) Adaptation of homotopy-perturbation method for numeric-analytic solution of system of ODEs. Phys Lett A 372:470–481
30. Chowdhury MSH, Hashim I, Momani S (2009) The multistage homotopy-perturbation method: A powerful scheme for handling the Lorenz system. Chaos Solitons Fractals 40:1929–1937
31. Yu Y, Li H-X (2009) Application of the multistage homotopy-perturbation method to solve a class of hyperchaotic systems. Chaos Solitons Fractals 42:2330–2337
32. Hashim I, Chowdhury MSH, Mawa S (2008) On multistage homotopy-perturbation method applied to nonlinear biochemical reaction model. Chaos Solitons Fractals 36:823–827
33. Hock W, Schittkowski K (1981) Test examples for nonlinear programming codes. Springer, Berlin
34. Schittkowski K (1987) More test examples for nonlinear programming codes. Springer, Berlin

Part III
Fractional Calculus in Mathematics and Physics

Chapter 13
On the Hadamard Type Fractional Differential System

Ziqing Gong, Deliang Qian, Changpin Li, and Peng Guo

1 Introduction

In recent decades, the fractional differential equations has been paid more and more attention, which mostly involve the Riemann–Liouville fractional calculus or the Caputo one [1–6]. The Hadamard calculus (differentiation and integration) has not been mentioned so much as other kinds of fractional derivative, even if it has been presented many years before [7].

In the following, the definitions of the Hadamard derivative and integral are introduced [8].

Definition 13.1. The Hadamard fractional integral of order $\alpha \in R^+$ of function $f(x)$, $\forall x > 1$, is defined by:

$$_H D_{1,x}^{-\alpha} f(x) = \frac{1}{\Gamma(\alpha)} \int_1^x \left(\ln \frac{x}{t} \right)^{\alpha-1} f(t) \frac{dt}{t}, \tag{13.1}$$

where $\Gamma(\cdot)$ is the Euler Gamma function.

Definition 13.2. The Hadamard derivative of order $\alpha \in [n-1, n)$, $n \in Z^+$, of function $f(x)$ is given as follows:

$$_H D_{1,x}^{\alpha} f(x) = \frac{1}{\Gamma(n-\alpha)} \left(x \frac{d}{dx} \right)^n \int_1^x \left(\ln \frac{x}{t} \right)^{n-\alpha-1} f(t) \frac{dt}{t}. \tag{13.2}$$

Z. Gong • C. Li (✉) • P. Guo
Department of Mathematics, Shanghai University, Shanghai 200444, P. R. China
e-mail: ziqinggong@126.com; lcp@shu.edu.cn; pengguo7909@yahoo.com.cn

D. Qian
Department of Mathematics, Zhongyuan University of Technology, Zhengzhou 450007, P. R. China
e-mail: deliangqian@126.com

D. Baleanu et al. (eds.), *Fractional Dynamics and Control*,
DOI 10.1007/978-1-4614-0457-6_13, © Springer Science+Business Media, LLC 2012

From the above definitions, the differences between Hadamard fractional derivative and the Riemann–Liouville fractional derivative are obvious, which include two aspects: firstly, no matter what the definitions of integral or derivative, the kernel in the Hadamard integral has the form of $\ln \frac{x}{t}$ instead of the form of $(x-t)$ which is involved in the Riemann–Liouville integral; secondly, the Hadamard derivative has the operator $\left(x\frac{d}{dx}\right)^n$, whose construction is well suited to the case of the half-axis and is invariant relative to dilation [9], while the Riemann–Liouville derivative has the operator $\left(\frac{d}{dx}\right)^n$.

Next, some of propositions with the Hadamard calculus are formed as follows.

Proposition 13.1. *If $0 < \alpha < 1$, the following relations hold*

(i) $_HD_{1,x}^{-\alpha}(\ln x)^{\beta-1} = \dfrac{\Gamma(\beta)}{\Gamma(\beta+\alpha)}(\ln x)^{\beta+\alpha-1}$;

(ii) $_HD_{1,x}^{\alpha}(\ln x)^{\beta-1} = \dfrac{\Gamma(\beta)}{\Gamma(\beta-\alpha)}(\ln x)^{\beta-\alpha-1}$.

Proof. Here we only prove (ii), (i) can be proved similar to (ii). Direct calculations yield

$$HD_{1,x}^{\alpha}(\ln x)^{\beta-1} = \left(x\frac{d}{dx}\right)\cdot\frac{1}{\Gamma(1-\alpha)}\int_1^x\left(\ln\frac{x}{t}\right)^{-\alpha}(\ln t)^{\beta-1}\frac{dt}{t}$$

$$= \left(x\frac{d}{dx}\right)\cdot\frac{(\ln x)^{\beta-\alpha}}{\Gamma(1-\alpha)}\int_1^x\left(1-\frac{\ln t}{\ln x}\right)^{-\alpha}\left(\frac{\ln t}{\ln x}\right)^{\beta-1}d\frac{\ln t}{\ln x}$$

$$= \left(x\frac{d}{dx}\right)\cdot\frac{(\ln x)^{\beta-\alpha}}{\Gamma(1-\alpha)}B(1-\alpha,\beta)$$

$$= \left(x\frac{d}{dx}\right)\cdot\frac{(\ln x)^{\beta-\alpha}}{\Gamma(1-\alpha)}\frac{\Gamma(1-\alpha)\Gamma(\beta)}{\Gamma(\beta-\alpha+1)}$$

$$= \frac{\Gamma(\beta)}{\Gamma(\beta-\alpha+1)}\cdot x\cdot\frac{d\left((\ln x)^{(\beta-\alpha)}\right)}{dx}$$

$$= \frac{\Gamma(\beta)}{\Gamma(\beta-\alpha)}(\ln x)^{(\beta-\alpha-1)}.$$

This completes the proof. □

The following results are available in [8].

Proposition 13.2. *If $\alpha \geq 0$ and $\beta = 1$, for any $j = [\alpha]+1$, the following relations hold*

(i) $\left(_HD_{1,t}^{\alpha}1\right)(x) = \dfrac{1}{\Gamma(1-\alpha)}(\ln x)^{-\alpha}$;

(ii) $\left(_HD_{1,t}^{\alpha}(\ln t)^{\alpha-j}\right)(x) = 0$.

Next, we will introduce the weighted space $C_{\gamma,\ln}[a,b]$, $C_{\delta,\gamma}^n[a,b]$ of the function f on the finite interval $[a,b]$, if $\gamma \in C(0 \leq Re(\gamma) < 1)$, $n-1 < \alpha \leq n$, then

$$C_{\gamma,\ln}[a,b] := \left\{ f(x) : \ln(\tfrac{x}{a})^\gamma f(x) \in C[a,b], \|f\|_{C_\gamma} = \|(\ln \tfrac{x}{a})^\gamma f(x)\|_C \right\},$$

$$C_{0,\ln}[a,b] = C[a,b],$$

and

$$C_{\delta,\gamma}^n[a,b] := \left\{ g(x) : (\delta^n g)(x) \in C_{\gamma,\ln}[a,b], \right.$$

$$\left. \|g\|_{C_{\gamma,\ln}} = \sum_{k=0}^{n-1} \|\delta^k g\|_C + \|\delta^n g\|_{C_{\gamma,\ln}} \right\},$$

$$\delta = x\frac{d}{dx}.$$

Theorem 13.1. *Let $\alpha > 0$, $n = -[-\alpha]$ and $0 \leq \gamma < 1$. Let G be an open set in R and let $f : (a,b] \times G \longrightarrow R$ be a function such that: $f(x,y) \in C_{\gamma,\ln}[a,b]$ for any $y \in G$, then the following problem*

$$_HD_{a,t}^\alpha(x) = f(x,y(x)), (\alpha > 0), \tag{13.3}$$

$$_HD_{a,t}^{\alpha-k}(a+) = b_k, b_k \in R, (k = 1,\ldots,n, n = -[-\alpha]), \tag{13.4}$$

satisfies the following Volterra integral equation:

$$y(x) = \sum_{j=1}^n \frac{b_j}{\Gamma(\alpha-j+1)} \left(\ln \frac{t}{a} \right)^{\alpha-j} + \frac{1}{\Gamma(\alpha)} \int_a^x \left(\ln \frac{x}{t} \right)^{\alpha-1} f[t,y(t)]\frac{dt}{t}, (x > a > 0),$$

$$\tag{13.5}$$

i.e., $y(x) \in C_{n-\alpha,\ln}[a,b]$ satisfies the relations (13.3)–(13.4) if and only if it satisfies the Volterra integral equation (13.5).

In particular, if $0 < \alpha \leq 1$, the problem (13.3)–(13.4) is equivalent to the following equation:

$$y(x) = \frac{b}{\Gamma(\alpha)} \left(\ln \frac{t}{a} \right)^{\alpha-1} + \frac{1}{\Gamma(\alpha)} \int_a^x \left(\ln \frac{x}{t} \right)^{\alpha-1} f[t,y(t)]\frac{dt}{t}, (x > a > 0). \tag{13.6}$$

2 The Generalized Gronwall Inequality

The Gronwall inequality, which plays a very important role in classical differential systems, has been generalized by Ye and Gao [10] which is used to fractional differential equations with Riemann–Liouville derivative. In this paper we further generalize the inequality. We firstly recall the classical Gronwall inequality which can be found in [11].

Theorem 13.2. *If*

$$x(t) \leq h(t) + \int_{t_0}^{t} k(s)x(s)ds, \, t \in [t_0, T),$$

where all the functions involved are continuous on $[t_0, T)$, $T \leq \infty$, *and* $k(t) \geq 0$, *then* $x(t)$ *satisfies*

$$x(t) \leq h(t) + \int_{t_0}^{t} h(s)k(s)exp\left[\int_{s}^{t} k(u)du\right]ds, \, t \in [t_0, T).$$

If, in addition, $h(t)$ *is nondecreasing, then*

$$x(t) \leq h(t)exp\left(\int_{t_0}^{t} k(s)ds\right), \, t \in [t_0, T).$$

The generalized Gronwall inequality corresponding to the Riemann–Liouville type fractional differential system is introduced as follows which is presented in Ye and Gao [10].

Theorem 13.3. *Suppose* $\alpha > 0$, $a(t)$ *is a nonnegative function and locally integrable on* $0 \leq t < T$ *(some* $T \leq +\infty$) *and* $g(t)$ *is a nonnegative, nondecreasing, continuous function defined on* $0 \leq t < T$, $g(t) \leq M$ *(constant), and suppose* $u(t)$ *is nonnegative and locally integrable on* $0 \leq t < T$ *with*

$$u(t) \leq a(t) + g(t) \int_{0}^{t} (t-s)^{\alpha-1}u(s)ds,$$

on the interval. Then

$$u(t) \leq a(t) + \int_{0}^{t}\left[\sum_{n=1}^{\infty} \frac{(g(t)\Gamma(\alpha))^n}{\Gamma(n\alpha)}(t-s)^{n\alpha-1}a(s)\right]ds, \, 0 \leq t < T.$$

This inequality can be used to estimate the bound of the Lyapunov exponents for both the Riemann–Liouville fractional differential systems and the Caputo ones [5]. In the following, we derive another inequality which can be regarded as a modification of Theorem 3.

Theorem 13.4. *Suppose* $\alpha > 0$, $a(t)$ *and* $u(t)$ *are nonnegative functions and locally integrable on* $1 \leq t < T$ $(\leq +\infty)$, *and* $g(t)$ *is a nonnegative, nondecreasing, continuous function defined on* $1 \leq t < T$, $g(t) \leq M$ *(constant). If the following inequality*

$$u(t) \leq a(t) + g(t) \int_{1}^{t} \left(\ln\frac{t}{s}\right)^{\alpha-1} u(s)\frac{ds}{s}, \, 1 \leq t < T, \tag{13.7}$$

holds. Then

$$u(t) \leq a(t) + \int_1^t \left[\sum_{n=1}^\infty \frac{(g(t)\Gamma(\alpha))^n}{\Gamma(n\alpha)} \left(\ln \frac{t}{s} \right)^{n\alpha-1} a(s) \right] \frac{ds}{s}, \quad 1 \leq t < T. \qquad (13.8)$$

Proof. Let

$$B\phi(t) = g(t) \int_1^t \left(\ln \frac{t}{s} \right)^{n\alpha-1} \phi(s) \frac{ds}{s}.$$

Then

$$u(t) \leq a(t) + Bu(t).$$

Iterating the inequality, one has

$$u(t) \leq \sum_{k=0}^{n-1} B^k a(t) + B^n u(t).$$

In the following, we prove

$$B^n u(t) \leq \int_1^t \frac{(g(t)\Gamma(\alpha))^n}{\Gamma(n\alpha)} \left(\ln \frac{t}{s} \right)^{n\alpha-1} u(s) \frac{ds}{s}, \qquad (13.9)$$

and $B^n u(t) \to +\infty$ for each $t \in (1, T)$.

Obviously, (13.9) holds when $n = 1$. Suppose it holds for $n = k$. Let $n = k+1$, then one has

$$B^{k+1} u(t) = B\left(B^k u(t) \right) \leq g(t) \int_1^t \left(\ln \frac{t}{s} \right)^{\alpha-1} \left[\int_1^s \frac{(g(t)\Gamma(\alpha))^n}{\Gamma(n\alpha)} \left(\ln \frac{s}{\tau} \right)^{k\alpha-1} u(\tau) \frac{d\tau}{\tau} \right] \frac{ds}{s}.$$

Under the condition that $g(t)$ is nondecreasing, one obtains

$$B^{k+1} u(t) \leq (g(t))^{k+1} \int_1^t \left(\ln \frac{t}{s} \right)^{\alpha-1} \left[\int_1^s \frac{(\Gamma(\alpha))^n}{\Gamma(n\alpha)} \left(\ln \frac{s}{\tau} \right)^{k\alpha-1} u(\tau) \frac{d\tau}{\tau} \right] \frac{ds}{s}.$$

By interchanging the order of integration, we get

$$B^{k+1} u(t) \leq (g(t))^{k+1} \int_1^t \left[\int_\tau^t \frac{(\Gamma(\alpha))^k}{\Gamma(k\alpha)} \left(\ln \frac{t}{s} \right)^{\alpha-1} \left(\ln \frac{s}{\tau} \right)^{k\alpha-1} \frac{ds}{s} \right] u(\tau) \frac{d\tau}{\tau}$$

$$= \int_1^t \frac{(g(t)\Gamma(\alpha))^{k+1}}{\Gamma((k+1)\alpha)} \left(\ln \frac{t}{s} \right)^{(k+1)\alpha-1} u(s) \frac{ds}{s},$$

where the integral

$$\int_\tau^t \left(\ln\frac{t}{s}\right)^{\alpha-1} \left(\ln\frac{s}{\tau}\right)^{k\alpha-1} \frac{ds}{s} = \left(\ln\frac{t}{\tau}\right)^{k\alpha+\alpha-1} \int_0^1 (1-z)^{\alpha-1} z^{k\alpha-1} dz$$

$$= \left(\ln\frac{t}{\tau}\right)^{(k+1)\alpha-1} B(k\alpha,\alpha)$$

$$= \frac{\Gamma(\alpha)\Gamma(k\alpha)}{\Gamma((k+1)\alpha)} \left(\ln\frac{t}{\tau}\right)^{(k+1)\alpha-1},$$

is obtained, where $\ln s = \ln\tau + z\ln\frac{t}{\tau}$ is used.

Therefore, (13.9) is true.

Moreover, since

$$B^n u(t) \le \int_1^t \frac{(M\Gamma(\alpha))^n}{\Gamma(n\alpha)} \left(\ln\frac{t}{s}\right)^{n\alpha-1} u(s)\frac{ds}{s} \to 0,$$

as $n \to +\infty$, for $t \in [1,T)$.

Hence this completes the proof.   □

Corollary 13.1. *Let $g(t) = b > 0$ in (13.7). The inequality (13.7) turns into the following form*

$$u(t) \le a(t) + b\int_1^t \left(\ln\frac{t}{s}\right)^{\alpha-1} u(s)\frac{ds}{s}.$$

Furthermore

$$u(t) \le a(t) + \int_1^t \left[\sum_{n=1}^\infty \frac{(b\Gamma(\alpha))^n}{\Gamma(n\alpha)} \left(\ln\frac{t}{s}\right)^{n\alpha-1} a(s)\right] \frac{ds}{s}, \quad (1 \le t < T).$$

Corollary 13.2. *Under the assumption of Theorem 4, suppose that $a(t)$ is a nondecreasing function on $[1,T)$. Then*

$$u(t) \le a(t)E_{\alpha,1}(g(t)\Gamma(\alpha)(\ln t)^\alpha),$$

where $E_{\alpha,1}$ is the Mittag-leffler function defined by

$$E_{\alpha,1} = \sum_{k=0}^\infty \frac{z^k}{\Gamma(k\alpha+1)}.$$

Proof. The assumptions imply

$$u(t) \leq a(t) \left[1 + \int_1^t \sum_{n=1}^{\infty} \frac{(g(t)\Gamma(\alpha))^n}{\Gamma(n\alpha)} \left(\ln \frac{t}{s} \right)^{n\alpha-1} \frac{ds}{s} \right]$$

$$= a(t) \sum_{n=0}^{\infty} \frac{(g(t)\Gamma(\alpha)\ln t)^n}{\Gamma(n\alpha+1)}$$

$$= a(t) E_\alpha(g(t)\Gamma(\alpha)(\ln t)^\alpha).$$

This ends the proof.   □

3   The Dependence of Solution on Parameters

As far as we are concerned, there have been some papers dedicated to study the
dependence of the solution on the order and the initial condition to the fractional
differential equation with Riemann–Liouville type and Caputo type derivative, while
quite few papers are contributed to the Hadamard type fractional differential system.
In this section, we study the dependence of the solution on the order and the initial
condition of the fractional differential equation with Hadamard fractional derivative.

Now we consider the following fractional system:

$$_HD_{1,t}^\alpha y(t) = f(t, y(t)), \tag{13.10}$$

$$_HD_{1,t}^{\alpha-1} y(t)|_{t=1} = \eta, \tag{13.11}$$

where $0 < \alpha < 1, 1 \leq t < T \leq +\infty, f : [1, T) \times R \to R$.

The existence and uniqueness of the initial value problem (13.10)–(13.11) have
been studied in [8], in which the dependence of a solution on initial conditions has
also been considered. Here, we investigate the dependence on both the initial value
conditions and the derivative order.

Obviously, the problem (13.10)–(13.11) can be changed into the Volterra integral
equation.

$$y(t) = \frac{\eta}{\Gamma(\alpha)} (\ln t)^{\alpha-1} + \frac{1}{\Gamma(\alpha)} \int_1^t (\ln t)^{\alpha-1} f(\tau, y(\tau)) \frac{d\tau}{\tau}. \tag{13.12}$$

In effect, the Volterra equation (13.12) is equivalent to the initial value problem
(13.10)–(13.11).

Theorem 13.5. *Let $\alpha > 0$ and $\delta > 0$ such that $0 < \alpha - \delta < \alpha \leq 1$. Also let the
function f be continuous and satisfy the Lipschitz condition with respect to the
second variable:*

$$|f(t, y) - f(t, z)| \leq L|y - z|,$$

for a constant L independent of t, y, z in R. For $1 \le t \le h < T$, assume that y and z are the solutions of the initial value problems (13.10)–(13.11) and

$$_H D_{1,t}^{\alpha-\delta} z(t) = f(t, z(t)), \tag{13.13}$$

$$_H D_{1,t}^{\alpha-\delta-1} z(t)|_{t=1} = \bar{\eta}, \tag{13.14}$$

respectively. Then, the following relation holds for $1 < t \le h$:

$$|z(t) - y(t)| \le A(t) + \int_1^t \left[\sum_{n=1}^{\infty} \left(\frac{L}{\Gamma(\alpha)} \Gamma(\alpha-\delta) \right)^n \frac{(\ln \frac{t}{s})^{n(\alpha-\delta)-1}}{\Gamma(n(\alpha-\delta))} A(s) \right] \frac{ds}{s},$$

where

$$A(t) = \left| \frac{\bar{\eta}}{\Gamma(\alpha-\delta)} (\ln t)^{\alpha-\delta-1} - \frac{\eta}{\Gamma(\alpha)} (\ln t)^{\alpha-1} \right| + \left| \frac{(\ln t)^{\alpha-\delta}}{(\alpha-\delta)\Gamma(\alpha)} - \frac{(\ln t)^{\alpha}}{\Gamma(\alpha+1)} \right| \cdot \|f\|$$

$$+ \left| \frac{(\ln t)^{\alpha-\delta}}{\alpha-\delta} \left[\frac{1}{\Gamma(\alpha-\delta)} - \frac{1}{\Gamma(\alpha)} \right] \right| \cdot \|f\|,$$

and

$$\|f\| = \max_{1 \le t \le h} |f(t, y)|.$$

Proof. The solutions of the initial value problem (13.10)–(13.11) and (13.13)–(13.14) are as follows:

$$y(t) = \frac{\eta}{\Gamma(\alpha)} (\ln t)^{\alpha-1} + \frac{1}{\Gamma(\alpha)} \int_1^t (\ln t)^{\alpha-1} f(\tau, y(\tau)) \frac{d\tau}{\tau},$$

and

$$z(t) = \frac{\bar{\eta}}{\Gamma(\alpha-\delta)} (\ln t)^{\alpha-\delta-1} + \frac{1}{\Gamma(\alpha-\delta)} \int_1^t (\ln t)^{\alpha-\delta-1} f(\tau, z(\tau)) \frac{d\tau}{\tau}.$$

So we have

$$|z(t) - y(t)| \le \left| \frac{\bar{\eta}}{\Gamma(\alpha-\delta)} (\ln t)^{\alpha-\delta-1} - \frac{\eta}{\Gamma(\alpha)} (\ln t)^{\alpha-1} \right|$$

$$+ \left| \frac{1}{\Gamma(\alpha-\delta)} \int_1^t (\ln t)^{\alpha-\delta-1} f(\tau, z(\tau)) \frac{d\tau}{\tau} - \frac{1}{\Gamma(\alpha)} \int_1^t (\ln t)^{\alpha-\delta-1} f(\tau, z(\tau)) \frac{d\tau}{\tau} \right|$$

$$+ \left| \frac{1}{\Gamma(\alpha)} \int_1^t (\ln t)^{\alpha-\delta-1} f(\tau, z(\tau)) \frac{d\tau}{\tau} - \frac{1}{\Gamma(\alpha)} \int_1^t (\ln t)^{\alpha-\delta-1} f(\tau, y(\tau)) \frac{d\tau}{\tau} \right|$$

$$+ \left| \frac{1}{\Gamma(\alpha)} \int_1^t (\ln t)^{\alpha-\delta-1} f(\tau, y(\tau)) \frac{d\tau}{\tau} - \frac{1}{\Gamma(\alpha)} \int_1^t (\ln t)^{\alpha-1} f(\tau, y(\tau)) \frac{d\tau}{\tau} \right|$$

$$\le A(t) + \frac{1}{\Gamma(\alpha)} \int_1^t (\ln t)^{\alpha-\delta-1} L |z(\tau) - y(\tau)| \frac{d\tau}{\tau},$$

where

$$A(t) = \left| \frac{\bar{\eta}}{\Gamma(\alpha-\delta)} (\ln t)^{\alpha-\delta-1} - \frac{\eta}{\Gamma(\alpha)} (\ln t)^{\alpha-1} \right| + \left| \frac{(\ln t)^{\alpha-\delta}}{(\alpha-\delta)\Gamma(\alpha)} - \frac{(\ln t)^{\alpha}}{\Gamma(\alpha+1)} \right| \cdot \|f\|$$

$$+ \left| \frac{(\ln t)^{\alpha-\delta}}{\alpha-\delta} \left[\frac{1}{\Gamma(\alpha-\delta)} - \frac{1}{\Gamma(\alpha)} \right] \right| \cdot \|f\|.$$

Applying Theorem 1 to the above inequality yields:

$$|z(t) - y(t)| \le A(t) + \int_1^t \left[\sum_{n=1}^{\infty} \left(\frac{L}{\Gamma(\alpha)} \Gamma(\alpha-\delta) \right)^n \frac{(\ln \frac{t}{s})^{n(\alpha-\delta)-1}}{\Gamma(n(\alpha-\delta))} A(s) \right] \frac{ds}{s}.$$

The proof is finished.   □

Next, we give an example to discuss the approximate solution of the Hadamard fractional differential equation.

$$_HD_{1,t}^{1-\delta} x(t) = x(t), \tag{13.15}$$

$$_HD_{1,t}^{-\delta} x(t)|_{t=1} = 1, \tag{13.16}$$

where $1 \le t < T \le +\infty$, $\delta \in R^+$ is small enough.

For the above question, we need not get its asymptotic solution. We can find its approximate solution quickly in the other way. Now we consider the simple problem as follows:

$$_HD_{1,t}^1 y(t) = y(t), \tag{13.17}$$

$$_HD_{1,t}^0 y(t)|_{t=1} = 1. \tag{13.18}$$

Combining the corresponding evaluation and Theorem 5, one has

$$A(t) = \left| \frac{1}{\Gamma(1-\delta)} (\ln t)^{-\delta} - 1 \right| + \left| \frac{(\ln t)^{1-\delta}}{1-\delta} - \ln t \right| \cdot \|x\| + \left| \frac{(\ln t)^{1-\delta}}{1-\delta} \left[\frac{1}{\Gamma(1-\delta)} - 1 \right] \right| \cdot \|y\|.$$

When $\delta \longrightarrow 0$ and $t \in [1, T)$, we get $A(t) \longrightarrow 0$.

Actually, $\delta \longrightarrow 0$ and $t \in [1, T)$, one has

$$|x(t) - y(t)| = |e^{\ln t} - (\ln t)^{\delta} e^{(\ln t)^{1-\delta}}| \longrightarrow 0.$$

The example shows that the Hadamard differential equation is dependent on both the initial value conditions and the order of derivative.

4   Estimation of the Bound of the Lyapunov Exponents

Recently, Li, Chen and Li, Xia have obtained the bound of the Lyapunov exponents of the discrete-time system, the ordinary differential system respectively. For details, see [12, 13]. Also, Li, et al. firstly introduced the Lyapunov exponents for the fractional differential systems with Riemann–Liouville derivative and Caputo derivative, and determined the bounds of their Lyapunov exponents [5]. In this paper, we use the modified Gronwall inequality to derive the bound of the Lyapunov exponents of the fractional differential system with Hadamard derivative.

Theorem 13.6. *The following fractional differential system with Hadamard derivative*

$$\begin{cases} {}_H D_{t_0,t}^{\alpha} x(t) = f(x,t), \\ (x,t) \in \Omega \times (t_0,+\infty) \subset R^n \times (t_0,+\infty),\ \alpha \in (0,1), t_0 > 0, \\ {}_H D_{t_0,t}^{\alpha-1} x(t)|_{t=t_0} = x_0, \end{cases} \qquad (13.19)$$

has its first variation equation

$$\begin{cases} {}_H D_{t_0,t}^{\alpha} \Phi(t) = f_x(x,t)\Phi(t), \\ (x,t) \in \Omega \times (t_0,+\infty) \subset R^n \times (t_0,+\infty),\ \alpha \in (0,1), t_0 > 0, \\ \Phi(t_0) = I, \end{cases} \qquad (13.20)$$

where I is an identity matrix and

$$\Phi(t) = \frac{\partial}{\partial s}\phi(t;x_0 + s\Phi(t))|_{s=0} = D_x\phi(t_0;x_0),$$

$\phi(t_0;x_0)$ *is the fundamental solution to the system.*

Proof. The proof is similar to that in [5], we omit the details here.   □

Definition 13.3. Let $u_k(t),\ k = 1,2,\ldots,n$ be the eigenvalues of $\Phi(t)$ of system (13.20), which satisfy $|u_1(t)| \le |u_2(t)| \le \cdots \le |u_n(t)|$. Then the Lyapunov exponents l_k of the trajectory $x(t)$ solving (13.20) are defined by:

$$l_k = \limsup_{t\to\infty} \frac{1}{t} \ln |u_k(t)|,\quad k = 1,2,\ldots,n.$$

These exponents $l_k, k = 1,2,\ldots,n$, are real numbers. The existence of the limit for the classical differential system was established [14]. For the fractional differential system, it still holds. Obviously, Φ is not invertible when $u_1(t) = 0$, which implies $l_1 = -\infty$. But this case does not happen in general. Hence, we always assume that $u_1(t)$ is not (identically) equal to zero. Therefore, Φ is always supposed to be invertible.

Next, we estimate the bound of the Lyapunov exponents for the fractional differential systems with Hadamard derivative. But firstly, let's take a look at the following lemma [15].

Lemma 13.1. *If $0 < \alpha < 2$, β is an arbitrary complex number, u is an arbitrary real number such that $\frac{\pi\alpha}{2} < u < \min\{\pi, \pi\alpha\}$, then for an arbitrary integer $p \geq 1$ the following expansion holds*

$$E_{\alpha,\beta}(z) = \frac{1}{\alpha} z^{(1-\beta)/\alpha} e^{z^{1/\alpha}} - \sum_{k=1}^{p} \frac{z^{-k}}{\Gamma(\beta - k\alpha)} + O(|z|^{-1-p}), \quad |z| \to \infty, \ |\arg(z)| \leq u.$$

By Lemma 1, we can directly obtain the asymptotic expansion of the Mittag-Leffler function

$$E_{\alpha,\alpha}(K(\ln t)^\alpha) \approx \frac{e^{K^{\frac{1}{\alpha}}}}{\alpha} K^{\frac{1}{\alpha}-1} (\ln t)^{1-\alpha} t, \quad t \to +\infty,$$

where K is a positive constant.

Integrating system (13.19) gives

$$\Phi(t) = \frac{\left(\ln \frac{t}{t_0}\right)^{\alpha-1}}{\Gamma(\alpha)} I + \frac{1}{\Gamma(\alpha)} \int_{t_0}^{t} \left(\ln \frac{t}{\tau}\right)^{\alpha-1} f_x(x,\tau)\Phi(\tau) \frac{d\tau}{\tau}.$$

Taking the matrix norm of both sides of the above equation leads to

$$\|\Phi(t)\| \leq \frac{\left(\ln \frac{t}{t_0}\right)^{\alpha-1}}{\Gamma(\alpha)} + \frac{M}{\Gamma(\alpha)} \int_{t_0}^{t} \left(\ln \frac{t}{\tau}\right)^{\alpha-1} \|\Phi(\tau)\| \frac{d\tau}{\tau},$$

where the constant M is assumed the bound of $\|f_x(x,t)\|$.

Applying Corollary 2 to the above integral inequality brings about

$$\|\Phi(t)\| \leq \left(\ln \frac{t}{t_0}\right)^{\alpha-1} E_{\alpha,\alpha}\left(M\left(\ln \frac{t}{t_0}\right)^\alpha\right).$$

By the fact that the spectral radius of a given matrix is not bigger than its norm, we have

$$|u_n(t)| \leq \|\Phi(t)\| \leq \left(\ln \frac{t}{t_0}\right)^{\alpha-1} E_{\alpha,\alpha}\left(M\left(\ln \frac{t}{t_0}\right)^\alpha\right).$$

Using the definition of the Lyapunov exponents and applying Lemma 1, one gets

$$l_n = \lim_{t \to +\infty} \sup \frac{1}{t} \ln |u_n(t)| \le \lim_{t \to +\infty} \sup \frac{1}{t} \ln \| \varPhi(t) \|$$

$$\le \lim_{t \to +\infty} \sup \frac{1}{t} \ln \left(\left(\ln \frac{t}{t_0} \right)^{\alpha-1} E_{\alpha,\alpha} \left(M \left(\ln \frac{t}{t_0} \right)^{\alpha} \right) \right)$$

$$= \lim_{t \to +\infty} \sup \frac{1}{t} \ln \left(\frac{e^{K^{\frac{1}{\alpha}}}}{\alpha} K^{\frac{1}{\alpha}-1} \left(\ln \frac{t}{t_0} \right)^{1-\alpha} \frac{t}{t_0} \right)$$

$$= 0.$$

So the Lyapunov exponents of systems (13.19) satisfy

$$-\infty < l_1 \le \cdots \le l_n \le 0.$$

Therefore we eventually derive the upper bound of the Lyapunov exponents for the fractional differential systems with Hadamard derivative and the upper bound is zero, which means that generally the fractional differential system with Hadamard derivative has no chaotic attractor in the sense of the definition 3. We do not know whether or not such a system is chaotic in the other sense. Such a problem is still open. We hope the studies in this respect will appear elsewhere.

Acknowledgements This work was partially supported by the National Natural Science Foundation of China under Grant no. 10872119 and Shanghai Leading Academic Discipline Project under Grant no. S30104.

References

1. Li CP, Deng WH (2007) Remarks on fractional derivatives. Appl Math Comput 187(2): 777–784
2. Li CP, Dao XH, Guo P (2009) Fractional derivatives in complex plane. Nonlinear Anal-Theor 71(5–6):1857–1869
3. Li CP, Zhao ZG (2009) Asymptotical stability analysis of linear fractional differential systems. J Shanghai Univ (Engl Ed) 13(3):197–206
4. Qian DL, Li CP, Agarwal RP, Wong PJY (2010) Stability analysis of fractional differential system with Riemann–Liouville derivative. Math Comput Model 52(5–6):862–874
5. Li CP, Gong ZQ, Qian DL, Chen YQ (2010) On the bound of the Lyapunov exponents for the fractional differential systems. Chaos 20(1):013127
6. Miller KS, Ross B (1993) An introduction to the fractional calculus and fractional differential equations. Wiley, New York
7. Hadamard J (1892) Essai sur létude des fonctions données par leur développement de Taylor. J Math Pures Appl 8(Ser. 4):101–186
8. Kilbas AA, Srivastava HM, Trujillo JJ (2006) Theory and applications of fractional differential equations. Elsevier, Amersterdam
9. Samko SG, Kilbas AA, Marichev OI (1993) Fractional integrals and derivatives: theory and applications. Gordon and Breach Science Publishers, Switzerland

10. Ye HP, Gao JM (2007) A generalized Gronwall inequality and its application to a fractional differential equation. J Math Anal Appl 328:1075–1081
11. Corduneanu C (1971) Principle of differential and integral equations. Allyn and Bacon, Boston
12. Li CP, Chen GR (2004) Estimating the Lyapunov exponents of discrete systems. Chaos 14(2):343–346
13. Li CP, Xia X (2004) On the bound of the Lyapunov exponents for continuous systems. Chaos 14(3):557–561
14. Oseledec VI (1968) A multiplicative ergodic theorem: Liapunov characteristic numbers for dynamical systems. Trans Mosc Math Soc 19:197–231
15. Podlubny I (1999) Fractional differential equations. Academic, New York

Chapter 14
Robust Synchronization and Parameter Identification of a Unified Fractional-Order Chaotic System

E.G. Razmjou, A. Ranjbar, Z. Rahmani, and R. Ghaderi

1 Introduction

Chaotic systems have held researchers' interest in the past decades. Some nonlinear systems can model various natural and man-made systems, and are known to have great sensitivity to initial conditions. This means two system starting trajectories from their arbitrary and almost the same initial states could evolve in dramatically different fashions, and soon become uncorrelated and unpredictable. In recent years, a new direction of chaos research has emerged, in which fractional-order calculus is applied to dynamic systems [19].

Fractional calculus is in essence as an extension of the ordinary calculus, with almost 300-year-old history. In spite of the long history, the application of the fractional calculus to physics and engineering is just a recent focus of interest [15]. It has been found that the behavior of many physical systems can be properly described by using the fractional-order system theory. For example, heat conduction [8], quantum evolution of complex systems [9], and diffusion waves [6] are known systems governed by the fractional-order equations. In fact, real world process generally or most likely is fractional-order system [16]. More recently, there is a new trend to investigate the control and dynamics of fractional-order dynamical systems.

Ahmad and Sprott [1] have shown that nonlinear chaotic systems can still show chaos when their models become fractional. Ahmad and Harba [2] investigated chaos control for fractional-chaotic systems, where controllers have been designed using "backstepping" method of nonlinear control design. Li and Chen [12] found that chaos exists in the fractional-order Chen system with order less than three. Linear feedback control of chaos in this system is studied. In [11] chaos synchronization of fractional-order chaotic systems is studied.

E.G. Razmjou • A. Ranjbar (✉) • Z. Rahmani • R. Ghaderi
Intelligent System Research Group, Babol University of Technology,
Faculty of Electrical and Computer Engineering, P.O. Box 47135-484, Babol, Iran
e-mail: ehsan.razmju@gmail.com; a.ranjbar@nit.ac.ir; rahmaniz@gmail.com; r_ghaderi@nit.ac.ir

D. Baleanu et al. (eds.), *Fractional Dynamics and Control*,
DOI 10.1007/978-1-4614-0457-6_14, © Springer Science+Business Media, LLC 2012

A unified chaotic system is a chaotic system, which depends on a parameter $\alpha \in [0, 1]$. If $0 \leq \alpha < 0.8$, the unified chaotic system reduces to the generalized Lorenz chaotic system; the unified chaotic system is reduced to the Lü chaotic system when $\alpha = 0.8$. The choice $0.8 < \alpha \leq 1$ makes the unified chaotic system the generalized Chen chaotic system. Several researchers have focused on control and synchronization of the unified chaotic system. Chen and Lu [4] considered that the parameter of the two unified chaotic systems is unknown and an adaptive controller is used to achieve synchronization based on the Lyapunov stability theory. Chen et al. [5] investigated the stabilization and synchronization of the unified chaotic system via an impulsive control method. Lu et al. [13] used linear feedback and adaptive control to synchronize an identical unified chaotic system with only one input controller. Ucar et al. [17] used a nonlinear active controller to synchronize two coupled unified chaotic systems with three control inputs. Wang and Liu [18] proved that the unified chaotic system is equivalent to a passive one which becomes asymptotically stabilized at equilibrium points. Wang and Song [20] studied the synchronization problem of two identical unified chaotic systems using three different methods. They used a linear feedback controller, a nonlinear feedback method, and an impulsive controller to synchronize the systems. In [21] based on the sliding mode theory, synchronization of two identical unified chaotic is discussed.

In this chapter adaptive sliding mode control will be designed to synchronize two fractional-order unified chaotic systems. This will be done when parameters are unknown and need to be identified, especially when initial conditions of master and slave systems are different. This is also done when the slave system is perturbed by the uncertainties in the dynamic. Also Unlike many well-known methods of the sliding mode control, no knowledge on the bound of uncertainty and disturbance is required.

The chapter is organized as follows: Section 2 describes the unified system. Fractional-order adaptive controller is proposed to synchronize and identify parameters of the two unified systems in Sect. 3. Simulation study is given in Sect. 4, to illustrate the effectiveness of the proposed controller. The chapter will be concluded in Sect. 5.

2 System Description

Unified chaotic system is a system whose behavior incorporates the behavior of the chaotic Lorenz, Chen, and the Lü systems. The unified chaotic system is governed by the following set of ordinary differential equations:

$$\begin{cases} \frac{dx}{dt} = (25\alpha + 10)(y - x) \\ \frac{dy}{dt} = (28 - 35\alpha)x - xz + (29\alpha - 1)y. \\ \frac{dz}{dt} = xy - \frac{8+\alpha}{3}z \end{cases} \qquad (14.1)$$

The states of the system (6) are x, y, and z and the key parameter of the system is α which takes values in the range [0, 1] to become chaotic. When $\alpha = 0.01$, the unified chaotic system represents the Lorenz chaotic attractor. It represents the Lü chaotic attractor when $\alpha = 0.8$. Similarly, when $\alpha = 1$, it represents the Chen chaotic attractor [3]. Moreover, for $\alpha \in [0,0.8]$, system (6) is called the general Lorenz system. Ultimately, system (6) is called the general Chen system when $\alpha \in [0.8, 1]$ [7].

The work will be expanded when a fractional order of (14.2) is considered. It means the standard derivatives in (14.1) are replaced by fractional derivatives, which are as follows:

$$\begin{cases} \frac{d^q x}{dt^q} = (25\alpha + 10)(y - x) \\ \frac{d^q y}{dt^q} = (28 - 35\alpha)x - xz + (29\alpha - 1)y, \\ \frac{d^q z}{dt^q} = xy - \frac{8+\alpha}{3}z \end{cases} \qquad (14.2)$$

where q, as the fractional order, is subjected to $0 < q \leq 1$.

The chaos in fractional-order unified systems (Chen, Lü and Lorenz-like) for $q = 0.9, 0.95, 0.99$ is shown in [14]. From fractional-order unified chaotic system in (14.2) a generalized type can be given as follows:

$$\begin{cases} \frac{d^q x}{dt^q} = a(y - x) \\ \frac{d^q y}{dt^q} = bx - xz + cy. \\ \frac{d^q z}{dt^q} = xy - dz \end{cases} \qquad (14.3)$$

2.1 Fractional-Order Chen System

From [14], the fractional-order Chen system is represented by:

$$\begin{cases} \frac{d^q x_1}{dt^q} = a_1(x_2 - x_1) \\ \frac{d^q x_2}{dt^q} = (c_1 - a_1)x_1 - x_1 x_3 + c_1 x_2. \\ \frac{d^q x_3}{dt^q} = x_1 x_2 - b_1 x_3 \end{cases} \qquad (14.4)$$

The fractional-order Chen system as master is represented from (14.4), where x_1, x_2, and x_3 are the states and a_1, b_1, and c_1 are unknown constant parameters of the master dynamic. A similar uncertain driven slave system may be written as:

$$\begin{cases} \frac{d^q y_1}{dt^q} = a_2(t)(y_2 - y_1) + \Delta f_1(y_1, y_2, y_3) + u_1 \\ \frac{d^q y_2}{dt^q} = (c_2(t) - a_2(t))y_1 - y_1 y_3 + c_2(t)y_2 + \Delta f_2(y_1, y_2, y_3) + u_2. \\ \frac{d^q y_3}{dt^q} = y_1 y_2 - b_2(t)y_3 + \Delta f_3(y_1, y_2, y_3) + u_3 \end{cases} \qquad (14.5)$$

$a_2(t)$, $b_2(t)$, and $c_2(t)$ are time dependent unknown parameters, which must be identified through the behavior of the master. $\Delta f_i(y_1, y_2, y_3)$ ($i = 1, 2, 3$) are uncertain terms, representing the unknown part of dynamic. The uncertainty is assumed upper bounded by a positive constant σ as $|\Delta f_i(y_1, y_2, y_3)| \leq \sigma$.

Note that the slave dynamic contains three individual input control signals. The control will be designed such that the master and the slave are synchronized after starting from different initial conditions. The error will be defined between the states of the master in (14.4) and the slave systems in (14.5), which is as follows:

$$\begin{cases} \frac{d^q e_1}{dt^q} = a_2(t)(e_2 - e_1) + \tilde{a}(x_2 - x_1) + \Delta f_1(y_1, y_2, y_3) + u_1 \\ \frac{d^q e_2}{dt^q} = c_2(t)(e_1 + e_2) + \tilde{c}(x_1 + x_2) - y_1 e_3 - e_1 x_3 + a_2(t)e_1 - x_1\tilde{a} + \Delta f_2(y_1, y_2, y_3) + u_2, \\ \frac{d^q e_3}{dt^q} = y_1 e_2 + e_1 x_2 - b_2(t)e_3 - x_3\tilde{b} + \Delta f_3(y_1, y_2, y_3) + u_3 \end{cases}$$

(14.6)

where $e_1 = y_1 - x_1$, $e_2 = y_2 - x_2$, and $e_3 = y_3 - x_3$ are the states error. Likewise the tilda-term shows the deviation of parameters from their nominal values as:

$$\tilde{a} = a_2(t) - a_1, \quad \tilde{b} = b_2(t) - b_1, \quad \tilde{c} = c_2(t) - c_1.$$

2.2 Fractional Order Lü System

From [14], the fractional-order Lü system is expressed by:

$$\begin{cases} \frac{d^q x_1}{dt^q} = a_1(x_2 - x_1) \\ \frac{d^q x_2}{dt^q} = -x_1 x_3 + c_1 x_2. \\ \frac{d^q x_3}{dt^q} = x_1 x_2 - b_1 x_3 \end{cases}$$

(14.7)

The master fractional-order Lü system is represented in (14.7), whilst the forced uncertain slave system may be written as:

$$\begin{cases} \frac{d^q y_1}{dt^q} = a_2(t)(y_2 - y_1) + \Delta f_1(y_1, y_2, y_3) + u_1 \\ \frac{d^q y_2}{dt^q} = -y_1 y_3 + c_2(t)y_2 + \Delta f_2(y_1, y_2, y_3) + u_2. \\ \frac{d^q y_3}{dt^q} = y_1 y_2 - b_2(t)y_3 + \Delta f_3(y_1, y_2, y_3) + u_3 \end{cases}$$

(14.8)

Deduction of equations in (14.7) from (14.8) yields the error dynamic by:

$$\begin{cases} \frac{d^q e_1}{dt^q} = a_2(t)(e_2 - e_1) + \tilde{a}(x_2 - x_1) + \Delta f_1(y_1, y_2, y_3) + u_1 \\ \frac{d^q e_2}{dt^q} = -y_1 e_3 - e_1 x_3 + c_2(t)e_2 + x_2\tilde{c} + \Delta f_2(y_1, y_2, y_3) + u_2. \\ \frac{d^q e_3}{dt^q} = y_1 e_2 + e_1 x_2 - b_2(t)e_3 - x_3\tilde{b} + \Delta f_3(y_1, y_2, y_3) + u_3 \end{cases}$$

(14.9)

In this chapter, the goal is to design an adaptive sliding mode controller such that the resultant error and the parameter identification approach zero. This means a robust synchronization will be achieved when:

$$\lim_{t \to \infty} |e(t)| = \lim_{t \to \infty} |y(t) - x(t)| = 0$$

and

$$\lim_{t \to \infty} |\tilde{a}| = \lim_{t \to \infty} |a_2(t) - a_1| = 0$$

$$\lim_{t \to \infty} |\tilde{b}| = \lim_{t \to \infty} |b_2(t) - b_1| = 0$$

$$\lim_{t \to \infty} |\tilde{c}| = \lim_{t \to \infty} |c_2(t) - c_1| = 0.$$

3   Sliding Mode Controller

3.1   Design of the Controller for the Fractional-Order Chen System

A primary step in designing the sliding mode controller is to choose a sliding surface. An appropriate switching surface with integral operation is proposed such that the sliding motion on the manifold achieves desired properties. However, a sliding surface may be defined in the form of:

$$S(t) = \int_0^t (e_1(\tau) + e_2(\tau) + e_3(\tau))d\tau + D^{q-1}(e_1 + e_2 + e_3). \tag{14.10}$$

Since the dynamic is of the fractional, a similar fractional dynamic surface is suggested. Due to complexity of the current synchronization task, an integral dynamic term is dedicated to be included in the surface. This will be shown providing a faster synchronization with less error. The situation $S(t) = 0$ proves a stable dynamic for $e(t)$. Our aim is to design a controller to enable the system reaching the sliding surface in a finite time. To ensure the occurrence of the sliding motion, a control law and the adaptation mechanism are proposed by:

$$\begin{cases} u_1 = -e_1 - a_2(t)(e_2 - e_1) - \eta k \, \mathrm{sgn}(S) \\ u_2 = -e_2 - c_2(t)(e_2 + e_1) + a_2(t)e_1 + y_1 e_3 + e_1 x_3 - \eta k \, \mathrm{sgn}(S) \\ u_3 = -e_3 - y_1 e_2 - e_1 x_2 + b_2(t)e_3 - \eta k \, \mathrm{sgn}(S) \end{cases} \tag{14.11}$$

$$\begin{cases} \dot{a}_2(t) = -S(x_2 - 2x_1) \\ \dot{b}_2(t) = S x_3 \\ \dot{c}_2(t) = -S(x_2 + x_1) \end{cases} \tag{14.12}$$

where $\eta > 1$ and k is the reaching gain, achieved by the following adaptive law:

$$\dot{k} = \frac{\gamma}{3}|S|, \quad k(0) = \hat{k} > 0 \tag{14.13}$$

showing γ is a positive constant number.

Lemma 14.1 (Barbalat lemma, [10]). *If $\omega : R \to R$ is a uniformly continuous function for $t \geq 0$ and if $\lim\limits_{t\to\infty} \int_0^t \omega(\lambda)d\lambda$ exists and is finite, then:*

$$\lim_{t\to\infty} \omega(t) = 0.$$

Statement 1. Consider the error dynamic (6) with unknown parameters and disturbance uncertainties. This system is controlled by the adaptive sliding mode controller (14.11) together with the adaptation mechanism (14.12). Consequently the state error trajectory converges to the sliding surface $S(t) = 0$.

Proof. Consider the following Lyapunov function as:

$$V = \frac{1}{2}S^2 + \frac{1}{2}(\tilde{a}^2 + \tilde{b}^2 + \tilde{c}^2) + \frac{3}{2\gamma}(\sigma - k)^2,$$

where σ is an upper bound of the uncertainty. Then, the appropriate first derivative is obtained by:

$$\dot{V} = S\dot{S} + \dot{a}_2(t)\tilde{a} + \dot{b}_2(t)\tilde{b} + \dot{c}_2(t)\tilde{c} + \frac{3}{\gamma}(\sigma - k)(-3\dot{k})$$

$$= S\begin{bmatrix} e_1 + e_2 + e_3 + a_2(t)(e_2 - e_1) + \tilde{a}(x_2 - x_1) \\ +\Delta f_1(y_1 + y_2 + y_3) + u_1 + c_2(e_2 + e_1) \\ +\tilde{c}(x_2 + x_1) - a_2 e_1 - x_1\tilde{a} - y_1 e_3 - e_1 x_3 \\ +\Delta f_2(y_1 + y_2 + y_3) + u_2 + y_1 e_2 + x_2 e_1 - b_2 e_3 - x_3\tilde{b} \\ +\Delta f_3(y_1 + y_2 + y_3) + u_3 \end{bmatrix}$$

$$+\dot{a}_2(t)\tilde{a} + \dot{b}_2(t)\tilde{b} + \dot{c}_2(t)\tilde{c} + \frac{3}{\gamma}(\sigma - k)(-3\dot{k}). \tag{14.14}$$

Substitution of (14.11) and (14.12) into (14.14) achieves the derivative of the Lyapunov function as:

$$\dot{V} = S\begin{bmatrix} -3\eta k \, \mathrm{sgn}(S) + \Delta f_1(y_1, y_2, y_3) + \Delta f_2(y_1, y_2, y_3) \\ +\Delta f_3(y_1, y_2, y_3) \end{bmatrix}$$

$$+\frac{3}{\gamma}(\sigma - k)(-3\dot{k}) \leq 3\sigma|S| - 3\eta k|S| + \frac{3}{\gamma}(\sigma - k)(-3\dot{k}). \tag{14.15}$$

From (14.13) and (14.15) we achieve:

$$\dot{V} = S\dot{S} \leq 3k|S|(1 - \eta). \tag{14.16}$$

From $\eta > 1$ and $k, k(0) > 0$, the derivative of Lyapunov function accordingly yields:

$$\dot{V} \leq 3k|S|(1 - \eta) = -\omega(t) \leq 0, \qquad (14.17)$$

where $\omega(t) = 3k|S|(\eta - 1)$. Using (14.11) and (14.12) concludes that the reaching condition $\dot{V} \leq 0$ is always maintained. Since $\dot{V}$ is negative semi-definite, the origin in the error dynamic is not an asymptotically stable point. On the other hand as $\dot{V} \leq 0$ then $S \in L_\infty$ and $\tilde{a}, \tilde{b}, \tilde{c} \in L_\infty$, accordingly $V(t) \in L_\infty$ (i.e. $S, \tilde{a}, \tilde{b}, \tilde{c}, V(t)$ are bounded). Then we have:

$$\int_0^t \omega(\lambda)d\lambda \leq \int_0^t -\dot{V}d\lambda = V(0) - V(t) \leq V(0).$$

As t approaches infinity, the above integral is always less than or equal to $V(0)$. Since $V(0)$ is positive and finite, $\lim_{t \to \infty} \int_0^t \omega(\lambda)d\lambda$ exists and is finite. Thus, according to the Barbalat's lemma, we obtain:

$$\lim_{t \to \infty} \omega(t) = \lim_{t \to \infty} 3k|S|(\eta - 1) = 0. \qquad (14.18)$$

Since $\eta > 1$, (14.18) implies $S = 0$. Hence the proof is completely achieved. ∎

3.2 Design of the Controller for the Fractional-Order Lü system

The same switching surface is used for the Lü system when the following proposed control law and the adaptation mechanism are used by:

$$\begin{cases} u_1 = -e_1 - a_2(t)(e_2 - e_1) - \eta k \, \text{sgn}(S) \\ u_2 = -e_2 - c_2(t)e_2 + a_2(t)e_1 + y_1 e_3 + e_1 x_3 - \eta k \, \text{sgn}(S) \\ u_3 = -e_3 - y_1 e_2 - e_1 x_2 + b_2(t)e_3 - \eta k \, \text{sgn}(S) \end{cases} \qquad (14.19)$$

$$\begin{cases} \dot{a}_2(t) = -S(x_2 - x_1) \\ \dot{b}_2(t) = Sx_3 \\ \dot{c}_2(t) = -Sx_2 \end{cases} \qquad (14.20)$$

Similar to the previous section, $\eta > 1$ and k as the reaching gain is achieved according to (14.13).

Statement 2. Consider the error dynamic in (14.9) with unknown parameters and disturbance uncertainties. The state error trajectory converges to the sliding surface $S(t) = 0$ if the sliding mode control law and the adaptation mechanism in (14.19) and (14.20) are applied.

Proof. Candidate the Lyapunov function as:

$$V = \frac{1}{2}S^2 + \frac{1}{2}(\tilde{a}^2 + \tilde{b}^2 + \tilde{c}^2) + \frac{3}{2\gamma}(\sigma - k)^2.$$

The time derivative of V is obtained by:

$$\dot{V} = S\dot{S} + \dot{a}_2(t)\tilde{a} + \dot{b}_2(t)\tilde{b} + \dot{c}_2(t)\tilde{c} + \frac{3}{\gamma}(\sigma - k)(-3\dot{k})$$

$$= S \begin{bmatrix} e_1 + e_2 + e_3 + a_2(t)(e_2 - e_1) + \tilde{a}(x_2 - x_1) \\ +\Delta f_1(y_1 + y_2 + y_3) + u_1 + c_2 e_2 \\ +\tilde{c}x_2 - y_1 e_3 - e_1 x_3 + \Delta f_2(y_1 + y_2 + y_3) + u_2 \\ +y_1 e_2 + x_2 e_1 - b_2 e_3 - x_3 \tilde{b} + \Delta f_3(y_1 + y_2 + y_3) + u_3 \end{bmatrix}$$

$$+\dot{a}_2(t)\tilde{a} + \dot{b}_2(t)\tilde{b} + \dot{c}_2(t)\tilde{c} + \frac{3}{\gamma}(\sigma - k)(-3\dot{k}). \tag{14.21}$$

Replacing (14.19), (14.20), and (14.13) into (14.21) achieves the derivative of the Lyapunov function as:

$$\dot{V} = S\dot{S} \le 3k|S|(1 - \eta). \tag{14.22}$$

Similar to the previous section the derivative of Lyapunov function remains negative. From inequality (14.22), it is concluded that there exists a finite time t_1 such that for all $t \ge t_1$, where the reach condition $\dot{V} \le 0$ is maintained. Similar to the previous section, using Barbalat's lemma provides $S = 0$ as $t \to \infty$. Thus the proof is completely achieved. ∎

4 The Simulation

A simulation has been carried out using SIMULINK®, where the order is set as $q = 0.95$. The Adams method is used to solve the system of differential equations during the simulation. Initial conditions of states of master and slave are, respectively, selected as (15, 10, 6) and (12, 8, 7). Meanwhile the master is perturbed by such an uncertainty term of:

$$\Delta f_i(y_1, y_2, y_3) = 0.5\sin\left(\sqrt{y_1^2 + y_2^2 + y_3^2}\right) \quad (i = 1, 2, 3).$$

Furthermore, the following adaptation law is appropriately chosen to update k:

$$\dot{k} = 2|S|, \quad k(0) = 10$$

together with the switching function in (14.10), which is shown in Fig. 14.3.

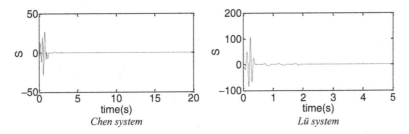

Fig. 14.1 Parameter identification of two unified systems when the master incorporates unknown parameters

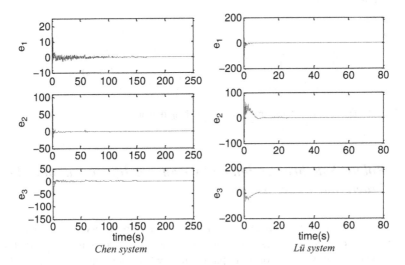

Fig. 14.2 Robust synchronization of fractional-order unified chaotic systems

4.1   Robust Synchronization and Parameter Identification of the Chen Fractional-Order System

In this section a robust synchronization together with a parameter identification of the Chen system are of concerned. Master system incorporates unknown constant parameters whilst the slave dynamic is perturbed by uncertainties. To achieve a robust synchronization and parameter identification, input controllers and the adaptation mechanism in (14.11) and (14.12) are, respectively, used. To obtain the Chen chaotic behavior, parameters in (14.4) is set to [14]:

$$a_1 = 40, \quad b_1 = 3, \quad c_1 = 28.$$

The result of the synchronization and parameter identification are, respectively, shown in Figs. 14.1 and 14.2, respectively.

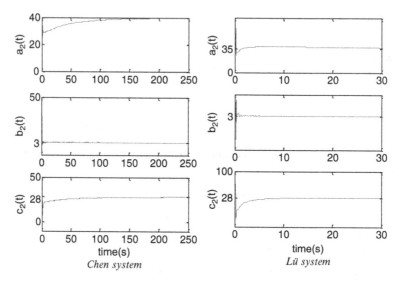

Fig. 14.3 Time response of the corresponding switching function $S(t)$

4.2 Robust Synchronization and Parameter Identification of Lü Fractional-Order System

Likewise a robust synchronization of Lü fractional-order system is considered here whilst some parameters are also identified. It is also assumed that the master involves unknown constant parameters when the slave is perturbed by such uncertainties. A robust synchronization and parameter identification will be achieved when input controllers and the adaptation mechanism are, respectively, used as in (14.19) and (14.20). The Lü system behaves chaotic when parameters in (14.7) are taken as [14]:

$$a_1 = 35, \quad b_1 = 3, \quad c_1 = 28.$$

The results of the synchronization and parameter identification are, respectively, shown in Figs. 14.1 and 14.2.

5 Conclusion

In this chapter, an adaptive sliding mode controller was used to synchronize a class of master–slave unified chaotic system through a Lyapunov based approach. This is achieved through using nonlinear inputs control when the system is also perturbed by the uncertainties. A novel switching surface is proposed to perform the task and to raise the convergence rate of the error in the closed-loop sliding mode control. This is achieved with no prior knowledge on the bound of uncertainty and disturbance is required. The states error converges to zero as time tends to infinity. The simulation result verifies the capability of the proposed adaptive control mechanism during the synchronization task through a simultaneous parameter identification scheme. The synchronization is made possible for two identical systems with different initial conditions. The result also shows the quality of that the proposed control scheme when it is found robust to bounded uncertainty.

References

1. Ahmad W, Sprott JC (2003) Chaos in fractional order system autonomous nonlinear systems. Chaos Solitons Fractals 16:339–351
2. Ahmad WM, Harba AM. (2003) On nonlinear control design for autonomous chaotic systems of integer and fractional orders. Chaos Solitons Fractals 18:693–701
3. Bowong S, Moukam Kakmeni FM, Dimi JL, Koina R (2006) Synchronizing chaotic dynamics with uncertainties using a predictable synchronization delay design. Commun Nonlinear Sci Numer Simul 11:973–987
4. Chen SH, Lu JH (2002) Synchronization of an uncertain unified chaotic system via adaptive control. Chaos Solitons Fractals 14(4):643–647
5. Chen S, Yang Q, Wang C (2004) Impulsive control and synchronization of unified chaotic systems. Chaos Solitons Fractals 20:751–758
6. EI-Sayed AMA (1996) Fractional order diffusion wave equation. Int J Theoret Phys 35:311–322
7. Femat R, Alvarez-Ramirez J, Fernandez-Anaya G (2000) Adaptive synchronization of high-order chaotic systems: a feedback with low-order parametrization. Phys D 139(3–4):231–246
8. Jenson VG, Jeffreys GV (1997) Mathematical methods in chemical engineering. Academic, New York
9. Kusnezov D, Bulgac A, Dang GD (1999) Quantum levy processes and fractional kinetics. Phys Rev Lett 82:1136–1139
10. Khalil HK (1992) Nonlinear systems. Macmillan, New York
11. Li C, Liao X, Yu J (2003) Synchronization of fractional order chaotic systems. Phys Rev 68:067203
12. Li C, Chen G (2004) Chaos in the fractional order Chen system and its control. Chaos Solitons Fractals 22:540–554
13. Lu J, Wu X, Han X, Lu J. (2004) Adaptive feedback synchronization of a unified chaotic system. Phys Lett A 329:327–333
14. Matouk AE (2009) Chaos synchronization between two different fractional systems of Lorenz family. Hindawi Publishing Corporation Mathematical Problems in Engineering, Article ID 572724, 11 p
15. Podlubny I (1999) Fractional differential equations. Academic, San Diego
16. Torvik PJ, Bagley RL (1984) On the appearance of the fractional derivative in the behavior of real materials. Trans ASME 51:294–298
17. Ucar A, Lonngren K, Bai E (2006) Synchronization of the unified chaotic systems via active control. Chaos Solitons Fractals 27:1292–1297
18. Wang F, Liu C (2007) Synchronization of unified chaotic system based on passive control. Phys D 225:55–60

Chapter 15
Fractional Cauchy Problems on Bounded Domains: Survey of Recent Results

Erkan Nane

1 Introduction

A celebrated paper of Einstein [8] established a mathematical link between random walks, the diffusion equation, and Brownian motion. The scaling limits of a simple random walk with mean zero, finite variance jumps yields a Brownian motion. The probability densities of the Brownian motion variables solve a diffusion equation, and hence we refer to the Brownian motion as the stochastic solution to the diffusion (heat) equation. The diffusion equation is the most familiar Cauchy problem. The general abstract Cauchy problem is $\partial_t u = Lu$, where $u(t)$ takes values in a Banach space and L is the generator of a continuous semigroup on that space, see [2]. If L generates a Markov process, then we call this Markov process a stochastic solution to the Cauchy problem $\partial_t u = Lu$, since its probability densities (or distributions) solve the Cauchy problem. This point of view has proven useful, for instance, in the modern theory of fractional calculus, since fractional derivatives are generators of certain (α-stable) stochastic processes, see [16].

Fractional derivatives are almost as old as their more familiar integer-order counterparts, see [23,28]. Fractional diffusion equations have recently been applied to problems in physics, finance, hydrology, and many other areas, see [11,13,22,30]. Fractional space derivatives are used to model anomalous diffusion or dispersion, where a particle plume spreads at a rate inconsistent with the classical Brownian motion model, and the plume may be asymmetric. When a fractional derivative replaces the second derivative in a diffusion or dispersion model, it leads to enhanced diffusion (also called superdiffusion). Fractional time derivatives are connected with anomalous subdiffusion, where a cloud of particles spreads more slowly than a classical diffusion. Fractional Cauchy problems replace the integer

E. Nane (✉)
Auburn University, Department of Mathematics and Statistics, 221 Parker Hall, Auburn, AL 36849, USA
e-mail: ezn0001@auburn.edu

D. Baleanu et al. (eds.), *Fractional Dynamics and Control*, DOI 10.1007/978-1-4614-0457-6_15, © Springer Science+Business Media, LLC 2012

time derivative by its fractional counterpart: $\partial_t^\beta u = Lu$. Here, $\partial_t^\beta g(t)$ indicates the Caputo fractional derivative in time, the inverse Laplace transform of $s^\beta \tilde{g}(s) - s^{\beta-1} g(0)$, where $\tilde{g}(s) = \int_0^\infty e^{-st} g(t) dt$ is the usual Laplace transform, see [6]. Nigmatullin [25] gave a physical derivation of the fractional Cauchy problem, when L is the generator of some continuous Markov process $\{Y(t)\}$ started at $x = 0$. The mathematical study of fractional Cauchy problems was initiated by Kochubei, Schneider and Wyss, see [12,13,31]. The existence and uniqueness of solutions was proved in [12, 13]. Fractional Cauchy problems were also invented independently by Zaslavsky [32] as a model for Hamiltonian chaos.

Stochastic solutions of fractional Cauchy problems are subordinated processes. If $X(t)$ is a stochastic solution to the Cauchy problem $\partial_t u = Au$, then under certain technical conditions, the subordinate process $X(E(t))$ is a stochastic solution to the fractional Cauchy problem $\partial_t^\beta u = Au$, see [3]. Here, $E(t)$ is the inverse or hitting time process to a stable subordinator $D(t)$ with index $\beta \in (0,1)$. That is, $E(t) = \inf\{x > 0 : D(x) > t\}$, and $D(t)$ is a Lévy process (continuous in probability with independent, stationary increments) whose smooth probability density $f_{D(1)}(t)$ has Laplace transform $e^{-s^\beta} = \tilde{f}_{D(1)}(s)$, see [29]. Just as Brownian motion is a scaling limit of a simple random walk, the stochastic solution to certain fractional Cauchy problems are scaling limits of continuous time random walks, in which the independent identically distributed (iid) jumps are separated by iid waiting times, see [19]. Fractional time derivatives arise from power law waiting times, where the probability of waiting longer than time, $t > 0$, falls off like $t^{-\beta}$ for large t value, see [16]. This is related to the fact that fractional derivatives are non-local operators defined by convolution with a power law, see [3].

In some applications, the waiting times between particle jumps evolve according to a more complicated process that cannot be adequately described by a single power law. Then, a waiting time model, that is, conditional power law leads to a distributed-order fractional derivative in time, defined by integrating the fractional derivative of order β against the probability distribution of the power law index, see [17]. The resulting distributed-order fractional Cauchy problem provides a more flexible model for anomalous sub-diffusion. The Lévy measure of a stable subordinator with index β is integrated against the power law index distribution to define a subordinator $W(t)$. Its inverse $E(t)$ produces a stochastic solution $X(E(t))$ of the distributed-order fractional Cauchy problem on $\mathbb{R}^d$, when $X(t)$ solves the original Cauchy problem $\partial_t u = Lu$.

2 Stochastic Solution of Heat Equation on Bounded Domains

Let D be a bounded domain in $\mathbb{R}^d$. We denote by $C^k(D), C^{k,\alpha}(D)$ and, $C^k(\bar{D})$ the space of k-times differentiable functions in D, the space of k-times differential functions with k-th derivative is Hölder's continuous of index α, and the space of functions that have all the derivatives up to order k extendable continuously up to the boundary ∂D of D, respectively. We refer to [20] for a detailed discussion of these spaces and concepts in this section.

Suppose that every point of ∂D is regular for D^C. The corresponding Markov process is a killed Brownian motion. We denote the eigenvalues and the eigenfunctions of the Laplacian $\Delta = \sum_{i=1}^{d} \partial_{x_i}^2$ by $\{\mu_n, \phi_n\}_{n=1}^{\infty}$, where $\phi_n \in C^{\infty}(D)$.

Remark 15.1. Eigenfunctions $\{\phi_n\}_{n=1}^{\infty}$ form an orthonormal basis of $L^2(D)$. In particular, the initial function f regarded as an element of $L^2(D)$ can be represented as

$$f(x) = \sum_{n=1}^{\infty} \bar{f}(n)\phi_n(x); \quad \bar{f}(n) = \int_D \phi_n(x)f(x)dx. \tag{15.1}$$

The corresponding heat kernel is given by

$$p_D(t,x,y) = \sum_{n=1}^{\infty} e^{-\mu_n t} \phi_n(x)\phi_n(y).$$

The series converges absolutely and uniformly on $[t_0, \infty) \times D \times D$ for all $t_0 > 0$. Let B be Brownian motion and let $\tau_D(X) = \inf\{\tau > 0 : X(\tau) \notin D\}$ be the first exit time of a process X from D. In this case, the semigroup given by

$$T_D(t)f(x) = \mathbb{E}_x[f(B(t))I(t < \tau_D(B))]$$

$$= \sum_{n=1}^{\infty} e^{-\mu_n t} \phi_n(x)\bar{f}(n) = \int_D p_D(t,x,y)f(y)dy \tag{15.2}$$

solves the heat equation in D with Dirichlet boundary conditions:

$$\partial_t u(t,x) = \Delta u(t,x), \quad x \in D, \, t > 0,$$

$$u(t,x) = 0, \quad x \in \partial D,$$

$$u(0,x) = f(x), \quad x \in D.$$

3   Fractional Cauchy Problem

Fractional derivatives in time are useful for physical models that involve sticking or trapping, see [19]. They are closely connected to random walk models with long waiting times between particle jumps, see [16]. The fractional derivatives are essentially convolutions with a power law. Various forms of the fractional derivative can be defined, depending on the domain of the power law kernel, and the way boundary points are handled, see [23, 28]. The Caputo fractional derivative invented by Caputo [6] is defined for $0 < \beta < 1$ as

$$\partial_t^{\beta} u(t,x) = \frac{1}{\Gamma(1-\beta)} \int_0^t \partial_r u(r,x) \frac{dr}{(t-r)^{\beta}}. \tag{15.3}$$

Its Laplace transform

$$\int_0^\infty e^{-st} \partial_t^\beta u(t,x)\, ds = s^\beta \tilde{u}(s,x) - s^{\beta-1} u(0,x) \tag{15.4}$$

incorporates the initial value in the same way as the first derivative. The Caputo derivative is useful for solving differential equations that involve a fractional time derivative, see [11, 26], because it naturally incorporates initial values.

Let $D \in \mathbb{R}^d$ be a bounded domain. In this section we will consider the fractional Cauchy problem:

$$\partial_t^\beta u(t,x) = \Delta u(t,x), \quad x \in D, \ t > 0;$$
$$u(t,x) = 0, \quad x \in \partial D, \ t > 0;$$
$$u(0,x) = f(x), \quad x \in D. \tag{15.5}$$

To obtain a solution, let $u(t,x) = G(t)F(x)$ be a solution of (15.5). Substituting in the PDE (15.5) leads to

$$F(x)\partial_t^\beta G(t) = G(t)\Delta F(x)$$

and now dividing both sides by $G(t)F(x)$, we obtain:

$$\frac{\partial_t^\beta G(t)}{G(t)} = \frac{\Delta F(x)}{F(x)} = -\mu.$$

That is,

$$\partial_t^\beta G(t) = -\mu G(t), \ t > 0; \tag{15.6}$$

$$\Delta F(x) = -\mu F(x) \ x \in D; \ F(x) = 0, \ x \in \partial D. \tag{15.7}$$

Eigenvalue problem (15.7) is solved by an infinite sequence of pairs (μ_n, ϕ_n), $n \geq 1$, where ϕ_n is a sequence of functions that form a complete orthonormal set in $L^2(D)$, $\mu_1 < \mu_2 \leq \mu_2 \leq \cdots$, and $\mu_n \to \infty$.

Using the μ_n determined by (15.7), we need to find a solution of (15.6) with $\mu = \mu_n$, which is the eigenvalue problem for the Caputo fractional derivative.

We next consider the eigenvalue problem for the Caputo fractional derivative of order $0 < \beta < 1$.

Lemma 15.1. *Let $\lambda > 0$. The unique solution of the eigenvalue problem*

$$\partial_t^\beta G(t) = -\lambda G(t), \ G(0) = 1 \tag{15.8}$$

is given by the Mittag-Leffler function:

$$G(t) = M_\beta(-\lambda t^\beta) = \sum_{n=0}^\infty \frac{(-\lambda t^\beta)^n}{\Gamma(1+\beta n)}. \tag{15.9}$$

For a detailed study of the Mittag-Leffler type functions we refer the reader to the tutorial paper by Gorenflo and Mainardi [10].

Therefore the solution to (15.6) is given by:

$$G(t) = G_0(n)M_\beta\left(-\mu t^\beta\right),$$

where $G_0(n) = \bar{f}(n)$ is selected to satisfy the initial condition f. Therefore using this lemma, we obtain a formal solution of the fractional Cauchy problem (15.5) as

$$u(t,x) = \sum_{n=1}^{\infty} \bar{f}(n)M_\beta\left(-\mu_n t^\beta\right)\phi_n(x). \tag{15.10}$$

Remark 15.2. The separation of variables technique works for a large class operators including uniformly elliptic operators L in divergence form, see Remark 15.5. For details, see [20]. Define a cube, i.e.,

$$D = \{x = (x_1, x_2, \ldots, x_d) : 0 < x_i < M \text{ for all } 1 \le i \le d\}.$$

The functions

$$\phi_n(x) = (2/M)^{d/2}\prod_{i=1}^{d}\sin(\pi n_i x_i/M)$$

parametrized by the multi-index of positive integers $n = \{n_1, n_2, \ldots, n_d\}$, form a complete orthonormal set of eigenfunctions of the Laplacian, with Dirichlet boundary conditions with corresponding eigenvalues

$$\mu_n = \pi^2 M^{-2}(n_1^2 + \cdots + n_d^2).$$

See, for example, Lemma 6.2.1 in [7]. In this case the boundary of the cube domain is not smooth.

3.1   Stochastic Solution

Fractional time derivatives emerge in anomalous diffusion models, when particles wait a long time between jumps. In the standard model, called a continuous time random walk (CTRW), a particle waits for a random time $J_n > 0$ and then takes a step of random size Y_n. Suppose the two sequences of i.i.d. random variables (J_n) and (Y_n) are independent. The particle arrives at location $X(n) = Y_1 + \cdots + Y_n$ at time $T(n) = J_1 + \cdots + J_n$. Since $N(t) = \max\{n \ge 0 : T(n) \le t\}$ is the number of jumps by time $t > 0$, the particle location at time t is $X(N(t))$. If $\mathbb{E}Y_n = 0$ and $\mathbb{E}[Y_n^2] < \infty$ then, as the time scale $c \to \infty$, the random walk of particle jumps has a scaling limit $c^{-1/2}X([ct]) \Rightarrow B(t)$, a standard Brownian motion. If $P(J_n > t) \sim ct^{-\beta}$ for some $0 < \beta < 1$ and $c > 0$, then the scaling limit $c^{-1/\beta}T([ct]) \Rightarrow D(t)$ is a strictly increasing stable Lévy process with index β, sometimes called a stable subordinator. The jump times $T(n)$ and the number of jumps $N(t)$ are inverses $\{N(t) \geq n\} = \{T(n) \leq t\}$, and it follows that the scaling limits are also inverses, see [16, Theorem 3.2]: $c^{-\beta}N(ct) \Rightarrow E(t)$, where

$$E(t) = \inf\{\tau : D(\tau) > t\}, \tag{15.11}$$

so that $\{E(t) \leq \tau\} = \{D(\tau) \geq t\}$. A continuous mapping argument in [16, Theorem 4.2] yields the CTRW scaling limit: Heuristically, since $N(ct) \approx c^{\beta}E(t)$, we have $c^{-\beta/2}X(N([ct])) \approx (c^{\beta})^{-1/2}X(c^{\beta}E(t)) \approx B(E(t))$, a time-changed Brownian motion. The density $u(t,x)$ of the process $B(E(t))$ solves a fractional Cauchy problem

$$\partial_t^{\beta} u(t,x) = \partial_x^2 u(t,x),$$

where the order of the fractional derivative equals the index of the stable subordinator. Roughly speaking, if the probability of waiting longer than time $t > 0$ between jumps falls off like $t^{-\beta}$, then the limiting particle density solves a diffusion equation that involves a fractional time derivative of the same order β.

The Laplace transform of $D(t)$ is given by:

$$\mathbb{E}(e^{-sD(t)}) = \int_0^{\infty} e^{-sx} f_{D(t)}(x)dx = e^{-ts^{\beta}}.$$

The inverse $E(t)$ of $D(t)$ has density

$$f_{E(t)}(l) = \partial_l P(E(t) \leq l) = \partial_l(1 - P(D(l) \leq t))$$

$$= -\partial_l \int_0^{\frac{t}{l^{1/\beta}}} f_{D(1)}(u)du$$

$$= (t/\beta)f_{D(1)}(tl^{-1/\beta})l^{-1-1/\beta}, \tag{15.12}$$

using the scaling property of the density $f_{D(t)}(l) = t^{-1/\beta}f_{D(1)}(lt^{-1/\beta})$, see [5].

Using the representation (15.12) and taking Laplace transforms we can show that the unique solution of the eigenvalue problem (15.8) is also given by

$$G(t) = \int_0^{\infty} \exp(-l\lambda)f_{E(t)}(l)dl = \mathbb{E}(\exp(-\lambda E(t))), \tag{15.13}$$

see [19] for the details.

Now using (15.2) and (15.13) we can express the solution to (15.5) as

$$
\begin{aligned}
u(t,x) &= \sum_{n=1}^{\infty} \bar{f}(n) M_\beta\left(-\mu_n t^\beta\right) \phi_n(x) \\
&= \sum_{n=1}^{\infty} \bar{f}(n) \left[\int_0^\infty \exp(-l\mu_n) f_{E(t)}(l) dl\right] \phi_n(x) \\
&= \int_0^\infty \left[\sum_{n=1}^{\infty} \bar{f}(n) \exp(-l\mu_n) \phi_n(x)\right] f_{E(t)}(l) dl \\
&= \int_0^\infty [T_D(l) f(x)] f_{E(t)}(l) dl \\
&= \mathbb{E}_x[f(B(E(t))) I(\tau_D(B) > E(t))] \\
&= \mathbb{E}_x[f(B(E(t))) I(\tau_D(B(E)) > t)].
\end{aligned}
\tag{15.14}
$$

Remark 15.3. Meerschaert et al. [20] established the conditions on the initial function f under which $u(t,x)$ is a classical solution (i.e., for each $t > 0$, $u(t,x) \in C^1(\bar{D}) \cap C^2(D)$ and for each $x \in D$, $u(t,x) \in C^1(0,\infty)$) of (15.5): that $\Delta f(x)$ has an eigenfunction expansion w.r.t. $\{\phi_n\}$, that is, absolutely and uniformly convergent. The analytic expression in $(0,M) \subset \mathbb{R}$ above is due to Agrawal [1].

4 Distributed-Order Fractional Cauchy Problems

Let μ be a finite measure with *supp* $\mu \subset (0,1)$. We consider the distributed order-time fractional derivative

$$
\mathbb{D}^{(\nu)} u(t,x) := \int_0^1 \partial_t^\beta u(t,x) \nu(d\beta), \quad \nu(d\beta) = \Gamma(1-\beta) \mu(d\beta). \tag{15.15}
$$

To ensure that $\mathbb{D}^{(\nu)}$ is well-defined, we impose the condition

$$
\int_0^1 \frac{1}{1-\beta} \mu(d\beta) < \infty \tag{15.16}
$$

as in [17, Eq. (3.3)]. Since $\Gamma(x) \sim 1/x$, as $x \to 0+$, this ensures that $\nu(d\beta)$ is a finite measure on $(0,1)$.

4.1 Eigenvalue Problem: Solution with Waiting Time Process

Stochastic solution to the distributed-order fractional Cauchy problem is obtained by considering a more flexible sequence of CTRW. At each scale $c > 0$, we are given i.i.d. waiting times (J_n^c) and i.i.d. jumps (Y_n^c). Assume the waiting times and jumps form triangular arrays whose row sums converge in distribution. Letting $X^c(n) = Y_1^c + \cdots + Y_n^c$ and $T^c(n) = J_1^c + \cdots + J_n^c$, we require that $X^c(cu) \Rightarrow A(t)$ and $T^c(cu) \Rightarrow W(t)$ as $c \to \infty$, where the limits $A(t)$ and $W(t)$ are independent Lévy processes. Letting $N_t^c = \max\{n \geq 0 : T^c(n) \leq t\}$, the CTRW scaling limit $X^c(N_t^c) \Rightarrow A(E_t^v)$, see [18, Theorem 2.1]. A power-law mixture model for waiting times was proposed in [17]: Take an i.i.d. sequence of mixing variables (B_i) with $0 < B_i < 1$ and assume $P\{J_i^c > u | B_i = \beta\} = c^{-1}u^{-\beta}$ for $u \geq c^{-1/\beta}$, so that the waiting times are power laws conditional on the mixing variables. The waiting time process $T^c(cu) \Rightarrow W(t)$ a nondecreasing Lévy process, or subordinator, with $\mathbb{E}[e^{-sW(t)}] = e^{-t\psi_W(s)}$ and Laplace exponent

$$\psi_W(s) = \int_0^\infty (e^{-sx} - 1)\phi_W(dx). \tag{15.17}$$

The Lévy measure

$$\phi_W(t, \infty) = \int_0^1 t^{-\beta}\mu(d\beta) = \int_0^1 \frac{t^{-\beta}}{\Gamma(1-\beta)} v(d\beta), \tag{15.18}$$

where μ is the distribution of the mixing variable B_i, see [17, Theorem 3.4 and Remark 5.1]. A computation in [17, Eq. (3.18)] using $\int_0^\infty (1 - e^{-st})\beta t^{-\beta-1}dt = \Gamma(1-\beta)s^\beta$ shows that:

$$\psi_W(s) = \int_0^1 s^\beta \Gamma(1-\beta)\mu(d\beta) = \int_0^1 s^\beta v(d\beta). \tag{15.19}$$

Then $c^{-1}N_t^c \Rightarrow E^v(t)$, the inverse subordinator, see [17, Theorem 3.10]. The general infinitely divisible Lévy process limit $A(t)$ forms a strongly continuous convolution semigroup with generator L (e.g., see [2]) and the corresponding CTRW scaling limit $A(E^v(t))$ is the stochastic solution to the distributed-order fractional Cauchy problem [17, Eq. (5.12)] defined by

$$\mathbb{D}^{(v)}u(t,x) = Lu(t,x). \tag{15.20}$$

Since $\phi_W(0, \infty) = \infty$ in (15.18), Theorem 3.1 in [18] implies that the inverse subordinator

$$E^v(t) = \inf\{x > 0 : W(x) > t\} \tag{15.21}$$

has density

$$g(t,x) = \int_0^t \phi_W(t - y, \infty)P_{W(x)}(dy). \tag{15.22}$$

This same condition ensures also that $E^v(t)$ is almost surely continuous, since $W(t)$ jumps in every interval, and hence is strictly increasing. Further, it follows from the definition (15.21) that $E^v(t)$ is monotone nondecreasing.

We say that a function is a **mild solution** to a pseudo-differential equation if its transform solves the corresponding equation in transform space. The next Lemma follows easily by taking Laplace transforms.

Lemma 15.2 ([21]). *For any $\lambda > 0$, $h(t,\lambda) = \int_0^\infty e^{-\lambda x} g(t,x)\,dx = \mathbb{E}[e^{-\lambda E^V(t)}]$ is a mild solution of*

$$\mathbb{D}^{(v)} h(t,\lambda) = -\lambda h(t,\lambda); \quad h(0,\lambda) = 1. \tag{15.23}$$

Kochubei [15] considered the following: Let $\rho(\alpha)$ be a right continuous non-decreasing step function on $(0,1)$. Assume that ρ has two sequences of jump points, β_n and v_n, $n = 0,1,2,\ldots$, where $\beta_n \to 0$, $v_n \to 1$, $\beta_0 = v_0 \in (0,1)$. Suppose also that the sequence $\{\beta_n\}$ is strictly decreasing and $\{v_n\}$ is strictly increasing. Let $\gamma_n^1 = (\rho(\beta_n) - \rho(\beta_n - 0))$ and $\gamma_n^2 = (\rho(v_n) - \rho(v_n - 0))$, and define the distributed order differential operator as:

$$\mathbb{D}^{(\rho)} u(t,x) = \sum_{n=0}^\infty \gamma_n^1 \partial_t^{\beta_n} u(t,x) + \sum_{n=0}^\infty \gamma_n^2 \partial_t^{v_n} u(t,x). \tag{15.24}$$

Since ρ is a finite measure we have

$$\sum_{n=1}^\infty \gamma_n^1 < \infty, \quad \sum_{n=1}^\infty \gamma_n^2 < \infty.$$

Here the corresponding subordinator is the sum of infinitely many independent stable subordinators;

$$W(t) = \sum_{n=0}^\infty (\gamma_n^1)^{1/\beta_n} \Gamma(1 - \beta_n) W^{\beta_n}(t) + \sum_{n=1}^\infty (\gamma_n^2)^{1/v_n} \Gamma(1 - v_n) W^{v_n}(t) \tag{15.25}$$

for independent stable subordinators $W^{\beta_n}(t)$, $W^{v_n}(t)$ for $n = 0,1,\cdots$.

Lemma 15.3. *Let $E^\rho(t) = \inf\{x > 0 : W(x) > t\}$. Then $h(t,\mu) = \mathbb{E}\left(e^{-\mu E^\rho(t)}\right)$ is the classical solution to the eigenvalue problem*

$$\mathbb{D}^{(\rho)} h(t,\mu) = -\mu h(t,\mu), \quad h(0,\mu) = 1.$$

Using inverse Laplace transforms Kochubei [15] established the following representation of $h(t,\mu)$:

$$h(t,\mu) = \frac{\mu}{\pi} \int_0^\infty r^{-1} e^{-tr} \frac{H_1(r)}{H_2(r)}\,dr, \tag{15.26}$$

where

$$H_1(r) = \sum_{n=0}^\infty \left[(\gamma_n^1) r^{\beta_n} \sin(\pi \beta_n) + (\gamma_n^2) r^{v_n} \sin(\pi v_n) \right]$$

$$H_2(r) = \left\{ \mu + \sum_{n=0}^{\infty} \left[\gamma_n^1 r^{\beta_n} \cos(\pi \beta_n) + \gamma_n^2 r^{v_n} \cos(\pi v_n) \right] \right\}^2$$

$$+ \left\{ \sum_{n=0}^{\infty} \left[\gamma_n^1 r^{\beta_n} \sin(\pi \beta_n) + \gamma_n^2 r^{v_n} \sin(\pi v_n) \right] \right\}^2. \qquad (15.27)$$

Let $D \subset \mathbb{R}^d$ be a bounded domain with $\partial D \in C^{1,\alpha}$ for some $0 < \alpha < 1$, and $D_\infty = (0,\infty) \times D$. We will write $u \in C^k(\bar{D})$ to mean that for each fixed $t > 0$, $u(t,\cdot) \in C^k(\bar{D})$, and $u \in C_b^k(\bar{D}_\infty)$ to mean that $u \in C^k(\bar{D}_\infty)$ and is bounded. Define

$$\mathcal{H}_\Delta(D_\infty) = \{u : D_\infty \to \mathbb{R} : \ \Delta u(t,x) \in C(D_\infty)\};$$

$$\mathcal{H}_\Delta^b(D_\infty) = \mathcal{H}_\Delta(D_\infty) \cap \{u : |\partial_t u(t,x)| \le k(t) g(x), \ g \in L^\infty(D), \ t > 0\},$$

for some functions k and b satisfying the condition

$$b(\lambda) \int_0^1 \int_0^t \frac{k(s)ds}{(t-s)^\beta} d\mu(\beta) < \infty, \qquad (15.28)$$

for $t, \lambda > 0$ and

$$k(t) \sum_{n=1}^{\infty} b(\lambda_n) \bar{f}(n) |\phi_n(x)| < \infty. \qquad (15.29)$$

Theorem 15.1. *Let $f \in C^1(\bar{D}) \cap C^2(D)$ for which the eigenfunction expansion (of Δf) with respect to the complete orthonormal basis $\{\phi_n : \ n \in \mathbb{N}\}$ converges uniformly and absolutely. Then the classical solution to the distributed-order fractional Cauchy problem*

$$\mathbb{D}^{(\rho)} u(t,x) = \Delta u(t,x), \ x \in D, \ t \ge 0;$$

$$u(t,x) = 0, \ x \in \partial D, \ t \ge 0;$$

$$u(0,x) = f(x), \ x \in D, \qquad (15.30)$$

for $u \in \mathcal{H}_\Delta^b(D_\infty) \cap C_b(\bar{D}_\infty) \cap C^1(\bar{D})$, with the distributed order fractional derivative $\mathbb{D}^{(\rho)}$ defined by (15.24), is given by

$$u(t,x) = \mathbb{E}_x[f(B(E^\rho(t))) I(\tau_D(B) > E^\rho(t))]$$

$$= \mathbb{E}_x[f(B(E^\rho(t))) I(\tau_D(B(E^\rho)) > t)]$$

$$= \int_0^\infty [T_D(l) f(x)] f_{E^\rho(t)}(l) dl$$

$$= \sum_0^\infty \bar{f}(n) \phi_n(x) h(t, \mu_n). \qquad (15.31)$$

In this case, $b(\lambda) = \lambda$, and $k(t)$ is given by $k(t) = Ct^{\beta_0-1}$, $0 < \beta_0 < 1$.

Proof. The proof is similar to the proof of Theorem 3.1 in [21]. We give the main parts of the proof here.

Denote the Laplace transform $t \to s$ of $u(t,x)$ by

$$\tilde{u}(s,x) = \int_0^\infty e^{-st} u(t,x)dt.$$

Since wet are working on a bounded domain, the Fourier transform methods in [19] are not useful. Instead, we will employ Hilbert space methods. Hence, given a complete orthonormal basis $\{\phi_n(x)\}$ on $L^2(D)$, we will call

$$\bar{u}(t,n) = \int_D \phi_n(x)u(t,x)dx;$$

$$\hat{u}(s,n) = \int_D \phi_n(x) \int_0^\infty e^{-st}u(t,x)dtdx$$

$$= \int_D \phi_n(x)\tilde{u}(s,x)dx$$

$$= \int_0^\infty e^{-st}\bar{u}(t,x)dt \quad \text{(when Fubini Thm. holds)}, \qquad (15.32)$$

respectively the ϕ_n and the ϕ_n-Laplace transforms. Since $\{\phi_n\}$ is a complete orthonormal basis for $L^2(D)$, we can invert the ϕ_n-transform to obtain

$$u(t,x) = \sum_n \bar{u}(t,n)\psi_n(x)$$

for any $t > 0$, where the above series converges in the L^2 sense (e.g., see [27, Proposition 10.8.27]).

Assume that $u(t,x)$ solves (15.31). Using Green's second identity, we obtain

$$\int_D [u\Delta\phi_n - \phi_n\Delta u]dx = \int_{\partial D} \left[u\frac{\partial\phi_n}{\partial\theta} - \phi_n\frac{\partial u}{\partial\theta} \right] ds = 0,$$

since $u|_{\partial D} = 0 = \phi_n|_{\partial D}$, $u \in C^1(\bar{D})$ by assumption, and $\phi_n \in C^1(\bar{D})$ by [9, Theorem 8.29]. Hence, the ϕ_n-transform of Δu is

$$\int_D \phi_n(x)\Delta u(t,x)dx = -\mu_n \int_D u(t,x)\phi_n(x)dx$$

$$= -\mu_n\bar{u}(t,n), \qquad (15.33)$$

as ϕ_n is the eigenfunction of the Laplacian corresponding to eigenvalue μ_n.

The fact that the operator $\mathbb{D}^{(\rho)}$ commutes with the ϕ_n-transform follows from (15.28).

Taking the ϕ_n-transform of (15.31) we obtain that

$$\mathbb{D}^{(\rho)}\bar{u}(t,n) = -\mu_n\bar{u}(t,n). \tag{15.34}$$

From Lemma 15.3 we get the solution

$$\bar{u}(t,n) = \bar{f}(n)h(t,\mu_n) = \bar{f}(n)\mathbb{E}\left(e^{-\mu_n E^\rho(t)}\right).$$

Now inverting the ϕ_n-transform gives

$$u(t,x) = \sum_0^\infty \bar{f}(n)\phi_n(x)h(t,\mu_n).$$

The stochastic representation uses Lemma 15.3 and the stochastic representation of the killed semigroup of Brownian motion (15.2).

We use the representation (15.26) to establish the fact that the solution is a classical solution. The details of the proof can be seen from the proof of the main results in [20, 21]. □

Remark 15.4. Let $v(d\beta) = p(\beta)d\beta$ for some $p \in C^1(0,1)$, and $0 < \beta_0 < \beta_1 < 1$ be such that

$$C(\beta_0,\beta_1,p) = \int_{\beta_0}^{\beta_1} \sin(\beta\pi)\Gamma(1-\beta)p(\beta)d\beta > 0. \tag{15.35}$$

Then

$$h(t,\lambda) = \mathbb{E}\left[e^{-\lambda E^v(t)}\right] = \frac{\lambda}{\pi}\int_0^\infty r^{-1}e^{-tr}\frac{\Phi_1(r)}{\Phi_2(r)}dr, \tag{15.36}$$

where

$$\Phi_1(r) = \int_0^1 r^\beta \sin(\beta\pi)\Gamma(1-\beta)p(\beta)d\beta$$

$$\Phi_2(r) = \left[\int_0^1 r^\beta \cos(\beta\pi)\Gamma(1-\beta)p(\beta)d\beta + \lambda\right]^2$$
$$+ \left[\int_0^1 r^\beta \sin(\beta\pi)\Gamma(1-\beta)p(\beta)d\beta\right]^2.$$

In this case we have $|\partial_t h(t,\lambda)| \le \lambda k(t)$, where

$$k(t) = [C(\beta_0,\beta_1,p)\pi]^{-1}\left[\Gamma(1-\beta_1)t^{\beta_1-1} + \Gamma(1-\beta_0)t^{\beta_0-1}\right]. \tag{15.37}$$

Hence, $h(t,\lambda)$ is a classical solution to (15.23). The representation (15.36) is due to Kochubei [14], which follows by inverting the Laplace transform of (15.23).

For $u \in \mathcal{H}_\Delta^b(D_\infty) \cap C_b(\bar{D}_\infty) \cap C^1(\bar{D})$ for k given by (15.37), Meerschaert et al. [21] shows that the solution to (15.30), where $\mathbb{D}^{(\rho)}$ replaced with the more general $\mathbb{D}^{(v)}$, is a strong (classical) solution for $f \in C^1(\bar{D}) \cap C^2(D)$ for which Δf has an absolutely and uniformly convergent eigenfunction expansion w.r.t $\{\phi_n\}$. Naber [24] studied the distributed-order fractional Cauchy problem in $D = (0, M) \subset \mathbb{R}$.

Remark 15.5. The methods of this paper also apply to the Cauchy problems that are obtained by replacing Laplacian with uniformly elliptic operator in divergence form defined on C^2 functions by

$$Lu = \sum_{i,j=1}^{d} \frac{\partial \left(a_{ij}(x)(\partial u/\partial x_i) \right)}{\partial x_j} \tag{15.38}$$

with $a_{ij}(x) = a_{ji}(x)$ and, for some $\lambda > 0$,

$$\lambda \sum_{i=1}^{n} y_i^2 \le \sum_{i,j=1}^{n} a_{ij}(x) y_i y_j \le \lambda^{-1} \sum_{i=1}^{n} y_i^2, \quad \forall y \in \mathbb{R}^d. \tag{15.39}$$

If X_t is a solution to $dX_t = \sigma(X_t)dB_t + b(X_t)dt$, $X_0 = x_0$, where σ is a $d \times d$ matrix, and B_t is a Brownian motion, then X_t is associated with the operator L with $a = \sigma\sigma^T$, see Chapters 1 and 5 in Bass [4]. Define the first exit time as $\tau_D(X) = \inf\{t \ge 0 : X_t \notin D\}$. The semigroup defined by $T_D(t)f(x) = \mathbb{E}_x[f(X_t)I(\tau_D(X) > t)]$ has generator L with Dirichlet boundary conditions, which follows by an application of the Itô formula.

References

1. Agrawal OP (2002) Solution for a fractional diffusion-wave equation defined in a bounded domain. Fractional order calculus and its applications. Nonlinear Dynam 29:145–155
2. Arendt W, Batty C, Hieber M, Neubrander F (2001) Vector-valued Laplace transforms and Cauchy problems. Monographs in Mathematics, 2nd edn. Vol 96, Birkhäuser, Basel, p 539
3. Baeumer B, Meerschaert MM (2001) Stochastic solutions for fractional Cauchy problems. Fract Calc Appl Anal 4:481–500
4. Bass RF (1998) Diffusions and elliptic operators. Springer-Verlag, New York
5. Bertoin J (1996) Lévy processes. Cambridge University Press, Cambridge
6. Caputo M (1967) Linear models of dissipation whose Q is almost frequency independent, Part II. Geophys J R Astr Soc 13:529–539
7. Davies EB (1995) Spectral theory and differential operators. Cambridge studies in advance mathematics, vol 42. Cambridge University Press, Cambridge
8. Einstein A (1906) On the theory of the Brownian movement. Ann Phys-Berlin 4:371–381
9. Gilbarg D, Trudinger NS (2001) Elliptic partial differential equations of second order. Reprint of the 1998 edn. Springer, New York
10. Gorenflo R, Mainardi F (1997) Fractional claculus: Integral and differential equations of fractional order. In: Capinteri A, Mainardi F (eds) Fractals and fractional calculus in continuum mechanics, Springer-Verlag, New York, pp 223–276
11. Gorenflo R, Mainardi F (2003) Fractional diffusion processes: Probability distribution and continuous time random walk. Lecture Notes Phys 621:148–166
12. Kochubei AN (1989) A Cauchy problem for evolution equations of fractional order. Diff Equat 25:967–974

13. Kochubei AN (1990) Fractional-order diffusion. Diff Equat 26:485–492
14. Kochubei AN (2008) Distributed order calculus and equations of ultraslow diffusion. J Math Anal Appl 340:252–281
15. Kochubei AN (2008) Distributed order calculus: an operator-theoretic interpretation. Ukraïn Mat Zh 60:478–486
16. Meerschaert MM, Scheffler HP (2004) Limit theorems for continuous time random walks with infinite mean waiting times. J Appl Probab 41:623–638
17. Meerschaert MM, Scheffler HP (2006) Stochastic model for ultraslow diffusion. Stoch Proc Appl 116:1215–1235
18. Meerschaert MM, Scheffler HP (2008) Triangular array limits for continuous time random walks. Stoch Proc Appl 118:1606–1633
19. Meerschaert MM, Benson DA, Scheffler HP, Baeumer B (2002) Stochastic solution of space–time fractional diffusion equations. Phys Rev E 65:1103–1106
20. Meerschaert MM, Nane E, Vellaisamy P (2009) Fractional Cauchy problems on bounded domains. Ann Probab 37:979–1007
21. Meerschaert MM, Nane E, Vellaisamy P (2011) Distributed-order fractional diffusions on bounded domains. J Math Anal Appl 379:216–228
22. Metzler R, Klafter J (2004) The restaurant at the end of the random walk: recent developments in the description of anomalous transport by fractional dynamics. J Phys A 37:161–208
23. Miller K, Ross B (1993) An introduction to the fractional calculus and fractional differential equations. Wiley, New York
24. Naber M (2004) Distributed order fractional sub-diffusion. Fractals 12:23–32
25. Nigmatullin RR (1986) The realization of the generalized transfer in a medium with fractal geometry. Phys Status Solidi B 133:425–430
26. Podlubny I (1999) Fractional differential equations. Academic Press, San Diego
27. Royden HL (1968) Real analysis, 2nd edn. MacMillan, New York
28. Samko S, Kilbas A, Marichev O (1993) Fractional integrals and derivatives: theory and applications. Gordon and Breach, London
29. Sato KI (1999) Lévy processes and infinitely divisible distributions. Cambridge University Press, Cambridge
30. Scalas E (2004) Five years of continuous-time random walks in econophysics. In: Namatame A (ed) Proc of WEHIA 2004, Kyoto, pp 3–16
31. Schneider WR, Wyss W (1989) Fractional diffusion and wave equations. J Math Phys 30:134–144
32. Zaslavsky G (1994) Fractional kinetic equation for Hamiltonian chaos. Phys D 76:110–122

Chapter 16
Fractional Analogous Models in Mechanics and Gravity Theories

Dumitru Baleanu and Sergiu I. Vacaru

1 Introduction

We can construct analogous fractional models of geometries and physical theories in explicit form if we use fractional derivatives resulting in zero for actions on constants (for instance, for the Caputo fractional derivative). This is important for elaborating geometric models of theories with fractional calculus even (performing corresponding nonholonomic deformations) we may prefer to work with another type of fractional derivatives.

In this chapter, we outline some key constructions for analogous classical and quantum fractional theories [1–6] when methods of nonholonomic and Lagrange–Finsler geometry are generalized to fractional dimensions.[1]

An important consequence of such geometric approaches is that using analogous and bi-Hamilton models (see integer dimension constructions [7, 7, 8]) and related solitonic systems, we can study analytically and numerically, as well to try to construct some analogous mechanical and gravitational systems, with the aim to mimic a nonlinear/fractional nonholonomic dynamics/evolution and even to provide certain schemes of quantization, like in the "fractional" Fedosov approach [4, 8].

---

[1]We recommend readers to consult in advance the above cited papers on details, notation conventions, and bibliography.

D. Baleanu (✉)
Department of Mathematics and Computer Sciences, Cankaya University, 06530, Ankara, Turkey

Institute of Space Sciences, P. O. Box, MG-23, R 76900, Magurele–Bucharest, Romania
e-mail: dumitru@cankaya.edu.tr

S.I. Vacaru
Science Department, University "Al. I. Cuza" Iaşi, 54, Lascar Catargi street,
Iaşi, Romania, 700107
e-mail: sergiu.vacaru@uaic.ro

D. Baleanu et al. (eds.), *Fractional Dynamics and Control*,
DOI 10.1007/978-1-4614-0457-6_16, © Springer Science+Business Media, LLC 2012

This work is organized in the form:

In Sect. 2, we remember the most important formulas on Caputo fractional derivatives and nonlinear connections. Section 3 is devoted to fractional Lagrange–Finsler geometries. There are presented the main constructions for analogous fractional gravity in Sect. 4.

2 Caputo Fractional Derivatives and N-Connections

We provide some important formulas on fractional calculus for nonholonomic manifold elaborated in [1–3, 5]. Our geometric arena consists from an abstract fractional manifold $\overset{\alpha}{V}$ (we shall use also the term "fractional space" as an equivalent one enabled with certain fundamental geometric structures) with prescribed nonholonomic distribution modeling both the fractional calculus and the non-integrable dynamics of interactions.

The fractional left, respectively, right Caputo derivatives are denoted

$$_{1x}\overset{\alpha}{\partial}_x f(x) := \frac{1}{\Gamma(s-\alpha)} \int\limits_{1x}^{x} (x-x')^{s-\alpha-1} \left(\frac{\partial}{\partial x'}\right)^s f(x')dx';$$

$$_x\overset{\alpha}{\partial}_{2x} f(x) := \frac{1}{\Gamma(s-\alpha)} \int\limits_{x}^{2x} (x'-x)^{s-\alpha-1} \left(-\frac{\partial}{\partial x'}\right)^s f(x')dx'. \qquad (16.1)$$

Using such operators, we can construct the fractional absolute differential $\overset{\alpha}{d} := (dx^j)^\alpha\ _0\overset{\alpha}{\partial}_j$ when $\overset{\alpha}{d}x^j = (dx^j)^\alpha \frac{(x^j)^{1-\alpha}}{\Gamma(2-\alpha)}$, where we consider $_1x^i = 0$.

We denote a fractional tangent bundle in the form $\underline{T}^{\alpha}M$ for $\alpha \in (0,1)$, associated with a manifold M of necessary smooth class and integer $\dim M = n$.[2] Locally, both the integer and fractional local coordinates are written in the form $u^\beta = (x^j, y^a)$. A fractional frame basis $\overset{\alpha}{\underline{e}}_\beta = e^{\beta'}_\beta(u^\beta)\overset{\alpha}{\partial}_{\beta'}$ on $\underline{T}^{\alpha}M$ is connected via a vierlbein transform $e^{\beta'}_\beta(u^\beta)$ with a fractional local coordinate basis

$$\overset{\alpha}{\partial}_{\beta'} = \left(\overset{\alpha}{\partial}_{j'} =\ _{1x^{j'}}\overset{\alpha}{\partial}_{j'}, \overset{\alpha}{\partial}_{b'} =\ _{1y^{b'}}\overset{\alpha}{\partial}_{b'}\right), \qquad (16.2)$$

---

[2]The symbol T is underlined to emphasize that we shall associate the approach with a fractional Caputo derivative.

for $j' = 1, 2, \ldots, n$ and $b' = n+1, n+2, \ldots, n+n$. The fractional co-bases are written as $\underline{e}^{\alpha\,\beta} = e^{\beta}_{\beta'}(u^{\beta}) \underline{d}u^{\alpha\,\beta'}$, where the fractional local coordinate co-basis is

$$\underline{d}u^{\alpha\,\beta'} = \left((dx^{i'})^{\alpha}, (dy^{a'})^{\alpha} \right). \tag{16.3}$$

It is possible to define a nonlinear connection (N-connection) $\overset{\alpha}{\mathbf{N}}$ for a fractional space $\overset{\alpha}{\mathbf{V}}$ by a nonholonomic distribution (Whitney sum) with conventional h- and v-subspaces, $\underline{h}\overset{\alpha}{\mathbf{V}}$ and $\underline{v}\overset{\alpha}{\mathbf{V}}$,

$$\underline{T}\overset{\alpha\,\alpha}{\mathbf{V}} = \underline{h}\overset{\alpha}{\mathbf{V}} \oplus \underline{v}\overset{\alpha}{\mathbf{V}}. \tag{16.4}$$

Locally, such a fractional N-connection is characterized by its local coefficients $\overset{\alpha}{\mathbf{N}} = \{{}^{\alpha}N_i^a\}$, when $\overset{\alpha}{\mathbf{N}} = {}^{\alpha}N_i^a(u)(dx^i)^{\alpha} \otimes \overset{\alpha}{\underline{\partial}}_a$.

On $\overset{\alpha}{\mathbf{V}}$, it is convenient to work with N-adapted fractional (co) frames,

$$ {}^{\alpha}\mathbf{e}_{\beta} = \left[{}^{\alpha}\mathbf{e}_j = \overset{\alpha}{\underline{\partial}}_j - {}^{\alpha}N_j^a \overset{\alpha}{\underline{\partial}}_a, \; {}^{\alpha}\mathbf{e}_b = \overset{\alpha}{\underline{\partial}}_b \right], \tag{16.5}$$

$$ {}^{\alpha}\mathbf{e}^{\beta} = \left[{}^{\alpha}\mathbf{e}^j = (dx^j)^{\alpha}, \; {}^{\alpha}\mathbf{e}^b = (dy^b)^{\alpha} + {}^{\alpha}N_k^b(dx^k)^{\alpha} \right]. \tag{16.6}$$

A fractional metric structure (d-metric) $\overset{\alpha}{\mathbf{g}} = \{{}^{\alpha}g_{\alpha\beta}\} = \left[{}^{\alpha}g_{kj}, {}^{\alpha}g_{cb} \right]$ on $\overset{\alpha}{\mathbf{V}}$ can be represented in different equivalent forms,

$$\begin{aligned} \overset{\alpha}{\mathbf{g}} &= {}^{\alpha}g_{\gamma\beta}(u)(du^{\gamma})^{\alpha} \otimes (du^{\beta})^{\alpha} = \eta_{k'j'} \, {}^{\alpha}\mathbf{e}^{k'} \otimes {}^{\alpha}\mathbf{e}^{j'} + \eta_{c'b'} \, {}^{\alpha}\mathbf{e}^{c'} \otimes {}^{\alpha}\mathbf{e}^{b'}, \\ &= {}^{\alpha}g_{kj}(x,y) \, {}^{\alpha}\mathbf{e}^{k} \otimes {}^{\alpha}\mathbf{e}^{j} + {}^{\alpha}g_{cb}(x,y) \, {}^{\alpha}\mathbf{e}^{c} \otimes {}^{\alpha}\mathbf{e}^{b} \end{aligned} \tag{16.7}$$

where matrices $\eta_{k'j'} = diag[\pm 1, \pm 1, \ldots, \pm 1]$ and $\eta_{a'b'} = diag[\pm 1, \pm 1, \ldots, \pm 1]$, for the signature of a "prime" spacetime $\mathbf{V}$, are obtained by frame transforms $\eta_{k'j'} = e^{k}_{k'} e^{j}_{j'} \, {}^{\alpha}g_{kj}$ and $\eta_{a'b'} = e^{a}_{a'} e^{b}_{b'} \, {}^{\alpha}g_{ab}$.

We can adapt geometric objects on $\overset{\alpha}{\mathbf{V}}$ with respect to a given structure $\overset{\alpha}{\mathbf{N}}$, calling them as distinguished objects (d-objects). For instance, a distinguished connection (d-connection) $\overset{\alpha}{\mathbf{D}}$ on $\overset{\alpha}{\mathbf{V}}$ is defined as a linear connection preserving under parallel transports the Whitney sum (16.4). There is an associated N-adapted differential 1-form

$$ {}^{\alpha}\mathbf{\Gamma}^{\tau}_{\beta} = {}^{\alpha}\Gamma^{\tau}_{\beta\gamma} \, {}^{\alpha}\mathbf{e}^{\gamma}, \tag{16.8}$$

parametrizing the coefficients (with respect to (19.2) and (19.1)) in the form ${}^{\alpha}\Gamma^{\gamma}_{\tau\beta} = \left({}^{\alpha}L^i_{jk}, {}^{\alpha}L^a_{bk}, {}^{\alpha}C^i_{jc}, {}^{\alpha}C^a_{bc} \right).$

The absolute fractional differential $^{\alpha}\mathbf{d} = \,_{1x}d_x + \,_{1y}d_y$ acts on fractional differential forms in N-adapted form. This is a fractional distinguished operator, d-operator; the value $^{\alpha}\mathbf{d} := \,^{\alpha}e^{\beta}\,^{\alpha}e_{\beta}$ splits into exterior h-/ v-derivatives, $_{1x}d_x : = (dx^i)^{\alpha}\,_{1x}\underline{\partial}_i = \,^{\alpha}e^j\,^{\alpha}e_j$ and $_{1y}d_y : = (dy^a)^{\alpha}\,_{1x}\underline{\partial}_a = \,^{\alpha}e^b\,^{\alpha}e_b$. Using such differentials, we compute the torsion and curvature (as fractional two d-forms derived for (16.8)) of $^{\alpha}\mathbf{D} = \{^{\alpha}\Gamma^{\tau}_{\beta\gamma}\}$,

$$^{\alpha}\mathcal{T}^{\tau} \doteqdot \,^{\alpha}\mathbf{D}\,^{\alpha}\mathbf{e}^{\tau} = \,^{\alpha}\mathbf{d}\,^{\alpha}\mathbf{e}^{\tau} + \,^{\alpha}\Gamma^{\tau}_{\beta} \wedge \,^{\alpha}\mathbf{e}^{\beta} \text{ and}$$

$$^{\alpha}\mathcal{R}^{\tau}_{\beta} \doteqdot \,^{\alpha}\mathbf{D}\,^{\alpha}\Gamma^{\tau}_{\beta} = \,^{\alpha}d\,^{\alpha}\Gamma^{\tau}_{\beta} - \,^{\alpha}\Gamma^{\gamma}_{\beta} \wedge \,^{\alpha}\Gamma^{\tau}_{\gamma} = \,^{\alpha}\mathbf{R}^{\tau}_{\beta\gamma\delta}\,^{\alpha}\mathbf{e}^{\gamma} \wedge \,^{\alpha}\mathbf{e}^{\delta}.$$

Contracting respectively the indices, we can compute the fractional Ricci tensor $^{\alpha}\mathcal{R}ic = \{^{\alpha}\mathbf{R}_{\alpha\beta} \doteqdot \,^{\alpha}\mathbf{R}^{\tau}_{\alpha\beta\tau}\}$ with components

$$^{\alpha}R_{ij} \doteqdot \,^{\alpha}R^k_{ijk}, \quad ^{\alpha}R_{ia} \doteqdot - \,^{\alpha}R^k_{ika}, \quad ^{\alpha}R_{ai} \doteqdot \,^{\alpha}R^b_{aib}, \quad ^{\alpha}R_{ab} \doteqdot \,^{\alpha}R^c_{abc}$$

and the scalar curvature of $^{\alpha}\mathbf{D}$,

$$^{\alpha}_s R \doteqdot \,^{\alpha}\mathbf{g}^{\tau\beta}\,^{\alpha}\mathbf{R}_{\tau\beta} = \,^{\alpha}R + \,^{\alpha}S, \quad ^{\alpha}R = \,^{\alpha}g^{ij}\,^{\alpha}R_{ij}, \quad ^{\alpha}S = \,^{\alpha}g^{ab}\,^{\alpha}R_{ab},$$

with $^{\alpha}\mathbf{g}^{\tau\beta}$ being the inverse coefficients to a d-metric (19.3).

The Einstein tensor of any metric compatible $^{\alpha}\mathbf{D}$, when $^{\alpha}\mathbf{D}_{\tau}\,^{\alpha}\mathbf{g}^{\tau\beta} = 0$, is defined $^{\alpha}\mathcal{E}ns = \{^{\alpha}\mathbf{G}_{\alpha\beta}\}$, where

$$^{\alpha}\mathbf{G}_{\alpha\beta} := \,^{\alpha}\mathbf{R}_{\alpha\beta} - \frac{1}{2}\,^{\alpha}\mathbf{g}_{\alpha\beta}\,^{\alpha}_s\mathbf{R}. \qquad (16.9)$$

The regular fractional mechanics defined by a fractional Lagrangian $^{\alpha}L$ can be equivalently encoded into canonical geometric data $(_L^{\alpha}\mathbf{N}, _L^{\alpha}\mathbf{g}, ^{\alpha}\mathbf{D})$, where we put the label L to emphasize that such geometric objects are induced by a fractional Lagrangian as we provided in [1–3, 5]. We also note that it is possible to "arrange" on $\mathbf{V}$ such nonholonomic distributions when a d-connection $_0^{\alpha}\mathbf{D} = \{_0^{\alpha}\widetilde{\Gamma}^{\gamma}_{\alpha'\beta'}\}$ is described by constant matrix coefficients, see details in [7, 8], for integer dimensions, and [5], for fractional dimensions.

3 Fractional Lagrange–Finsler Geometry

A Lagrange space $L^n = (M, L)$, of integer dimension n, is defined by a Lagrange fundamental function $L(x, y)$, i.e., a regular real function $L : TM \to \mathbb{R}$, for which the Hessian $_L g_{ij} = (1/2)\partial^2 L/\partial y^i \partial y^j$ is not degenerate.

We say that a Lagrange space L^n is a Finsler space F^n if and only if its fundamental function L is positive and two homogeneous with respect to variables y^i, i.e. $L = F^2$. For simplicity, we shall work with Lagrange spaces and their fractional generalizations, considering the Finsler ones to consist of a more particular, homogeneous, subclass.

Definition 16.1. A (target) fractional Lagrange space $\overset{\alpha}{\underline{L}}^n = (\overset{\alpha}{\underline{M}}, \overset{\alpha}{L})$ of fractional dimension $\alpha \in (0, 1)$, for a regular real function $\overset{\alpha}{L} : \underline{T}M \to \mathbb{R}$, when the fractional Hessian is

$$_L \overset{\alpha}{g}_{ij} = \frac{1}{4}\left(\overset{\alpha}{\partial_i}\overset{\alpha}{\partial_j} + \overset{\alpha}{\partial_j}\overset{\alpha}{\partial_i} \right) \overset{\alpha}{L} \neq 0. \tag{16.10}$$

In our further constructions, we shall use the coefficients $_L \overset{\alpha}{g}^{ij}$ being inverse to $_L \overset{\alpha}{g}_{ij}$ (16.10).[3] Any $\overset{\alpha}{\underline{L}}^n$ can be associated with a prime "integer" Lagrange space L^n.

The concept of nonlinear connection (N-connection) on $\overset{\alpha}{\underline{L}}^n$ can be introduced similarly to that on nonholonomic fractional manifold [1, 2] considering the fractional tangent bundle $\overset{\alpha}{\underline{T}}M$.

Definition 16.2. An N-connection $\overset{\alpha}{N}$ on $\overset{\alpha}{\underline{T}}M$ is defined by a nonholonomic distribution (Whitney sum) with conventional h- and v-subspaces, $\underline{h}\overset{\alpha}{\underline{T}}M$ and $\underline{v}\overset{\alpha}{\underline{T}}M$, when

$$\overset{\alpha}{\underline{T}}\overset{\alpha}{\underline{T}}M = \underline{h}\overset{\alpha}{\underline{T}}M \oplus \underline{v}\overset{\alpha}{\underline{T}}M. \tag{16.11}$$

Locally, a fractional N-connection is defined by a set of coefficients, $\overset{\alpha}{N} = \{^\alpha N_i^a\}$ computed as $\overset{\alpha}{N} = {}^\alpha N_i^a(u)(dx^i)^\alpha \otimes \overset{\alpha}{\partial_a}$, see formulas (19.15) and (16.3).

Let us consider values $y^k(\tau) = dx^k(\tau)/d\tau$, for $x(\tau)$ parametrizing smooth curves on a manifold M with $\tau \in [0, 1]$. The fractional analogs of such configurations are determined by changing $d/d\tau$ into the fractional Caputo derivative $\overset{\alpha}{\partial_\tau} = {}_{1\tau}\overset{\alpha}{\partial_\tau}$ when $^\alpha y^k(\tau) = \overset{\alpha}{\partial_\tau}x^k(\tau)$. For simplicity, we shall omit the label α for $y \in \underline{T}M$ if that will not result in ambiguities and/or we shall not associate with it an explicit fractional derivative along a curve.

---

[3] We shall put a left label L to certain geometric objects if it is necessary to emphasize that they are induced by Lagrange generating function. Nevertheless, such labels will be omitted (to simplify the notations) if that will not result in ambiguities.

By straightforward computations, following the same scheme as in [7] but with fractional derivatives and integrals, we prove:

Theorem 16.1. *Any $\overset{\alpha}{L}$ defines the fundamental geometric objects determining canonically a nonholonomic fractional Riemann–Cartan geometry on $\overset{\alpha}{T}M$ being satisfied the properties:*

1. *The fractional Euler–Lagrange equations, $\overset{\alpha}{\partial}_\tau({}_{1}y^i\overset{\alpha}{\partial}_i\overset{\alpha}{L}) - {}_{1}x^i\overset{\alpha}{\partial}_i\overset{\alpha}{L} = 0$, are equivalent to the fractional "nonlinear geodesic" (equivalently, semi-spray) equations*
$$\left(\overset{\alpha}{\partial}_\tau\right)^2 x^k + 2\overset{\alpha}{G}{}^k(x, {}^\alpha y) = 0, \text{ where}$$
$$\overset{\alpha}{G}{}^k = \tfrac{1}{4}\,{}_{L}g^{kj}\left[y^j\;{}_{1}y^i\;\overset{\alpha}{\partial}_j\left({}_{1}x^i\overset{\alpha}{\partial}_i\overset{\alpha}{L}\right) - {}_{1}x^i\overset{\alpha}{\partial}_i\overset{\alpha}{L}\right] \text{ defines the canonical N-connection}$$
$${}^{\alpha}_{L}N^a_j = {}_{1}y^j\,\overset{\alpha}{\partial}_j\overset{\alpha}{G}{}^k(x, {}^\alpha y).$$

2. *There is a canonical (Sasaki type) metric structure,*
$${}_{L}\overset{\alpha}{\mathbf{g}} = {}^{\alpha}_{L}g_{kj}(x,y)\,{}^\alpha e^k \otimes {}^\alpha e^j + {}^{\alpha}_{L}g_{cb}(x,y)\,{}^{\alpha}_{L}e^c \otimes {}^{\alpha}_{L}e^b,$$
where the frame structure (defined linearly by ${}^{\alpha}_{L}N^a_j$) is ${}^{\alpha}_{L}\mathbf{e}_v = ({}^{\alpha}_{L}\mathbf{e}_i, e_a)$.

3. *There is a canonical metrical distinguished connection*
$${}^{\alpha}_{c}\mathbf{D} = (h\,{}^{\alpha}_{c}D, v\,{}^{\alpha}_{c}D) = \left\{{}^{\alpha}_{c}\Gamma^\gamma_{\alpha\beta} = ({}^\alpha\widehat{L}{}^i_{jk},\;{}^\alpha\widehat{C}{}^i_{jc})\right\},$$

(in brief, d-connection), which is a linear connection preserving under parallelism the splitting (16.11) and metric compatible, i.e. ${}^{\alpha}_{c}\mathbf{D}\left({}_{L}\overset{\alpha}{\mathbf{g}}\right) = 0$, ${}^{\alpha}_{c}\Gamma^i_j =$
${}^{\alpha}_{c}\Gamma^i_{j\gamma}\,{}^{\alpha}_{L}e^\gamma = \widehat{L}{}^i_{jk}e^k + \widehat{C}{}^i_{jc}\,{}^{\alpha}_{L}e^c$, for $\widehat{L}{}^i_{jk} = \widehat{L}{}^a_{bk}, \widehat{C}{}^i_{jc} = \widehat{C}{}^a_{bc}$ in ${}^{\alpha}_{c}\Gamma^a_b = {}^{\alpha}_{c}\Gamma^a_{b\gamma}\,{}^{\alpha}_{L}e^\gamma =$
$\widehat{L}{}^a_{bk}e^k + \widehat{C}{}^a_{bc}\,{}^{\alpha}_{L}e^c$,

$${}^\alpha\widehat{L}{}^i_{jk} = \frac{1}{2}\,{}^\alpha g^{ir}\left({}^{\alpha}_{L}\mathbf{e}_k\,{}^{\alpha}_{L}g_{jr} + {}^{\alpha}_{L}\mathbf{e}_j\,{}^{\alpha}_{L}g_{kr} - {}^{\alpha}_{L}\mathbf{e}_r\,{}^{\alpha}_{L}g_{jk}\right),$$

$${}^\alpha\widehat{C}{}^a_{bc} = \frac{1}{2}\,{}^\alpha g^{ad}\left({}^\alpha e_c\,{}^{\alpha}_{L}g_{bd} + {}^\alpha e_c\,{}^{\alpha}_{L}g_{cd} - {}^\alpha e_d\,{}^{\alpha}_{L}g_{bc}\right)$$

are just the generalized Christoffel indices.[4]

Finally, in this section, we note that:

Remark 16.1. We note that ${}^{\alpha}_{c}\mathbf{D}$ is with nonholonomically induced torsion structure defined by 2-forms

$${}^{\alpha}_{L}\mathcal{T}^i = \widehat{C}{}^i_{jc}\,{}^\alpha e^i \wedge {}^{\alpha}_{L}e^c,$$

$${}^{\alpha}_{L}\mathcal{T}^a = -\frac{1}{2}\,{}_{L}\Omega^a_{ij}\,{}^\alpha e^i \wedge {}^\alpha e^j + \left({}^\alpha e_b\,{}^{\alpha}_{L}N^a_i - {}^\alpha\widehat{L}{}^a_{bi}\right){}^\alpha e^i \wedge {}^{\alpha}_{L}e^b$$

---

[4]For integer dimensions, we contract "horizontal" and "vertical" indices following the rule: $i = 1$ is $a = n+1$; $i = 2$ is $a = n+2$; ... $i = n$ is $a = n+n$.

computed from the fractional version of Cartan's structure equations

$$d\,{}^{\alpha}e^i - {}^{\alpha}e^k \wedge {}^{\alpha}_{c}\Gamma^i_k = -{}^{\alpha}_{L}\mathcal{T}^i, \ d\,{}^{\alpha}_{L}e^a - {}^{\alpha}_{L}e^b \wedge {}^{\alpha}_{c}\Gamma^a_b = -{}^{\alpha}_{L}\mathcal{T}^a,$$

$$d\,{}^{\alpha}_{c}\Gamma^i_j - {}^{\alpha}_{c}\Gamma^k_j \wedge {}^{\alpha}_{c}\Gamma^i_k = -{}^{\alpha}_{L}\mathcal{R}^i_j$$

in which the curvature 2-form is denoted ${}^{\alpha}_{L}\mathcal{R}^i_j$.

For any d-connection on ${}^{\alpha}\underline{T}M$, we can compute, respectively, the N-adapted coefficients of ${}^{\alpha}\mathcal{T}^\tau = \{{}^{\alpha}\Gamma^\tau_{\beta\gamma}\}$ and ${}^{\alpha}\mathcal{R}^\tau_\beta = \{{}^{\alpha}\mathbf{R}^\tau_{\beta\gamma\delta}\}$ as it is explained for general fractional nonholonomic manifolds in [1,2].

4 Analogous Fractional Gravity

Let us consider a "prime" nonholonomic manifold $\mathbf{V}$ is of integer dimension dim $\mathbf{V} = n + m, n \geq 2, m \geq 1$.[5] Its fractional extension ${}^{\alpha}\mathbf{V}$ is modelled by a quadruple $(\mathbf{V}, {}^{\alpha}\mathbf{N}, {}^{\alpha}\mathbf{d}, {}^{\alpha}\mathbf{I})$, where ${}^{\alpha}\mathbf{N}$ is a nonholonomic distribution stating a nonlinear connection (N-connection) structure. The fractional differential structure ${}^{\alpha}\mathbf{d}$ is determined by Caputo fractional derivative (19.13) following formulas (19.15) and (16.3).

For any frame and co-frame (dual) structures, ${}^{\alpha}e_{\alpha'} = ({}^{\alpha}e_{i'}, {}^{\alpha}e_{a'})$ and ${}^{\alpha}e^{\beta'} = ({}^{\alpha}e^{i'}, {}^{\alpha}e^{a'})$ on ${}^{\alpha}\mathbf{V}$, we can consider frame transforms

$$ {}^{\alpha}e_\alpha = A_\alpha^{\alpha'}(x,y)\,{}^{\alpha}e_{\alpha'} \ \text{and} \ {}^{\alpha}e^\beta = A^\beta_{\beta'}(x,y)\,{}^{\alpha}e^{\beta'}. \tag{16.12}$$

A subclass of frame transforms (16.12), for fixed "prime" and "target" frame structures, is called N-adapted if such nonholonomic transformations preserve the splitting defined by a N-connection structure $\mathbf{N} = \{N_i^a\}$.

Under (in general, nonholonomic) frame transforms, the metric coefficients of any metric structure ${}^{\alpha}\mathbf{g}$ on ${}^{\alpha}\mathbf{V}$ are recomputed following the formulas:

$$ {}^{\alpha}g_{\alpha\beta}(x,y) = A_\alpha^{\alpha'}(x,y)\,A^{\beta'}_\beta(x,y)\,{}^{\alpha}g_{\alpha'\beta'}(x,y). $$

---

[5]A nonholonomic manifold is a manifold endowed with a non-integrable (equivalently, nonholonomic, or anholonomic) distribution. There are three useful (for our considerations) examples when (1) $\mathbf{V}$ is a (pseudo) Riemannian manifold; (2) $\mathbf{V} = E(M)$, or (3) $\mathbf{V} = TM$, for a vector, or tangent, bundle on a base manifold M. We also emphasize that in this chapter we follow the conventions from [1,2,7] when left indices are used as labels and right indices may be abstract ones or running certain values.

For any fixed $\overset{\alpha}{\mathbf{g}}$ and $\overset{\alpha}{\mathbf{N}}$, there are N-adapted frame transforms when

$$\overset{\alpha}{\mathbf{g}} = {}^{\alpha}g_{ij}(x,y)\ {}^{\alpha}\mathbf{e}^i \otimes {}^{\alpha}\mathbf{e}^j + {}^{\alpha}h_{ab}(x,y)\ {}^{\alpha}\mathbf{e}^a \otimes {}^{\alpha}\mathbf{e}^b,$$

$$= {}^{\alpha}g_{i'j'}(x,y)\ {}^{\alpha}\mathbf{e}^{i'} \otimes {}^{\alpha}\mathbf{e}^{j'} + {}^{\alpha}h_{a'b'}(x,y)\ {}^{\alpha}\mathbf{e}^{a'} \otimes {}^{\alpha}\mathbf{e}^{b'},$$

where ${}^{\alpha}\mathbf{e}^a$ and ${}^{\alpha}\mathbf{e}^{a'}$ are elongated following formulas (19.2), respectively, by ${}^{\alpha}N^a_j$ and

$$
{}^{\alpha}N^{a'}_{j'} = A_a{}^{a'}(x,y)A^j{}_{j'}(x,y)\ {}^{\alpha}N^a_j(x,y), \tag{16.13}
$$

or, inversely, ${}^{\alpha}N^a_j = A_{a'}{}^a(x,y)A^{j'}{}_j(x,y)\ {}^{\alpha}N^{a'}_{j'}(x,y)$, with prescribed ${}^{\alpha}N^{a'}_{j'}$.

We preserve the N-connection splitting for any frame transform (16.12) when ${}^{\alpha}g_{i'j'} = A^i{}_{i'}A^j{}_{j'}\ {}^{\alpha}g_{ij}$, ${}^{\alpha}h_{a'b'} = A^a{}_{a'}A^b{}_{b'}\ {}^{\alpha}h_{ab}$, for $A_i{}^{i'}$ constrained to get holonomic ${}^{\alpha}\mathbf{e}^{i'} = A_i{}^{i'}\ {}^{\alpha}\mathbf{e}^i$, i.e. $[\ {}^{\alpha}\mathbf{e}^{i'},\ {}^{\alpha}\mathbf{e}^{j'}] = 0$ and ${}^{\alpha}\mathbf{e}^{a'} = dy^{a'} +\ {}^{\alpha}N^{a'}_{j'}dx^{j'}$, for certain $x^{i'} = x^{i'}(x^i,y^a)$ and $y^{a'} = y^{a'}(x^i,y^a)$, with ${}^{\alpha}N^{a'}_{j'}$ computed following formulas (16.13). Such conditions can be satisfied by prescribing from the very beginning a nonholonomic distribution of necessary type. The constructions can be equivalently inverted, when ${}^{\alpha}g_{\alpha\beta}$ and ${}^{\alpha}N^a_i$ are computed from ${}^{\alpha}g_{\alpha'\beta'}$ and ${}^{\alpha}N^{a'}_{i'}$, if both the metric and N-connection splitting structures are fixed on $\overset{\alpha}{\mathbf{V}}$.

An unified approach to Einstein–Lagrange/Finsler gravity for arbitrary integer and noninteger dimensions is possible for the fractional canonical d-connection ${}^{\alpha}\widehat{\mathbf{D}}$. The fractional gravitational field equations are formulated for the Einstein d-tensor (19.5), following the same principle of constructing the matter source ${}^{\alpha}\Upsilon_{\beta\delta}$ as in general relativity but for fractional metrics and d-connections, ${}^{\alpha}\widehat{\mathbf{E}}_{\beta\delta} =\ {}^{\alpha}\Upsilon_{\beta\delta}$. Such a system of integro-differential equations for generalized connections can be restricted to fractional nonholonomic configurations for ${}^{\alpha}\mathbf{V}$ if we impose the additional constraints

$$
{}^{\alpha}\widehat{L}^c_{aj} =\ {}^{\alpha}\mathbf{e}_a(\ {}^{\alpha}N^c_j),\quad {}^{\alpha}\widehat{C}^i_{jb} = 0,\quad {}^{\alpha}\Omega^a_{ji} = 0. \tag{16.14}
$$

There are no theoretical or experimental evidences that for fractional dimensions we must impose conditions of type (19.7) but they have certain physical motivation if we develop models which in integer limits result in the general relativity theory.

Acknowledgements This chapter summarizes the results presented in our corresponding talk at Conference "New Trends in Nanotechnology and Nonlinear Dynamical Systems", 25–27 July, 2010, Çankaya University, Ankara, Turkey.

References

1. Vacaru S Fractional nonholonomic Ricci flows, arXiv: 1004.0625
2. Vacaru S Fractional dynamics from Einstein gravity, general solutions, and black holes, arXiv: 1004.0628

3. Baleanu D, Vacaru S (2010) Fractional almost Kahler–Lagrange geometry. Nonlinear Dyn. published online in: arXiv: 1006.5535
4. Baleanu D, Vacaru S (2011) Fedosov quantization of fractional Lagrange spaces. Int J Theor Phys 50:233–243
5. Baleanu D, Vacaru S (2011) Constant curvature coefficients and exact solutions in fractional gravity and geometric mechanics. Centr Eur J Phys 9 arXiv: 1007.2864
6. Baleanu D, Vacaru S (2011) Fractional curve flows and solitonic hierarchies in gravity and geometric mechanics, J Math Phys 52: 053514
7. Vacaru S (2008) Finsler and Lagrange geometries in Einstein and string gravity. Int J Geom Meth Mod Phys 5:473–451
8. Vacaru S (2010) Einstein gravity as a nonholonomic almost Kähler geometry, Lagrange–Finsler variables, and deformation quantization. J Geom Phys 60:1289–13051
9. Vacaru S (2010) Curve flows and solitonic hierarchies generated by Einstein metrics. Acta Appl Math 110:73–107
10. Anco S, Vacaru S (2009) Curve flows in Lagrange–Finsler geometry, bi-Hamiltonian structures and solitons. J Geom Phys 59:79–103

Chapter 17
Schrödinger Equation in Fractional Space

Sami I. Muslih and Om P. Agrawal

1 Introduction

The concept of a "fractal" and a fractional dimensional space was introduced by Mandelbrot [1]. Historically, the first example of the fractional physical object was the Brownian motion [2]. In quantum physics, the first successful attempt of applying fractality concept was Feynman path integral approach [3]. Feynman and Hibbs [4] reformulated the non-relativistic quantum mechanics as a path integral over a Brownian path.

Several applications of fractional dimensional space could be cited. In 1970s, Stillinger [5] described a procedure for integration on a fractional space of dimension D (where D is a noninteger number), and generalized the Laplace operator $\Delta_D \psi(\mathbf{r})$ in this space as,

$$\Delta_D = \frac{\partial^2}{\partial r^2} + \frac{D-1}{r}\frac{\partial}{\partial r} + \frac{1}{r^2 \sin^{D-2}\theta}\frac{\partial}{\partial \theta}\sin^{D-2}\theta\frac{\partial}{\partial \theta}.$$

Many investigations into low-dimensional semiconductors [6–8] have used this Laplacian to solve the Schrödinger equation for hydrogen-like atoms for anisotropic solids to obtain energy bounds and the optical spectra as a function of fractional spatial dimension D. Recent progress includes the description of a single coordinate momentum operator in this fractional dimensional space based on generalized Wigner relations [9, 10] presenting realization of parastatics [11]. In some applications, a fractional dimension appears as an explicit parameter when

S.I. Muslih
Al-Azhar University-Gaza, Palestine
e-mail: smuslih@ictp.it

O.P. Agrawal (✉)
Southern Illinois University, Carbondale, Illinois, USA
e-mail: om@engr.siu.edu

D. Baleanu et al. (eds.), *Fractional Dynamics and Control*,
DOI 10.1007/978-1-4614-0457-6_17, © Springer Science+Business Media, LLC 2012

the physical problem is formulated in α dimensions in such a way that D maybe extended to noninteger values, as in Wilson's study of quantum field theory models in less than four dimensions [12], or in the approach to quantum mechanics by Stillinger [5]. It is worthwhile to mention that the experimental measurement of dimension D of our real world is given by $D = (3 \pm 10^{-6})$ [5, 12]. The fractional value of D agrees with the experimental physical observations that in general relativity, gravitational fields are understood to be geometric perturbations (curvatures) in our space time [13], rather than entities residing within a flat space time. Besides, Zeilinger and Svozil [14] noted that the current discrepancy between theoretical and experimental values of the anomalous magnetic moment of the electron could be resolved if we take the dimension D of our space as $D = 3 - (5.3 \pm 2.5) \times 10^{-7}$.

The formalism from [5] has been applied to problems such as excitons [8, 15–21], magnetoexcitons [22], impurities [19], semiconductor heterostructures [23], polarons [24], and superconductivity [25], often successfully mirroring computational results in specified problems. Also, the putative fractional dimension may be viewed as an effective dimension of compactified higher dimensions or as a manifestation of a nontrivial microscopic lattice structure of space [26].

In fractional dimensional model, one can study the energy spectrum for hydrogen-like atom-fractioan "Bohr atom" by solving the relevant Schrödinger equation in a noninteger dimensional space.

2 Fractional Schrödinger Equation

Recently, Laskin [2] showed that the path integral over Lévy trajectories leads to a fractional Schrödinger equation which can be written as

$$i\hbar \frac{\partial \psi(\mathbf{r},t)}{\partial t} = C_\alpha(-\hbar^2 \Delta)^{\alpha/2} \psi(\mathbf{r},t) + V(r,t)\psi(\mathbf{r},t), \qquad (17.1)$$

where $\hbar$ is the Plank's constant, $\mathbf{r}$ is the position vector and r its magnitude, Δ is the Laplacian operator, C_α is a constant of dimensions $erg^{1-\alpha}.cm^\alpha.sec^{-\alpha}$, $V(r,t)$ is the potential field, $\psi(\mathbf{r},t)$ is the wave function, and $(-\hbar^2 \Delta)^{\alpha/2}$ is the 3D quantum Riesz fractional derivative [2,27]. In the standard case, when $\alpha = 2$, we have $C_2 = 1/(2m)$, and (17.1) reduces to the standard Schrödinger equation. Since a Lévy trajectory is a trajectory in a fractional dimension space, we call (17.1) as the fractional Schrödinger equation in a fractional dimension space.

When $V(r,t) = V(r)$, i.e., the potential is not a function of time, using the method of separation of variables one can show that the solution of (17.1) is

$$\psi(\mathbf{r},t) = e^{(-i/\hbar)Et}\phi(\mathbf{r}), \qquad (17.2)$$

and the following equation is satisfied:

$$C_\alpha(-\hbar^2 \Delta)^{\alpha/2}\phi(\mathbf{r}) + V(r)\phi(\mathbf{r}) = E\phi(\mathbf{r}), \qquad (17.3)$$

where E is a number. Equation (17.3) is the time-independent fractional Schrödinger equation. For this case, the probability density is

$$\rho = |\psi(\mathbf{r},t)^2| = |\phi(\mathbf{r})^2| \tag{17.4}$$

i.e., the probability density is independent of time. Furthermore, (17.3) can be written as

$$H_\alpha \psi(\mathbf{r},t) = E \psi(\mathbf{r},t), \tag{17.5}$$

which is an eigenvalue equation. Here, H_α is the Hamiltonian operator given as

$$H_\alpha = C_\alpha (-\hbar^2 \Delta)^{\alpha/2} + V(r), \tag{17.6}$$

and E is the energy eigenvalue. In other words, the system has a well-defined energy and it is a stationary system. These results are the same as those for an ordinary Schrödinger equation with time-independent Hamiltonian [29]. Using (17.5) and (17.6), we have

$$H_\alpha = C_\alpha p^\alpha + V(r) = E_{\text{kin}} + V(r) = E, \tag{17.7}$$

where $\mathbf{p}$ $(= -i\hbar\nabla)$ is the momentum operator, p is the magnitude of $\mathbf{p}$, ∇ is the gradient operator, and E_{kin} $(= C_\alpha \mathbf{p}^\alpha)$ is the kinetic energy of the system. Thus, unless $\alpha = 2$, the total energy E of an ordinary and a fractional systems are not the same.

In the discussion to follow, we shall use these equations to find the radius of the fractional Bohr atom and the frequency of the radiated wave when an electron jumps from one energy level to another. This requires the virial theorem for fractional atom which is discussed next.

3 The Virial Theorem for Fractional Bohr Atom

To develop the virial theorem to be used for a stationary fractional dynamic system, consider a particle at a position $\mathbf{r}$ under the influence of a central force with a potential $V(r)$. The total Hamiltonian of the system is given by (see (17.7)),

$$H = C_\alpha p^\alpha + V(r). \tag{17.8}$$

Note that for a central force, the potential depends on r only. For this case, the Hamilton's equations lead to

$$\dot{p}_s = -\frac{\partial H}{\partial s} = -\frac{\partial V}{\partial s}, \quad s = x,y,z \tag{17.9}$$

$$\dot{s} = \frac{\partial H}{\partial p} = \alpha C_\alpha p^{\alpha-2} p_s, \quad s = x,y,z, \tag{17.10}$$

where s and p_s, $s = x, y, z$, are the components of $\mathbf{r}$ and $\mathbf{p}$ along the direction s. In deriving (17.9) and (17.10), we have used the fact that $p^2 = p_x^2 + p_y^2 + p_z^2$ and $V(r) = V(x, y, z)$.

We now define the scalar virial G as

$$G = \mathbf{p} \cdot \mathbf{r} = xp_x + yp_y + zp_z, \qquad (17.11)$$

where "$\cdot$" represents the dot product. Taking total time derivative of virial G and using (17.9) and (17.10), we obtain

$$\frac{dG}{dt} = \alpha C_\alpha p^\alpha + \left(-\frac{\partial V}{\partial r}\right) r = \alpha E_{\text{kin}} - \frac{\partial V}{\partial r} r. \qquad (17.12)$$

Here we have used the fact that $\nabla V \cdot \mathbf{r} = r(\partial V/\partial r)$. The average value dG/dt over an interval τ is given as

$$\frac{\overline{dG}}{dt} = \frac{1}{\tau} \int_0^\tau \frac{dG}{dt} dt = \frac{1}{\tau}[G(\tau) - G(0)] = \alpha \overline{E_{\text{kin}}} - \frac{\overline{\partial V}}{\partial r} r, \qquad (17.13)$$

where a bar on "*," i.e. $\overline{*}$, represents the average value of "*." The last part of (17.13) follows from (17.12). Assuming that $G(\tau)$ remains bounded and letting the limit of τ go to infinity, we obtain $\overline{dG}/dt = 0$, which leads to

$$\alpha \overline{E_{\text{kin}}} = \frac{\overline{\partial V}}{\partial r} r. \qquad (17.14)$$

Note that if $V(r)$ is of the form, C/r, we obtain $\alpha \overline{E_{\text{kin}}} = -\overline{V}$, which is the same as that given in [2]. Equation (17.14) will be used to obtain the radii of the fractional Bohr atom and the frequency of the radiated wave which are discussed next.

4   Radii and Frequency of the Fractional Bohr Atom in Fractional Dimensional Space

The potential energy $V(\mathbf{r})$ of a hydrogen-like fractional Bohr atom in a fractional dimensional space is [29]

$$V(\mathbf{r}) = -\frac{Ze^2 k_\varepsilon}{|\mathbf{r}|^{\varepsilon+1}}, \qquad 0 \le \varepsilon < 1, \qquad (17.15)$$

where Z is the atomic number, e is the charge of the electron, ε is a parameter, and k_ε is a space constant given as

$$k_\varepsilon = \frac{\Gamma\left(\frac{\varepsilon+3}{2}\right)}{2\pi^{(\varepsilon+3)/2}(\varepsilon+1)\varepsilon_0}. \qquad (17.16)$$

Here ε_0 is the dielectric coefficient of the free space, and ε is a constant which may account for various conditions discussed below. In the special case, when $\varepsilon = 0$, k_1 has a value of $1/(4\pi\varepsilon_0)$, and (17.15) reduces to standard potential energy term. Note that the expression for the potential energy considered here is different from that considered in [2]. Apart from the derivations given in [29], (17.15) is motivated by the fact that according to experimental measurements the central force differs slightly from the inverse square law. In addition, ε in (17.15) may account for the rotation of the nucleus, motion of the electron in fractional dimension, or simply a new potential function.

Substituting (17.15) into (17.14), we obtain

$$E_{\text{kin}} = -\frac{1+\varepsilon}{\alpha}V. \tag{17.17}$$

To obtain the radius of the fractional hydrogen-like atom, according to Bohr's postulate [30], we have

$$pa_n = n\hbar, \quad (n = 1, 2, 3, \ldots), \tag{17.18}$$

where, n is the quantum number and a_n is the radius of the nth electron circular orbit. Using (17.15), (17.17), (17.18), and the fact that $E_{\text{kin}} = C_\alpha p^\alpha$, we obtain

$$a_n = A_0 n^{\frac{\alpha}{\alpha-\varepsilon-1}}, \tag{17.19}$$

where the fractional Bohr radius A_0 (also known as the radius of the ground orbit) is given as

$$A_0 = \left[\frac{\alpha C_\alpha \hbar^\alpha}{Z e^2 k_\varepsilon(\varepsilon+1)}\right]^{\frac{1}{\alpha-\varepsilon-1}}. \tag{17.20}$$

The energy spectrum can be obtained from the total energy $\overline{E}$ given by

$$\overline{E} = \overline{E_{\text{kin}}} + \overline{V} = \frac{(\varepsilon-\alpha+1)}{(1+\varepsilon)}\overline{E_{\text{kin}}}. \tag{17.21}$$

Thus, the nth energy level of the fractional hydrogen-like atom is given by

$$E_{n,\alpha,\varepsilon} = B_{0,\alpha,\varepsilon} n^{\frac{-\alpha(\varepsilon+1)}{\alpha-\varepsilon-1}}, \tag{17.22}$$

where B_0 is defined as

$$B_{0,\alpha,\varepsilon} = \frac{(\varepsilon-\alpha+1)}{[\alpha^\alpha C_\alpha \hbar^\alpha(\varepsilon+1)]^{\frac{\varepsilon+1}{\alpha-\varepsilon-1}}}\left((Z e^2 k_\varepsilon)^\alpha\right)^{\frac{1}{\alpha-\varepsilon-1}}. \tag{17.23}$$

Equation (17.22) generalizes the well-known energy spectrum of the standard quantum mechanical hydrogen-like atom. In the special case, when $\alpha = 2$, $\varepsilon = 0$ and $C_2 = 1/(2m)$, the expression for $E_{n,\alpha,\varepsilon}$ reduces to

$$E_{n,2,0} = -\frac{Z^2 \mu e^4}{32\pi^2 \varepsilon_0^2 \hbar^2 n^2}, \tag{17.24}$$

which is the same as energy levels of the standard hydrogen atom (see [28]). However, to compare the results given here and in [28], note that one must replace here k_ε with 1 and the electron mass m with the reduced mass μ.

Now, the frequency of radiation associated with the transition of electrons from one allowed orbit corresponding to $n_1 = k$ to another corresponding to $n_2 = n$ is given by

$$
\omega_{k,n,\alpha,\varepsilon} = \left(\frac{E_{k,\alpha,\varepsilon} - E_{n,\alpha,\varepsilon}}{\hbar} \right),
$$

$$
= \frac{(\alpha - \varepsilon - 1)E_0}{\hbar} \left(\frac{1}{n^{\frac{\alpha(\varepsilon+1)}{\alpha-\varepsilon-1}}} - \frac{1}{k^{\frac{\alpha(\varepsilon+1)}{\alpha-\varepsilon-1}}} \right). \qquad (17.25)
$$

This expression reduces to the standard frequency shifts for $\alpha = 2$ and $\varepsilon = 0$ as

$$
\omega_{k,n} = \frac{E_0}{\hbar} \left(\frac{1}{n^2} - \frac{1}{k^2} \right). \qquad (17.26)
$$

5 Conclusions

The complexity of calculations involving the fractional dynamics has been illustrated by solving the fractional Schrodinger equation in a noninteger dimensional space. We have obtained the energy eigenvalues and the fractional Bohr radius for the hydrogen-like atom. One should notice that as a special case $\varepsilon = 0$, we obtain the same energy spectrum as obtained in [2].

Acknowledgements Dr. Sami I. Muslih would like to sincerely thank the Institute of International Education (Fulbright Scholar Program), New York, NY, and the Department of Mechanical Engineering and Energy Processes (MEEP) and the Dean of Graduate Studies at Southern Illinois University, Carbondale (SIUC), IL, for providing him the financial support and the necessary facilities during his stay at SIUC.

References

1. Mandelbrot BB (1983) The fractal geometry of nature. W.H. Freeman, New York
2. Laskin N (2002) Fractional schrodinger equation. Phys Rev E 66:056108–056115
3. Feynman RP (1972) Statistical mechanics. Benjamin. Reading, Mass
4. Feynman RP, Hibbs AR (1965) Quantum mechanics and path integrals. McGraw-Hill, New York
5. Stillinger FH (1977) Axiomatic basis for spaces with noninteger dimension. J Math Phys 18:1224–1234
6. He X (1990) Anisotropy and isotropy: A model of fraction-dimensional space. Solid State Commun 75:111–114

7. He X (1991) Excitons in anisotropic solids: The model of fractional-dimensional space. Phys Rev B 43:2063–2069
8. He X (1990) Fractional dimensionality and fractional derivative spectra of interband optical transitions. Phys Rev B 42:11751–11756
9. Matos-Abiague A (2001) Bose-like oscillator in fractional-dimensional space. J Phys A Math Gen 34:3125–3128
10. Matos-Abiague A (2001) Deformation of quantum mechanics in fractional-dimensional space. J Phys A Math Gen 34:11059–11071
11. Jing SJ (1998) A new kind of deformed calculus and parabosonic coordinate representation. Phys A Math Gen 31:6347–6354
12. Willson KG (1973) Quantum field - theory models in less than 4 dimensions. Phys Rev D 7:2911–2926
13. Misner CW, Thorne KS, Wheeler JA (1973) Gravitation. Freeman, San Francisco
14. Zeilinger A, Svozil K (1985) Measuring the dimension of space-time. Phys Rev Lett 54:2553–2555
15. Lefebvre P, Christol P, Mathieu H (1995) Universal formulation of excitonic linear absorption spectra in all semiconductor microstructures. Superlattice Microstuct 17:19–21
16. Mathieu H, Lefebvre P, Christol P (1992) Simple analytical method for calculating exciton binding energies in semiconductor quantum wells. Phys Rev B 46:4092–4101
17. Lefebvre P, Christol P, Mathieu H (1992) Excitons in semiconductor superlattices: Heuristic description of the transfer between Wannier-like and Frenkel-like regimes. Phys Rev B 46:13603–13606
18. Lefebvre P, Christol P, Mathieu H (1993) Unified formulation of excitonic absorption spectra of semiconductor quantum wells, superlattices, and quantum wires. Phys Rev B 46:17308–17315
19. Thilaagam A (1997) Stark shifts of excitonic complexes in quantum wells. Phys Rev B 56:4665–4670
20. Thilaagam A (1997) Exciton-phonon interaction in fractional dimensional space. Phys Rev B 56:9798–9804
21. Thilaagam A (1999) Pauli blocking effects in quantum wells. Phys Rev B 59:3027–3032
22. Reyes-Gomez E, Matos-Abiague A, Perdomo-Leiva CA, Dios-Leyva M, de, Oliveira LE (2000) Excitons and shallow impurities in $GaAs - Ga_{1-x}Al_xAs$ semiconductor heterostructures within a fractional- dimensional space approach: Magnetic-field effects. Phys Rev B. 61:13104–13114
23. Mikhailov ID, Betancur FJ, Escorcia RA, Sierrra-Ortega J (2003) Shallow donors in semiconductor heterostructures: Fractal dimension approach and the variational principle. Phys Rev B 67:115317
24. Matos-Abiague A (2002) Polaron effect in $GaAs - Ga_{1-x}Al_xAs$ quantum wells: A fractional-dimensional space approach. Phys Rev B 65:165321
25. Bak Z (2003) Superconductivity in a system of fractional spectral dimension. Phys Rev B 68:064511
26. Schafer A, Muller B (1986) Bounds for the fractal dimension of space. J Phys A Math Gen 19:3891
27. Riesz M (1949) L'intégrale de Riemann-Liouville et le problème de Cauchy. Acta Mathematica 81:1–223
28. Ghatak AK, Lokanathan S (1975) Quantum mechanics. Macmillan Company of India Limited
29. Muslih SI, Baleanu D (2007) Fractional multipoles in fractional space. Nonlinear Anal Real World Appl 8:198–203
30. Bohr N (1913) On the constitution of atoms and molecules. I Phil Mag 26:1–25, 476–502, 857–875

Chapter 18
Solutions of Wave Equation in Fractional Dimensional Space

Sami I. Muslih and Om P. Agrawal

1 Introduction

In 1918 the Mathematician Felix Hausdorff introduced the notion of fractional dimension. This concept became very important especially after the revolutionary discovery of fractal geometry by Mandelbrot [1], where he used the concept of fractionality, and worked out the relations between fractional dimension and integer dimension by using a scaling method to describe irregular geometries of complex objects such as clouds, mountains, cost lines, and travel path of a lightning. The first fractional physical phenomenon, that was observed sometime ago and that is still a subject of many investigations, is the Brownian motion [1]. Feynman and Hibbs were first to formulate the non-relativistic quantum mechanics as an integral over Brownian paths, which is now known as the Feynman path integral [2]. The concept of fractals has been applied in many other areas of physics ranging from the dynamics of fluids in porous media to the resistivity networks in electronics [3–5].

The corner stone of fractal geometry is measuring the dimension of the fractional space, and several applications of fractional dimensional space could be cited. In 1970s, Stillinger [6] described a procedure for integration on a fractional space of dimension D (where D is a non-integer number), and generalized the Laplace operator $\Delta_D \psi(\mathbf{r})$ in this space as,

$$\Delta_D = \frac{\partial^2}{\partial r^2} + \frac{D-1}{r}\frac{\partial}{\partial r} + \frac{1}{r^2 \sin^{D-2}\theta}\frac{\partial}{\partial \theta}\sin^{D-2}\theta\frac{\partial}{\partial \theta}.$$

S.I. Muslih
Al-Azhar University-Gaza, Palestine
e-mail: smuslih@ictp.it

O.P. Agrawal (✉)
Southern Illinois University, Carbondale, Illinois, USA
e-mail: om@engr.siu.edu

D. Baleanu et al. (eds.), *Fractional Dynamics and Control*,
DOI 10.1007/978-1-4614-0457-6_18, © Springer Science+Business Media, LLC 2012

Many investigations into low-dimensional semi conductors [7, 8] have used this Laplacian to solve the Schrödinger equation for hydrogen like atoms for anisotropic solids to obtain energy bounds and the optical spectra as a function of fractional spatial dimension D. Recent progress in the area of quantum mechanics includes the description of a single coordinate momentum operator in a fractional dimensional space based on the generalized Wigner relations [9, 10] presenting realization of parastatics [11]. In some applications, the fractional dimensions appear as an explicit parameter when the physical problem is formulated in D dimensions in such a way that D maybe extended to non-integer values, as in Wilson's study of quantum field theory models in less than four dimensions [12]. It is worthwhile to mention that the experimental measurement of the dimension D of our real world is given by $D = (3 \pm 10^{-6})$ [6, 12]. The fractional value of D agrees with the experimental physical observations that in general relativity, gravitational fields are understood to be geometric perturbations (curvatures) in our space-time [13], rather than entities residing within a flat space-time. Zeilinger and Svozil [14] noted that the current discrepancy between theoretical and experimental values of the anomalous magnetic moment of the electron could be resolved if we take the dimension D of our space as $3 - (5.3 \pm 2.5) \times 10^{-7}$.

In the field of electromagnetic theory, the fractional concepts of space-time in space-time dimensions is slightly different from four lead to eliminate the logarithmic divergence of quantum electrodynamics [9]. Muslih and Agrawal [15] have studied the solutions of the Riesz potential in fractional dimensional space. Besides, the scalar Helmholtz equation (which is the time independent case of the most general wave equation) is one of the most basic and important equation encountered in mathematical treatment of various phenomena in many areas of physical sciences and engineering. The solutions of Helmholtz equation in n dimensions (n being a positive integer) was studied by using the fractional calculus [16]. Now we can ask the following important question: Is the dimension of the geometry of the field source exactly n, an integer? In fact, the geometries of the field sources (electric charge distribution , or current charge densities) are irregular and have fractional dimensions [1]. Using Mandelbrot's results about fractals, we will develop a new method to solve the wave equation in a fractional dimensional space. The starting point for solving the wave equation in fractals is to define the Fourier transform and its inverse in a fractional dimensional space to solve non-homogenous partial differential equation in fractional dimensional space. After doing this we obtain the potential solutions in terms of the fractional Green's function (Kernel) $G_D(\mathbf{r}, \mathbf{r}', t, t')$ which depend on the dimensionality parameter D. A compact form for the Green's function for fractal systems with dimensionality $0 < D \leq 3$ is also obtained. The treatment proposed gives a better understanding of building, designing and developing fractal devices with low dimensionality including fractional antennas, which are more reliable, and cost less than traditional high performance antennas. These devises include, dot thin rod antenna ($0 < D \leq 1$), and fractal antenna ($0 < D \leq 3$), which receive or transmit signals with high stability.

The chapter is organized as follows: In Sect. 2, the wave equation is presented. In Sect. 3, we introduce Fourier transform method in fractional dimensional space and its inverse transform to solve the wave equation. Section 4 deals with the evaluation of the Greens function for fractal systems with dimensionality $0 < D \leq 3$. It is shown that the required time of the propagation t_D is greater than or equal to $t' + \frac{|\mathbf{r} - \mathbf{r}'|}{c}$), and no wave could propagate with time less than $t' + \frac{|\mathbf{r} - \mathbf{r}'|}{c}$. In Sect. 5, we solve the problem of moving charged particle in D dimensional space. Section 6 contains our conclusions.

2 Wave Equation in Fractional Dimensional Space

In order to solve the wave equation for electromagnetic field, we start from the time varying fields, the electric field $\mathbf{E}$ and the magnetic field $\mathbf{B}$ and their corresponding coupled Maxwell's equations as

$$\nabla \cdot \mathbf{E} = \rho / \varepsilon_0, \qquad \nabla \times \mathbf{E} = -\frac{\partial \mathbf{B}}{\partial t}, \tag{18.1}$$

$$\nabla \cdot \mathbf{B} = 0, \qquad \nabla \times \mathbf{B} = \mu_0 \left(\mathbf{J} + \varepsilon_0 \frac{\partial \mathbf{E}}{\partial t} \right), \tag{18.2}$$

where ρ is source charge density and $\mathbf{J}$ is source current density. One of the methods to solve the coupled Maxwell's equation is to consider the solutions of the electric field and the magnetic field as

$$\mathbf{E} = \nabla \phi(\mathbf{r}, t) - \frac{\partial \mathbf{A}(\mathbf{r}, t)}{\partial t}, \tag{18.3}$$

$$\mathbf{B} = \nabla \times \mathbf{A}(\mathbf{r}, t), \tag{18.4}$$

where $\phi(\mathbf{r}, t)$ and $\mathbf{A}(\mathbf{r}, t)$ are the scalar and the vector potentials, respectively.

Simultaneous solution of Maxwell's equations leads to the following wave equation

$$\nabla^2 \psi(\mathbf{r}, t) - \frac{1}{c^2} \frac{\partial^2}{\partial t^2} \psi(\mathbf{r}, t) = -f(\mathbf{r}, t) \tag{18.5}$$

In (18.5), if $\psi(\mathbf{r}, t)$ is the scalar potential then $f(\mathbf{r}, t) = \frac{\rho(\mathbf{r}, t)}{\varepsilon_0}$, and if $\psi(\mathbf{r}, t)$ is the vector potential, then $f(\mathbf{r}, t) = \mu_0 \mathbf{J}(\mathbf{r}, t)$. Note that the equations in integer and fractional dimensional spaces are the same.

3 Fourier Transform Method in Fractional Dimensional Space

The Fourier transform method in fractional dimensional space of Gaussian integral over fractional volume element dV_D is defined by Stillinger in [6]. In this section, the Fourier transform $g(k)$ of a continues function $f(x)$ will be considered over fractional line element $d^\alpha x$, $0 < \alpha \leq 1$ by using the the Mandelbrot [1] fractional line element which is defined as

$$d^\alpha x = \frac{\pi^{\alpha/2} |x|^{\alpha-1}}{\Gamma(\alpha/2)} dx, \tag{18.6}$$

where $0 < \alpha \leq 1$. Then the Fourier transformation is defined as

$$g(k) = F(f(x)) = \int f(x) e^{ikx} d^\alpha x, \tag{18.7}$$

and the inverse Fourier transformation $f(x)$ is given by

$$f(x) = F^{-1}(g(k)) = \left(\frac{1}{2\pi}\right)^\alpha \int g(k) e^{-ikx} d^\alpha k. \tag{18.8}$$

The above definition can be generalized for N dimensional vector $\mathcal{R}^N$. The dimension of fractional space is given by $D = \alpha_1 + \alpha_2 + \cdots + \alpha_N$, $0 < \alpha_i \leq 1$

Theorem 18.1. *The generalized Dirac delta function in the α dimensional fractional space satisfies the following identity*

$$\delta^\alpha(x - x') = \left(\frac{1}{2\pi}\right)^\alpha \int e^{ik(x-x')} d^\alpha k. \tag{18.9}$$

Proof.

$$F^{-1}(F(f(x))) = f(x) = \int \left(f(x') d^\alpha(x') \left(\frac{1}{2\pi}\right)^\alpha \int e^{ik(x-x')} d^\alpha k \right), \tag{18.10}$$

$$= \int (f(x') d^\alpha(x') \delta^\alpha(x - x')). \tag{18.11}$$

$\square$

Theorem 18.2. *The generalized Dirac delta function in the α dimensional fractional space can be defined as*

$$\delta^\alpha(x) = \lim_{\varepsilon \to \infty} \varepsilon^\alpha e^{-\pi \varepsilon^2 x^2}. \tag{18.12}$$

Proof. From Theorem 1, we have

$$\int_{-\infty}^{\infty} e^{ikx} d^\alpha k = (2\pi)^\alpha \delta^\alpha(x). \tag{18.13}$$

Now, let us define the following integral [6, 12]

$$I(\lambda, q) = \int_{-\infty}^{\infty} e^{(-\lambda x^2 + qx)} d^{\alpha}x = \lambda^{-\alpha/2} \pi^{\alpha/2} e^{q^2/4\lambda}. \tag{18.14}$$

Take $q = ik$ and $x \rightarrow k$. This yields

$$\int_{-\infty}^{\infty} e^{(-\lambda k^2 + ikx)} d^{\alpha}k = \lambda^{-\alpha/2} \pi^{\alpha/2} e^{-x^2/4\lambda}. \tag{18.15}$$

Hence, we arrive to the value of Dirac delta function as

$$\delta^{\alpha}(x) = \left(\frac{1}{2\pi}\right)^{\alpha} \int_{-\infty}^{\infty} e^{ikx} d^{\alpha}k = \lim_{\lambda \to 0} \left(\int_{-\infty}^{\infty} e^{(-\lambda k^2 + ikx)} d^{\alpha}k\right),$$

$$= \left(\frac{1}{2\pi}\right)^{\alpha} \lim_{\lambda \to 0} \left(\lambda^{-\alpha/2} \pi^{\alpha/2} e^{-x^2/4\lambda}\right). \tag{18.16}$$

By taking $\lambda = \frac{1}{4\pi\varepsilon^2}$, we arrive at the proof of Theorem 2. □

Theorem 18.3. *The generalized Dirac delta function satisfies the following identities*

1-

$$\int_{-\infty}^{\infty} f(x)\delta^{\alpha}(x) d^{\alpha}x = f(0). \tag{18.17}$$

2-

$$\int_{-\infty}^{\infty} \delta^{\alpha}(x) d^{\alpha}x = 1. \tag{18.18}$$

Proof. Using the scaling method of Mandelbrot (18.6), we have

$$\int_{-\infty}^{\infty} \delta^{\alpha}(x)f(x) d^{\alpha}x = \frac{2\pi^{\alpha/2}}{\Gamma(\alpha/2)} \left(\int_{0}^{\infty} \delta^{\alpha}(x)f(x)x^{\alpha-1}dx\right). \tag{18.19}$$

Substitution of Dirac delta function defined in (18.12), we obtain

$$\int_{-\infty}^{\infty} \delta^{\alpha}(x)f(x) d^{\alpha}x = \frac{2\pi^{\alpha/2}}{\Gamma(\alpha/2)} \left(\int_{0}^{\infty} \lim_{\varepsilon \to \infty} \varepsilon^{\alpha} e^{-\pi\varepsilon^2 x^2} f(x)x^{\alpha-1}dx\right). \tag{18.20}$$

Using $z = \pi\varepsilon^2 x^2$, (18.20) can be written as

$$\int_{-\infty}^{\infty} \delta^{\alpha}(x)f(x) d^{\alpha}x = \frac{1}{\Gamma(\alpha/2)} \left(\lim_{\varepsilon \to \infty} f(0) \int_{0}^{\infty} e^{-z} z^{(\alpha/2-1)}dz\right) = f(0). \tag{18.21}$$

In particular, when $f(x) = 1$, we arrive at

$$\int_{-\infty}^{\infty} \delta^{\alpha}(x) d^{\alpha}x = 1. \tag{18.22}$$

$\square$

In general the above definitions can be generalized for N dimensional vectors in $\mathcal{R}^N$. The dimension of a fractional space is given by $D = \alpha_1 + \alpha_2 + \cdots + \alpha_N$, $0 < \alpha_i \leq 1$. In this case the generalized Dirac delta function in the D dimensional fractional space satisfies the following identity

$$\delta^D(\mathbf{r} - \mathbf{r}') = \left(\frac{1}{2\pi}\right)^D \int e^{\mathbf{k}\cdot(\mathbf{r}-\mathbf{r}')} d^D k,$$

$$= \delta^{\alpha_1}(x_1 - x_1')\delta^{\alpha_2}(x_2 - x_2')\ldots\delta^{\alpha_N}(x_N - x_N'). \tag{18.23}$$

Now, to solve the wave equation (18.5) we use the Fourier transform method in the fractional dimensional space defined previously, which leads to

$$F\psi(\mathbf{r},t) = \phi(\mathbf{k},\omega) = \int_{R^{D+1}} \psi(\mathbf{r},t) e^{i(\mathbf{k}\cdot\mathbf{r}-\omega t)} \, d^D r \, dt, \tag{18.24}$$

$$Ff(\mathbf{r},t) = g(\mathbf{k},\omega) = \int_{R^{D+1}} f(\mathbf{r},t) e^{i(\mathbf{k}\cdot\mathbf{r}-\omega t)} \, d^D r \, dt. \tag{18.25}$$

The inverse Fourier transform read as

$$\psi(\mathbf{r},t) = F^{-1}\phi(\mathbf{k},\omega) = \left(\frac{1}{2\pi}\right)^{D+1} \int_{k^{D+1}} \phi(\mathbf{k},\omega) e^{-i(\mathbf{k}\cdot\mathbf{r}-\omega t)} \, d^D k \, d\omega, \tag{18.26}$$

$$f(\mathbf{r},t) = F^{-1}g(\mathbf{k},\omega) = \left(\frac{1}{2\pi}\right)^{D+1} \int_{k^{D+1}} g(\mathbf{k},\omega) e^{-i(\mathbf{k}\cdot\mathbf{r}-\omega t)} \, d^D k \, d\omega. \tag{18.27}$$

Taking the Fourier transform of (18.5), we have

$$\phi(\mathbf{k},\omega) = \frac{g(\mathbf{k},\omega)}{k^2 - \frac{\omega^2}{c^2}}. \tag{18.28}$$

Hence, we obtain the solution for $\psi_D(\mathbf{r},t)$ as

$$\psi_D(\mathbf{r},t) = \int_{R^{D+1}} G_D(\mathbf{r},\mathbf{r}',t,t') f(\mathbf{r}',t') \, d^D r' \, dt', \tag{18.29}$$

where $G_D(\mathbf{r},\mathbf{r}',t,t')$ is the Green's function (*Kernel*) and is given by

$$G_D(\mathbf{r},\mathbf{r}',t,t') = \left(\frac{1}{2\pi}\right)^{D+1} \int \left(\left\{\int_{K^D} \frac{e^{i\mathbf{k}\cdot(\mathbf{r}-\mathbf{r}')}}{k^2 - \frac{\omega^2}{c^2}} d^D k\right\} e^{-i\omega(t-t')}\right) d\omega. \tag{18.30}$$

4   Evaluating the Green's Function

Thus, the problem is reduced to evaluating the fractional integral (18.30). Let us first perform the integration over ω, and let us consider the case where $(t-t') > 0$. Note that the Green function should satisfy the boundary condition and should converge for $(t-t') \to \infty$. This means that our complex contour should be in the lower half plane, with simple poles at $\omega = \pm c|k|$. Hence, we obtain

$$\int_{-\infty}^{\infty} \left(\frac{e^{-i\omega(t-t')}}{k^2 - \frac{\omega^2}{c^2}} \right) d\omega = \frac{2\pi}{c|k|} \sin c|k|(t-t'). \tag{18.31}$$

Inserting the result (18.31) in (18.30), we obtain the Green's function as

$$G_D(\mathbf{r},\mathbf{r}',t,t') = \left(\frac{1}{2\pi} \right)^{D+1} \int_{K^D} \left(\frac{2\pi}{c|k|} \sin(c|k|(t-t')) \right) e^{i\mathbf{k}\cdot(\mathbf{r}-\mathbf{r}')} d^D k. \tag{18.32}$$

Using the following transformation introduced in reference [17]

$$\int_{K^D} \varphi(\mathbf{k}) e^{i\mathbf{k}\cdot(\mathbf{r}-\mathbf{r}')} d^D k = (2\pi)^{D/2} \int_0^{\infty} \frac{J_{D/2-1}(\rho|\mathbf{r}-\mathbf{r}'|)}{(\rho|\mathbf{r}-\mathbf{r}'|)^{D/2-1}} \varphi(\rho)\rho^{D-1} d\rho, \tag{18.33}$$

where $J_{v(x)}$ is the Bessel function of the first kind. We obtain

$$G_D(\mathbf{r},\mathbf{r}',t,t') = \left(\frac{1}{2\pi} \right)^{D+1} \int_0^{\infty} \left(\frac{2\pi}{c\rho} \sin(c\rho(t-t')) \right)$$
$$\times (2\pi)^{D/2} \frac{J_{D/2-1}(\rho|\mathbf{r}-\mathbf{r}'|)}{(\rho|\mathbf{r}-\mathbf{r}'|)^{D/2-1}} \varphi(\rho)\rho^{D-1} d\rho, \tag{18.34}$$

The Green's function $G_D(\mathbf{r},\mathbf{r}',t,t')$ can be written in a compact form as

$$G_D(\mathbf{r},\mathbf{r}',t,t') = \left(\frac{1}{2\pi} \right)^{D/2} \frac{1}{c|\mathbf{r}-\mathbf{r}'|^{D/2-1}} \int_0^{\infty} \sin(c\rho(t-t'))\rho^{D/2-1}$$
$$\times J_{D/2-1}(\rho|\mathbf{r}-\mathbf{r}'|) d\rho. \tag{18.35}$$

To evaluate the Green's function we have to consider the following cases:

Case 1: $D = 3$. Then from (18.35) for $D = 3$ we have,

$$G_3(\mathbf{r},\mathbf{r}',t,t') = \frac{\delta\left(t' - t + \frac{|\mathbf{r}-\mathbf{r}'|}{c}\right)}{4\pi|\mathbf{r}-\mathbf{r}'|}, \tag{18.36}$$

which represent the retarded Greens function in four space-time dimensions [18].

This means that the wave signal arrives to the observer at time $t_3 = t' + \frac{|\mathbf{r} - \mathbf{r}'|}{c}$ where t_3 is the time measured in three dimensional space.

Case 2: In fractional dimensional space with $D \neq 3$, the Green's function (18.35) can be calculated with the aid of the following identity [19]

$$\int_0^\infty x^\nu \sin(ax) J_\nu(bx) \, dx = \begin{cases} \frac{\sqrt{\pi} 2^\nu b^\nu}{(a^2 - b^2)^{\nu + 1/2} \Gamma(1/2 - \nu)} & [0 < b < a, \quad -1 < Re\nu < 1/2] \\ 0 & [0 < a < b, \quad -1 < Re\nu < 1/2] \end{cases}$$

(18.37)

In (35), $\nu = D/2 - 1$, $a = c(t - t')$ and $b = |\mathbf{r} - \mathbf{r}'|$. For $0 < D < 3$ and $0 < |\mathbf{r} - \mathbf{r}'| < c(t - t')$, we have

$$G_D(\mathbf{r}, \mathbf{r}', t, t') = \frac{1}{2c\Gamma(\frac{3-D}{2}) \pi^{(\frac{D-1}{2})} (c^2(t - t')^2 - |\mathbf{r} - \mathbf{r}'|^2)^{(\frac{D-1}{2})}}.$$

(18.38)

Which means that the signal arrives to the observer at time $t_D > (t_3 = t' + \frac{|\mathbf{r} - \mathbf{r}'|}{c})$. This relates to the fact, that for fractals and in a fractional dimensional space, the measurement of the distance is greater than the distance $|\mathbf{r} - \mathbf{r}'|$ in exactly three dimensional space. So, the wave signal takes longer time t_D to arrive the observer.

For, $0 < D < 3$ and $0 < c(t - t') < |\mathbf{r} - \mathbf{r}'|$, then $G_D(\mathbf{r}, \mathbf{r}', t, t') = 0$. Which means no wave signal propagates in this case. In conclusion, the wave signal will propagate and arrive the observation point at least with time not less than t_3.

In all the above calculations, once we obtain the Greens function, we can plug its value in the potentials defined in (18.29) as

$$\psi_D(\mathbf{r}, t) = \int_{R^{D+1}} \frac{f(\mathbf{r}', t') \delta(t' - t - \frac{|\mathbf{r} - \mathbf{r}'|}{c})}{4\pi |\mathbf{r} - \mathbf{r}'|} d^D r' \, dt', \quad D = 3,$$

(18.39)

$$\psi_D(\mathbf{r}, t) = \int_{R^{D+1}} \left(\frac{f(\mathbf{r}', t')}{2c\Gamma(\frac{3-D}{2}) \pi^{(\frac{D-1}{2})} (c^2(t - t')^2 - |\mathbf{r} - \mathbf{r}'|^2)^{(\frac{D-1}{2})}} \right) d^D r' \, dt',$$

$$0 < D < 3, \quad 0 < |\mathbf{r} - \mathbf{r}'| < c(t - t').$$

(18.40)

This finally allow us to calculate the transmitted fields **E** and **B** for any fractal source.

5 Moving Point Charge

In this section, we consider a source as a point charge moving with velocity **v**. Let us first assume that this particle is moving in exactly three dimensions. The charge and the current densities for this system are

$$\rho(\mathbf{r}',t') = q\delta(\mathbf{r}' - \mathbf{r}_*(t')), \tag{18.41}$$

$$\mathbf{J}(\mathbf{r}',t') = q\mathbf{v}\delta(\mathbf{r}' - \mathbf{r}_*(t')). \tag{18.42}$$

According to (18.39), the potential function is obtained as

$$\phi(\mathbf{r},t) = \frac{1}{4\pi\varepsilon_0} \int \frac{q\delta(\mathbf{r}' - \mathbf{r}_*(t'))\delta(t' - t + 1/c|\mathbf{r} - \mathbf{r}'|)}{|\mathbf{r} - \mathbf{r}'|} d^3r' dt'. \tag{18.43}$$

Integrating over the spatial coordinates, we have

$$\phi(\mathbf{r},t) = \frac{q}{4\pi\varepsilon_0} \int_{-\infty}^{\infty} \frac{\delta(t' - t + R(t')/c)}{R(t')} dt', \tag{18.44}$$

where, $R(t') = |\mathbf{r} - \mathbf{r}_*(t')|$. Now the delta-function $\delta(t' - t + R(t')/c)$ can be expressed as

$$\delta(t' - t + R(t')) = \Sigma_i \frac{\delta(t' - t_i)}{|f'(t_i)|_{f(t_i)=0}}, \tag{18.45}$$

where

$$f(t') = t' + R(t')/c - t. \tag{18.46}$$

Now $f(t') = 0 = t' + R(t')/c - t, \Rightarrow t' = t - R(t')/c$ and

$$f'(t') = 1 + \frac{1}{2c}\left[(\mathbf{r} - \mathbf{r}_*(t')) \cdot (\mathbf{r} - \mathbf{r}_*(t'))\right]^{-1/2}\left[\frac{d}{dt}(\mathbf{r} - \mathbf{r}_*(t')) \cdot (\mathbf{r} - \mathbf{r}_*(t'))\right],$$

$$= 1 - \frac{\mathbf{v} \cdot \mathbf{R}(t')}{cR(t')}, \tag{18.47}$$

where, $\mathbf{v} = \frac{d\mathbf{r}_*(t')}{dt'}$. Hence

$$\delta(t' - t + R(t')/c) = \frac{\delta(t' - t_{ret})}{\left(1 - \frac{\mathbf{v} \cdot \mathbf{R}(t')}{cR(t')}\right)}, \tag{18.48}$$

where $t_{ret} = t + R(t')/c$. Now, the integral (18.44) can be written as

$$\phi(\mathbf{r},t) = \frac{1}{4\pi\varepsilon_0} \int \frac{\delta(t' - t_{ret})}{R(t')\left(1 - \frac{\mathbf{v} \cdot \mathbf{R}}{cR}\right)} dt'. \tag{18.49}$$

Integrating over t' we obtain

$$\phi(\mathbf{r},t) = \frac{q}{4\pi\varepsilon_0\left(1 - \frac{\boldsymbol{\beta} \cdot \mathbf{R}}{cR}\right)}\bigg|_{t_{ret}}, \tag{18.50}$$

where $\beta = \frac{v}{c}$. Similarly, the vector potential $\mathbf{A}$ is calculated as

$$\mathbf{A} = \frac{\beta}{c}\phi(\mathbf{r},t). \tag{18.51}$$

Now let us consider that the charged particle is moving in a fractal media along z axis with a fractional dimensionality $D = \alpha_x + \alpha_y + \alpha_z = 2 + \alpha \ (0 < \alpha < 1)$. The charge and the current densities for this system are

$$\rho(\mathbf{r}',t) = q\delta^\alpha(z' - vt')\delta(x')\delta(y'), \tag{18.52}$$

$$\mathbf{J}(\mathbf{r}',t) = q\mathbf{v}\delta^\alpha(z' - vt')\delta(x')\delta(y'). \tag{18.53}$$

Making use of (18.40), we obtain the potential $\phi(z,t)$ along z axis as

$$\phi(z,t) = \int_{c(t-t')<|\mathbf{r}-\mathbf{r}'|} \left(\frac{q\delta^\alpha(z' - vt')\delta(x')\delta(y')}{2c\Gamma(\frac{\alpha+1}{2})\pi^{(\frac{\alpha+1}{2})} (c^2(t-t')^2 - |\mathbf{r}-\mathbf{r}'|^2)^{(\frac{\alpha+1}{2})}} \right)$$
$$\times \, d^\alpha z' \, dx' \, dy' \, dt'. \tag{18.54}$$

Integrating over the spatial coordinates, we obtain

$$\phi(z,t) = \int_{c(t-t')<|\mathbf{r}-\mathbf{r}'|} \frac{q}{2c\Gamma(\frac{\alpha+1}{2})\pi^{(\frac{\alpha+1}{2})} (c^2(t-t')^2 - (z-vt')^2)^{(\frac{\alpha+1}{2})}} \, dt'. \tag{18.55}$$

Noting that, $c(t-t') < |\mathbf{r} - \mathbf{r}'|$ and $t > t'$, implies that $\frac{t'(c-v)}{(ct-z)} < 1$. Let $\frac{t'(c-v)}{(ct-z)} = x$. Hence the integral (18.55) can be written as

$$\phi(z,t) = \frac{q}{2c\Gamma(\frac{\alpha+1}{2})\pi^{(\frac{\alpha+1}{2})}} \frac{(ct+z)^{-\frac{1+\alpha}{2}}}{(c-v)(ct-z)^{(\frac{\alpha-1}{2})}} \int_0^1 dx(1-x)^{\frac{-(1+\alpha)}{2}} (1-ux)^{\frac{-(1+\alpha)}{2}}, \tag{18.56}$$

where $u = \frac{(c+v)(ct-z)}{(c-v)(ct+z)}$. Using the integral [19]

$$\int_0^1 dx x^{\lambda-1}(1-x)^{\mu-1}(1-vx)^{-\nu} = \beta(\lambda,\mu)_2F_1(\nu,\lambda;\lambda+\mu;v),$$

$$[Re\,\lambda > 0,\ Re\,\mu > 0,\ |v| < 1], \tag{18.57}$$

where $_2F_1(\nu,\lambda;\lambda+\mu;v)$ is the Gauss hypergeometric function, we arrive at the following expression

$$\phi(z,t) = \frac{q}{2c\Gamma(\frac{\alpha+1}{2})\pi^{(\frac{\alpha+1}{2})}} \frac{(ct+z)^{-\frac{1+\alpha}{2}}}{(c-v)(ct-z)^{(\frac{\alpha-1}{2})}}$$

$$\times \frac{\Gamma(\frac{1-\alpha}{2})}{\Gamma(\frac{3-\alpha}{2})} {}_2F_1\left(\frac{\alpha+1}{2},1;\frac{3-\alpha}{2};\frac{(c+v)(ct-z)}{(c-v)(ct+z)}\right). \qquad (18.58)$$

6   Conclusion

In this chapter we have given an application of time-dependent wave equation in fractional space and we solved the wave equation in fractional space by applying the Fourier transform method for nonhomogeneous partial differential equation. We obtained the retarded potentials and the values of the Green's function which depend on the dimensionality parameter D. We showed that the required time of the propagation $t_D \geq t' + \frac{|\mathbf{r}-\mathbf{r}'|}{c}$, and no wave could propagate with time less than $t' + \frac{|\mathbf{r}-\mathbf{r}'|}{c}$. The potential $\psi_3(\mathbf{r},t)$ for the exact three dimensional case $D = 3$ is recovered as a special case.

Acknowledgements One of the authors (S.M.) would like to sincerely thank the Institute of International Education (Fulbright Scholar Program), New York, NY, and the Department of Mechanical Engineering and Energy Processes (MEEP) and the Dean of Graduate Studies at Southern Illinois University, Carbondale (SIUC), IL, for providing him the financial support and the necessary facilities during his stay at SIUC.

References

1. Mandelbrot B (1983) The fractal geometry of nature. W.H. Freeman, New York
2. Feynman RP, Hibbs AR (1965) Quantum and path integrals. McGraw-Hill, New York
3. Fractals in physics. Proceedings of the VI Trieste International Symposium on Fractal Physics. ICTP, Trieste, Italy, July 9–12, 1985, North-Holland, Amsterdam (1986)
4. Carpinteri A, Mainardi F (1997) Fratals and fractional calculus in continum mechanics. Springer, New York
5. Zaslavsky GM (2002) Chaos, fractional kinetics, and anomalous transport. Phys Rep 371:461–580
6. Stillinger FH (1977) Axiomatic basis for spaces with noninteger dimension. J Math Phys 18:1224–1234
7. He X (1990) Anisotropy and isotropy: a model of fraction-dimensional space. Solid State Commun 75:111–114
8. He X (1991) Excitons in anisotropic solids: the model of fractional-dimensional space. Phys Rev B 43:2063–2069
9. Matos-Abiague A (2001) Bose-like oscillator in fractional-dimensional space. J Phys A Math Gen 34: 3125–3128
10. Matos-Abiague A (2001) Deformation of quantum mechanics in fractional-dimensional space. J Phys A Math Gen 34:11059–11071

11. Jing SJ (1998) A new kind of deformed calculus and parabosonic coordinate representation. J Phys A Math Gen 31:6347–6354
12. Willson KG (1973) Quantum field – theory models in less than 4 dimensions. Phys Rev D 7:2911–2926
13. Misner CW, Thorne KS, Wheeler JA (1973) Gravitation. Freeman, San Francisco
14. Zeilinger A, Svozil K (1985) Measuring the dimension of space-time. Phys Rev Lett 54:2553–2555
15. Muslih SI, Agrawal OP (2010) Riesz fractional derivatives and fractional dimensional space. Int J Theor Phys 49:270–275
16. Engheta N (1997) On the role of fractional calculus in electromagnetic theory. IEEE Antenna Propagation Mag 39:35–46
17. Muslih SI, Agrawal OP (2009) A scaling method and its applications to problems in fractional dimensional space. J Math Phys 50:123501–123511
18. Jackson JD (1999) Classical electrodynamics. Wiley, New York
19. Gradshleyn IS, Ryzhik IM (2007) Table of integrals and products. Academic, New York

Chapter 19
Fractional Exact Solutions and Solitons
in Gravity

Dumitru Baleanu and Sergiu I. Vacaru

1 Introduction

Recently, we extended the fractional calculus to Ricci flow theory, gravity and geometric mechanics, solitonic hierarchies, etc. [1–6]. In this work, we outline some basic geometric constructions related to fractional derivatives and integrals and their applications in modern physics and mechanics.

Our approach is also connected to a method when nonholonomic deformations of geometric structures[1] induce a canonical connection, adapted to a necessary type of nonlinear connection structure, for which the matrix coefficients of curvature are constant [7, 8]. For such an auxiliary connection, it is possible to define a biHamiltonian structure and derive the corresponding solitonic hierarchy.

The chapter is organized as given below: In Sect. 19.2, we outline the geometry of N-adapted fractional manifolds and provide an introduction to fractional gravity. In Sect. 19.3, we show how fractional gravitational field equations can be solved in a general form. Section 19.4 is devoted to the main theorem on fractional solitonic hierarchies corresponding to metrics and connections in fractional gravity. The appendix contains necessary definitions and formulas on Caputo fractional derivatives.

---

[1]Determined by a fundamental Lagrange/Finsler/Hamilton generating function (or, for instance, and Einstein metric).

D. Baleanu (✉)
Department of Mathematics and Computer Sciences, Cankaya University, 06530,
Ankara, Turkey

Institute of Space Sciences, P. O. Box, MG-23, R 76900, Magurele–Bucharest, Romania
e-mail: dumitru@cankaya.edu.tr

S.I. Vacaru
Science Department, University "Al. I. Cuza" Iaşi, 54, Lascar Catargi street,
Iaşi, Romania, 700107
e-mail: sergiu.vacaru@uaic.ro

D. Baleanu et al. (eds.), *Fractional Dynamics and Control*,
DOI 10.1007/978-1-4614-0457-6_19, © Springer Science+Business Media, LLC 2012

2   Fractional Nonholonomic Manifolds and Gravity

Let us consider a "prime" nonholonomic manifold $\mathbf{V}$ of integer dimension dim $\mathbf{V} = n + m, n \geq 2, m \geq 1$.[2] Its fractional extension $\overset{\alpha}{\mathbf{V}}$ is modeled by a quadruple $(\overset{\alpha}{\mathbf{V}}, \overset{\alpha}{\mathbf{N}}, \overset{\alpha}{\mathbf{d}}, \overset{\alpha}{\mathbf{I}})$, where $\overset{\alpha}{\mathbf{N}}$ is a nonholonomic distribution stating a nonlinear connection (N-connection) structure (for details, see Appendix 5). The fractional differential structure $\overset{\alpha}{\mathbf{d}}$ is determined by Caputo fractional derivative (19.13) following formulas (19.15). The noninteger integral structure $\overset{\alpha}{\mathbf{I}}$ is defined by rules of type (19.14).

A nonlinear connection (N-connection) $\overset{\alpha}{\mathbf{N}}$ for a fractional space $\overset{\alpha}{\mathbf{V}}$ is defined by a nonholonomic distribution (Whitney sum) with conventional h- and v-subspaces, $\underline{h}\overset{\alpha}{\mathbf{V}}$ and $\underline{v}\overset{\alpha}{\mathbf{V}}$, $\underline{T}\overset{\alpha}{\mathbf{V}} = \underline{h}\overset{\alpha}{\mathbf{V}} \oplus \underline{v}\overset{\alpha}{\mathbf{V}}$.

A fractional N-connection is defined by its local coefficients $\overset{\alpha}{\mathbf{N}} = \{ \,^{\alpha}N_i^a \}$ and $\overset{\alpha}{\mathbf{N}} = \,^{\alpha}N_i^a(u)(\mathrm{d}x^i)^{\alpha} \otimes \overset{\alpha}{\underline{\partial}}_a$. For a N-connection $\overset{\alpha}{\mathbf{N}}$, we can always construct a class of fractional (co)-frames (N-adapted) linearly depending on $\,^{\alpha}N_i^a$,

$$\,^{\alpha}\mathbf{e}_{\beta} = \left[\,^{\alpha}\mathbf{e}_j = \overset{\alpha}{\underline{\partial}}_j - \,^{\alpha}N_j^a \overset{\alpha}{\underline{\partial}}_a, \,^{\alpha}\mathbf{e}_b = \overset{\alpha}{\underline{\partial}}_b \right], \tag{19.1}$$

$$\,^{\alpha}\mathbf{e}^{\beta} = \left[\,^{\alpha}\mathbf{e}^j = (\mathrm{d}x^j)^{\alpha}, \,^{\alpha}\mathbf{e}^b = (\mathrm{d}y^b)^{\alpha} + \,^{\alpha}N_k^b(\mathrm{d}x^k)^{\alpha} \right]. \tag{19.2}$$

The nontrivial nonholonomy coefficients are computed $\,^{\alpha}W_{ib}^a = \overset{\alpha}{\partial}_b \,^{\alpha}N_i^a$ and $\,^{\alpha}W_{ij}^a = \,^{\alpha}\Omega_{ji}^a = \,^{\alpha}\mathbf{e}_i \,^{\alpha}N_j^a - \,^{\alpha}\mathbf{e}_j \,^{\alpha}N_i^a$, for $[\,^{\alpha}\mathbf{e}_{\alpha}, \,^{\alpha}\mathbf{e}_{\beta}] = \,^{\alpha}\mathbf{e}_{\alpha} \,^{\alpha}\mathbf{e}_{\beta} - \,^{\alpha}\mathbf{e}_{\beta} \,^{\alpha}\mathbf{e}_{\alpha} = \,^{\alpha}W_{\alpha\beta}^{\gamma} \,^{\alpha}\mathbf{e}_{\gamma}$. In the above formulas, the values $\,^{\alpha}\Omega_{ji}^a$ are called the coefficients of N-connection curvature. A nonholonomic manifold defined by a structure $\overset{\alpha}{\mathbf{N}}$ is called, in brief, a N-anholonomic fractional manifold.

We write a metric structure $\overset{\alpha}{\mathbf{g}} = \{ \,^{\alpha}g_{\underline{\alpha}\underline{\beta}} \}$ on $\overset{\alpha}{\mathbf{V}}$ in the form:

$$\overset{\alpha}{\mathbf{g}} = \,^{\alpha}g_{kj}(x,y) \,^{\alpha}\mathbf{e}^k \otimes \,^{\alpha}\mathbf{e}^j + \,^{\alpha}g_{cb}(x,y) \,^{\alpha}\mathbf{e}^c \otimes \,^{\alpha}\mathbf{e}^b$$

$$= \eta_{k'j'} \,^{\alpha}\mathbf{e}^{k'} \otimes \,^{\alpha}\mathbf{e}^{j'} + \eta_{c'b'} \,^{\alpha}\mathbf{e}^{c'} \otimes \,^{\alpha}\mathbf{e}^{b'}, \tag{19.3}$$

where matrices $\eta_{k'j'} = diag[\pm 1, \pm 1, ..., \pm 1]$ and $\eta_{a'b'} = diag[\pm 1, \pm 1, ..., \pm 1]$, for the signature of a "prime" spacetime $\mathbf{V}$, are obtained by frame transforms $\eta_{k'j'} = e^k_{\ k'} e^j_{\ j'} \,^{\alpha}g_{kj}$ and $\eta_{a'b'} = e^a_{\ a'} e^b_{\ b'} \,^{\alpha}g_{ab}$.

---

[2]A nonholonomic manifold is a manifold endowed with a nonintegrable (equivalently, nonholonomic, or anholonomic) distribution.

A distinguished connection (d-connection) $\overset{\alpha}{\mathbf{D}}$ on $\overset{\alpha}{\mathbf{V}}$ is defined as a linear connection preserving under parallel transports the Whitney. We can associate an N-adapted differential one-form $^{\alpha}\Gamma^{\tau}_{\beta} = {}^{\alpha}\Gamma^{\tau}_{\beta\gamma}\,{}^{\alpha}\mathbf{e}^{\gamma}$, parametrizing the coefficients (with respect to (19.2) and (19.1)) in the form $^{\alpha}\Gamma^{\gamma}_{\tau\beta} = \left({}^{\alpha}L^{i}_{jk},\ {}^{\alpha}L^{a}_{bk},\ {}^{\alpha}C^{i}_{jc},\ {}^{\alpha}C^{a}_{bc} \right)$. The absolute fractional differential $^{\alpha}\mathbf{d} = {}_{1x}\mathrm{d}_x + {}_{1y}\mathrm{d}_y$ acts on fractional differential forms in N-adapted form; the value $^{\alpha}\mathbf{d} := {}^{\alpha}\mathbf{e}^{\beta}\,{}^{\alpha}\mathbf{e}_{\beta}$ splits into exterior h-/v-derivatives, ${}_{1x}\mathrm{d}_x := (\mathrm{d}x^i)^{\alpha}\ {}_{1x}\overset{\alpha}{\underline{\partial}}_i = {}^{\alpha}e^j\,{}^{\alpha}\mathbf{e}_j$ and ${}_{1y}\mathrm{d}_y := (\mathrm{d}y^a)^{\alpha}\ {}_{1x}\overset{\alpha}{\underline{\partial}}_a = {}^{\alpha}e^b\,{}^{\alpha}e_b$.

The torsion and curvature of a fractional d-connection $\overset{\alpha}{\mathbf{D}} = \{{}^{\alpha}\Gamma^{\tau}_{\beta\gamma}\}$ can be defined and computed, respectively, as fractional two-forms,

$$^{\alpha}\mathcal{T}^{\tau} \doteq \overset{\alpha}{\mathbf{D}}\,{}^{\alpha}\mathbf{e}^{\tau} = {}^{\alpha}\mathbf{d}\,{}^{\alpha}\mathbf{e}^{\tau} + {}^{\alpha}\Gamma^{\tau}_{\beta} \wedge {}^{\alpha}\mathbf{e}^{\beta}\ \text{and}$$

$$^{\alpha}\mathcal{R}^{\tau}_{\beta} \doteq \overset{\alpha}{\mathbf{D}}\,{}^{\alpha}\Gamma^{\tau}_{\beta} = {}^{\alpha}d\,{}^{\alpha}\Gamma^{\tau}_{\beta} - {}^{\alpha}\Gamma^{\gamma}_{\beta} \wedge {}^{\alpha}\Gamma^{\tau}_{\gamma} = {}^{\alpha}\mathbf{R}^{\tau}_{\beta\gamma\delta}\,{}^{\alpha}\mathbf{e}^{\gamma} \wedge {}^{\alpha}\mathbf{e}^{\delta}. \quad (19.4)$$

There are two another important geometric objects: the fractional Ricci tensor $^{\alpha}\mathcal{R}ic = \{{}^{\alpha}\mathbf{R}_{\alpha\beta} \doteq {}^{\alpha}\mathbf{R}^{\tau}_{\alpha\beta\tau}\}$, with ${}^{\alpha}R_{ij} \doteq {}^{\alpha}R^{k}_{ijk}$, ${}^{\alpha}R_{ia} \doteq -{}^{\alpha}R^{k}_{ika}$, ${}^{\alpha}R_{ai} \doteq {}^{\alpha}R^{b}_{aib}$, ${}^{\alpha}R_{ab} \doteq {}^{\alpha}R^{c}_{abc}$, and the scalar curvature of fractional d-connection $\overset{\alpha}{\mathbf{D}}$, $^{\alpha}_{s}\mathbf{R} \doteq {}^{\alpha}\mathbf{g}^{\tau\beta}\,{}^{\alpha}\mathbf{R}_{\tau\beta} = {}^{\alpha}R + {}^{\alpha}S$, $^{\alpha}R = {}^{\alpha}g^{ij}\,{}^{\alpha}R_{ij}$, $^{\alpha}S = {}^{\alpha}g^{ab}\,{}^{\alpha}R_{ab}$, with $^{\alpha}\mathbf{g}^{\tau\beta}$ being the inverse coefficients to a d-metric (19.3). We can introduce the Einstein tensor $^{\alpha}\mathcal{E}ns = \{{}^{\alpha}\mathbf{G}_{\alpha\beta}\}$,

$$^{\alpha}\mathbf{G}_{\alpha\beta} := {}^{\alpha}\mathbf{R}_{\alpha\beta} - \frac{1}{2}\,{}^{\alpha}\mathbf{g}_{\alpha\beta}\,{}^{\alpha}_{s}\mathbf{R}. \quad (19.5)$$

For applications, we can consider more special classes of d-connections:

- There is a unique canonical metric compatible fractional d-connection $^{\alpha}\widehat{\mathbf{D}} = \left\{ {}^{\alpha}\widehat{\Gamma}^{\gamma}_{\alpha\beta} = \left({}^{\alpha}\widehat{L}^{i}_{jk},\ {}^{\alpha}\widehat{L}^{a}_{bk},\ {}^{\alpha}\widehat{C}^{i}_{jc},\ {}^{\alpha}\widehat{C}^{a}_{bc} \right) \right\}$, when $^{\alpha}\widehat{\mathbf{D}}\,({}^{\alpha}\mathbf{g}) = 0$, satisfying the conditions $^{\alpha}\widehat{T}^{i}_{jk} = 0$ and $^{\alpha}\widehat{T}^{a}_{bc} = 0$, but $^{\alpha}\widehat{T}^{i}_{ja}$, $^{\alpha}\widehat{T}^{a}_{ji}$, and $^{\alpha}\widehat{T}^{a}_{bi}$ are not zero. The N-adapted coefficients are given in explicit form in our works [1–6].
- The fractional Levi–Civita connection $^{\alpha}\nabla = \{{}^{\alpha}\Gamma^{\gamma}_{\alpha\beta}\}$ can be defined in standard form but for the fractional Caputo left derivatives acting on the coefficients of a fractional metric.

On spaces with nontrivial nonholonomic structure, it is preferred to work on $\overset{\alpha}{\mathbf{V}}$ with $^{\alpha}\widehat{\mathbf{D}} = \{{}^{\alpha}\widehat{\Gamma}^{\gamma}_{\tau\beta}\}$ instead of $^{\alpha}\nabla$ (the last one is not adapted to the N-connection splitting). The torsion $^{\alpha}\widehat{\mathcal{T}}^{\tau}$ (19.4) of $^{\alpha}\widehat{\mathbf{D}}$ is uniquely induced nonholonomically by off-diagonal coefficients of the d-metric (19.3).

With respect to N-adapted fractional bases (19.1) and (19.2), the coefficients of the fractional Levi–Civita and canonical d-connection satisfy the distorting relations

$^{\alpha}\Gamma^{\gamma}_{\alpha\beta} = {}^{\alpha}\widehat{\Gamma}^{\gamma}_{\alpha\beta} + {}^{\alpha}Z^{\gamma}_{\alpha\beta}$, where the N-adapted coefficients of distortion tensor $Z^{\gamma}_{\alpha\beta}$ are computed in [6].

An unified approach to Einstein–Lagrange/Finsler gravity for arbitrary integer and noninteger dimensions is possible for the fractional canonical d-connection $^{\alpha}\widehat{\mathbf{D}}$. The fractional gravitational field equations are formulated for the Einstein d-tensor (19.5), following the same principle of constructing the matter source $^{\alpha}\Upsilon_{\beta\delta}$ as in general relativity but for fractional metrics and d-connections,

$$^{\alpha}\widehat{\mathbf{E}}_{\beta\delta} = {}^{\alpha}\Upsilon_{\beta\delta}. \tag{19.6}$$

Such a system of integro-differential equations for generalized connections can be restricted to fractional nonholonomic configurations for $^{\alpha}\mathbf{V}$ if we impose the additional constraints:

$$^{\alpha}\widehat{L}^{c}_{aj} = {}^{\alpha}e_a({}^{\alpha}N^{c}_j), \quad {}^{\alpha}\widehat{C}^{i}_{jb} = 0, \quad {}^{\alpha}\Omega^{a}_{ji} = 0. \tag{19.7}$$

There are not theoretical or experimental evidences that for fractional dimensions we must impose conditions of type (19.7) but they have certain physical motivation if we develop models which in integer limits result in the general relativity theory.

3   Exact Solutions in Fractional Gravity

We studied in detail [2] what type of conditions must satisfy the coefficients of a metric (19.3) for generating exact solutions of the fractional Einstein equations (19.6). For simplicity, we can use a "prime" dimension splitting of type $2 + 2$ when coordinates are labeled in the form $u^{\beta} = (x^j, y^3 = v, y^4)$, for $i, j, ... = 1, 2$ and the metric ansatz has one Killing symmetry when the coefficients do not depend explicitly on variable y^4.

3.1   Separation of Equations for Fractional and Integer Dimensions

The solutions of equations can be constructed for a general source[3] $^{\alpha}\Upsilon^{\alpha}_{\beta} = diag[\,^{\alpha}\Upsilon_{\gamma}; \,^{\alpha}\Upsilon_1 = {}^{\alpha}\Upsilon_2 = {}^{\alpha}\Upsilon_2(x^k, v); \,^{\alpha}\Upsilon_3 = {}^{\alpha}\Upsilon_4 = {}^{\alpha}\Upsilon_4(x^k)]$. For such sources and ansatz with Killing symmetries for metrics, the Einstein equations (19.6) can be integrated in general form.

---

[3]Such parametrizations of energy-momentum tensors are quite general ones for various types of matter sources.

We can construct "non-Killing" solutions depending on all coordinates,

$$\alpha\mathbf{g} = {}^\alpha g_i(x^k)\,{}^\alpha dx^i \otimes {}^\alpha dx^i + {}^\alpha\omega^2(x^j,v,y^4)\,{}^\alpha h_a(x^k,v)\,{}^\alpha\mathbf{e}^a \otimes {}^\alpha\mathbf{e}^a,$$

$$\alpha\mathbf{e}^3 = {}^\alpha dy^3 + {}^\alpha w_i(x^k,v)\,{}^\alpha dx^i,\ \ {}^\alpha\mathbf{e}^4 = {}^\alpha dy^4 + {}^\alpha n_i(x^k,v)\,{}^\alpha dx^i, \qquad (19.8)$$

for any ${}^\alpha\omega$ for which ${}^\alpha\mathbf{e}_k\,{}^\alpha\omega = \overset{\alpha}{\underline{\partial}}_k\,{}^\alpha\omega + {}^\alpha w_k\,{}^\alpha\omega^* + {}^\alpha n_k\overset{\alpha}{\underline{\partial}}_{y^4}\,{}^\alpha\omega = 0$. Configurations with fractional Levi–Civita connection ${}^\alpha\nabla$, of type (19.7), can be extracted by imposing additional constraints

$$\alpha w_i^* = {}^\alpha\mathbf{e}_i \ln|\,{}^\alpha h_4|,\ \ {}^\alpha\mathbf{e}_k\,{}^\alpha w_i = {}^\alpha\mathbf{e}_i\,{}^\alpha w_k,\ \ {}^\alpha n_i^* = 0,\ \overset{\alpha}{\underline{\partial}}_i\,{}^\alpha n_k = \overset{\alpha}{\underline{\partial}}_k\,{}^\alpha n_i, \qquad (19.9)$$

where ${}^\alpha a^\bullet = \overset{\alpha}{\underline{\partial}}_1 a = {}_{1x^1}\overset{\alpha}{\underline{\partial}}_{x^1}{}^\alpha a$, ${}^\alpha a' = \overset{\alpha}{\underline{\partial}}_2 a = {}_{1x^2}\overset{\alpha}{\underline{\partial}}_{x^2}{}^\alpha a$, ${}^\alpha a^* = \overset{\alpha}{\underline{\partial}}_v a = {}_{1v}\overset{\alpha}{\underline{\partial}}_v{}^\alpha a$, being used the left Caputo fractional derivatives (19.15).

3.2 Solutions with ${}^\alpha h_{3,4}^* \neq 0$ and ${}^\alpha\Upsilon_{2,4} \neq 0$

For simplicity, we provide only a class of exact solution with metrics of type (19.8) when ${}^\alpha h_{3,4}^* \neq 0$ (in [2], there are all analyzed possibilities for coefficients[4]) We consider the ansatz:

$$\alpha\mathbf{g} = e^{{}^\alpha\psi(x^k)}\,{}^\alpha dx^i \otimes {}^\alpha dx^i + h_3(x^k,v)\,{}^\alpha\mathbf{e}^3 \otimes {}^\alpha\mathbf{e}^3 + h_4(x^k,v)\,{}^\alpha\mathbf{e}^4 \otimes {}^\alpha\mathbf{e}^4,$$

$$\alpha\mathbf{e}^3 = {}^\alpha dv + {}^\alpha w_i(x^k,v)\,{}^\alpha dx^i,\ \ {}^\alpha\mathbf{e}^4 = {}^\alpha dy^4 + {}^\alpha n_i(x^k,v)\,{}^\alpha dx^i \qquad (19.10)$$

We consider any nonconstant ${}^\alpha\phi = {}^\alpha\phi(x^i,v)$ as a generating function. We have to solve, respectively, the two-dimensional fractional Laplace equation, for ${}^\alpha g_1 = {}^\alpha g_2 = e^{{}^\alpha\psi(x^k)}$. Then we integrate on v, in order to determine ${}^\alpha h_3$, ${}^\alpha h_4$, and ${}^\alpha n_i$, and solve algebraic equations, for ${}^\alpha w_i$. We obtain (computing consequently for a chosen ${}^\alpha\phi(x^k,v)$):

$$\alpha g_1 = {}^\alpha g_2 = e^{{}^\alpha\psi(x^k)},\ \ {}^\alpha h_3 = \pm\frac{|\,{}^\alpha\phi^*(x^i,v)|}{{}^\alpha\Upsilon_2},$$

$$\alpha h_4 = {}^\alpha_0 h_4(x^k) \pm 2\,{}_{1v}\overset{\alpha}{I}_v\frac{(\exp[2\,{}^\alpha\phi(x^k,v)])^*}{{}^\alpha\Upsilon_2},$$

$$\alpha w_i = -\overset{\alpha}{\underline{\partial}}_i\,{}^\alpha\phi/\,{}^\alpha\phi^*,\ \ {}^\alpha n_i = {}^\alpha_1 n_k(x^i) + {}^\alpha_2 n_k(x^i)\,{}_{1v}\overset{\alpha}{I}_v\left[\,{}^\alpha h_3/(\sqrt{|\,{}^\alpha h_4|})^3\right], \qquad (19.11)$$

---

[4]By nonholonomic transforms, various classes of solutions can be transformed from one to another.

where $\,^{\alpha}_0 h_4(x^k)$, $\,^{\alpha}_1 n_k(x^i)$, and $\,^{\alpha}_2 n_k(x^i)$ are integration functions, and $\,_{1v}I_v^{\alpha}$ is the fractional integral on variables v and

$$\,^{\alpha}\phi = \ln\left|\frac{\,^{\alpha}h_4^*}{\sqrt{|\,^{\alpha}h_3\,^{\alpha}h_4|}}\right|, \quad \,^{\alpha}\gamma = \left(\ln|\,^{\alpha}h_4|^{3/2}/|\,^{\alpha}h_3|\right)^*,$$

$$\,^{\alpha}\alpha_i = \,^{\alpha}h_4^* \overset{\alpha}{\underline{\partial}}_k \,^{\alpha}\phi, \quad \,^{\alpha}\beta = \,^{\alpha}h_4^* \,^{\alpha}\phi^*. \tag{19.12}$$

For $\,^{\alpha}h_4^* \neq 0$; $\,^{\alpha}\Upsilon_2 \neq 0$, we have $\,^{\alpha}\phi^* \neq 0$. The exponent $e^{\,^{\alpha}\psi(x^k)}$ is the fractional analog of the "integer" exponential functions and is called the Mittag–Leffler function $E_\alpha[(x - \,^1 x)^\alpha]$. For $\,^{\alpha}\psi(x) = E_\alpha[(x - \,^1 x)^\alpha]$, we have $\overset{\alpha}{\underline{\partial}}_i E_\alpha = E_\alpha$.

We have to constrain the coefficients (19.11) to satisfy the conditions (19.9) in order to construct exact solutions for the Levi–Civita connection $\,^{\alpha}\nabla$. To select such classes of solutions, we can fix a nonholonomic distribution when $\,^{\alpha}_2 n_k(x^i) = 0$ and $\,^{\alpha}_1 n_k(x^i)$ are any functions satisfying the conditions $\overset{\alpha}{\underline{\partial}}_i \,^{\alpha}_1 n_k(xj) = \overset{\alpha}{\underline{\partial}}_k \,^{\alpha}_1 n_i(x^j)$. The constraints on $\,^{\alpha}\phi(x^k, v)$ are related to the N-connection coefficients $\,^{\alpha}w_i = -\overset{\alpha}{\underline{\partial}}_i \,^{\alpha}\phi / \,^{\alpha}\phi^*$ following relations:

$$(\,^{\alpha}w_i[\,^{\alpha}\phi])^* + \,^{\alpha}w_i[\,^{\alpha}\phi](\,^{\alpha}h_4[\,^{\alpha}\phi])^* + \overset{\alpha}{\underline{\partial}}_i \,^{\alpha}h_4[\,^{\alpha}\phi] = 0,$$

$$\overset{\alpha}{\underline{\partial}}_i \,^{\alpha}w_k[\,^{\alpha}\phi] = \overset{\alpha}{\underline{\partial}}_k \,^{\alpha}w_i[\,^{\alpha}\phi],$$

where, for instance, we denoted by $\,^{\alpha}h_4[\,^{\alpha}\phi]$ the functional dependence on $\,^{\alpha}\phi$. Such conditions are always satisfied for $\,^{\alpha}\phi = \,^{\alpha}\phi(v)$ or if $\,^{\alpha}\phi = const$ when $\,^{\alpha}w_i(x^k, v)$ can be any functions with zero $\,^{\alpha}\beta$ and $\,^{\alpha}\alpha_i$, see (19.12).

4 The Main Theorem on Fractional Solitonic Hierarchies

In [6–8], we proved that the geometric data for any fractional metric (in a model of fractional gravity or geometric mechanics) naturally define a N-adapted fractional biHamiltonian flow hierarchy inducing anholonomic fractional solitonic configurations.

Theorem 19.1. *For any N-anholonomic fractional manifold with prescribed fractional d-metric structure, there is a hierarchy of bi-Hamiltonian N-adapted fractional flows of curves $\gamma(\tau, \mathbf{l}) = h\gamma(\tau, \mathbf{l}) + v\gamma(\tau, \mathbf{l})$ described by geometric nonholonomic fractional map equations. The 0 fractional flows are defined as convective (traveling wave) maps*

$$\gamma_\tau = \gamma_\mathbf{l}, \text{ distinguished } (h\gamma)_\tau = (h\gamma)_{h\mathbf{X}} \text{ and } (v\gamma)_\tau = (v\gamma)_{v\mathbf{X}}.$$

There are fractional +1 flows defined as non-stretching mKdV maps

$$-(h\gamma)_\tau = {}^\alpha\mathbf{D}_{h\mathbf{X}}^2 (h\gamma)_{h\mathbf{X}} + \frac{3}{2} | {}^\alpha\mathbf{D}_{h\mathbf{X}} (h\gamma)_{h\mathbf{X}} |_{hg}^2 (h\gamma)_{h\mathbf{X}},$$

$$-(v\gamma)_\tau = {}^\alpha\mathbf{D}_{v\mathbf{X}}^2 (v\gamma)_{v\mathbf{X}} + \frac{3}{2} | {}^\alpha\mathbf{D}_{v\mathbf{X}} (v\gamma)_{v\mathbf{X}} |_{vg}^2 (v\gamma)_{v\mathbf{X}},$$

and fractional +2,... flows as higher order analogs. Finally, the fractional -1 flows are defined by the kernels of recursion fractional operators inducing non-stretching fractional maps ${}^\alpha\mathbf{D}_{h\mathbf{Y}} (h\gamma)_{h\mathbf{X}} = 0$ *and* ${}^\alpha\mathbf{D}_{v\mathbf{Y}} (v\gamma)_{v\mathbf{X}} = 0$.

5 Fractional Caputo N-Anholonomic Manifolds

The fractional left and right Caputo derivatives, respectively, are defined as:

$$_{1x}\overset{\alpha}{\underline{\partial}}_x f(x) := \frac{1}{\Gamma(s-\alpha)} \int_{1x}^{x} (x-x')^{s-\alpha-1} \left(\frac{\partial}{\partial x'}\right)^s f(x')dx';$$

$$_x\overset{\alpha}{\underline{\partial}}_{2x} f(x) := \frac{1}{\Gamma(s-\alpha)} \int_{x}^{2x} (x'-x)^{s-\alpha-1} \left(-\frac{\partial}{\partial x'}\right)^s f(x')dx'. \quad (19.13)$$

We can introduce $\overset{\alpha}{d} := (dx^j)^\alpha {}_0\overset{\alpha}{\underline{\partial}}_j$ for the fractional absolute differential, where $\overset{\alpha}{dx^j} = (dx^j)^\alpha \frac{(x^j)^{1-\alpha}}{\Gamma(2-\alpha)}$ if $_1x^i = 0$. Such formulas allow us to elaborate the concept of fractional tangent bundle $\underline{T}^\alpha M$, for $\alpha \in (0,1)$, associated to a manifold M of necessary smooth class and integer $\dim M = n$.[5]

Let us denote by $L_z(_1x, {}_2x)$ the set of those Lesbegue measurable functions f on $[_1x, {}_2x]$ when $\|f\|_z = (\int_{1x}^{2x} |f(x)|^z dx)^{1/z} < \infty$ and $C^z[_1x, {}_2x]$ be the space of functions which are z times continuously differentiable on this interval. For any real-valued function $f(x)$ defined on a closed interval $[_1x, {}_2x]$, there is a function $F(x) = {}_{1x}\overset{\alpha}{I}_x f(x)$ defined by the fractional Riemann–Liouville integral ${}_{1x}\overset{\alpha}{I}_x f(x) := \frac{1}{\Gamma(\alpha)} \int_{1x}^{x} (x-x')^{\alpha-1} f(x')dx'$, when $f(x) = {}_{1x}\overset{\alpha}{\underline{\partial}}_x F(x)$, for all $x \in [_1x, {}_2x]$, satisfies the conditions

---

[5]For simplicity, we may write both the integer and fractional local coordinates in the form $u^\beta = (x^j, y^a)$. We underlined the symbol T in order to emphasize that we shall associate the approach to a fractional Caputo derivative.

$$_{1x}\overset{\alpha}{\underline{\partial}}_x\left(\,_{1x}\overset{\alpha}{I}_xf(x)\right) = f(x),\ \alpha > 0,$$

$$_{1x}\overset{\alpha}{I}_x\left(\,_{1x}\overset{\alpha}{\underline{\partial}}_xF(x)\right) = F(x) - F(\,_1x),\ 0 < \alpha < 1. \tag{19.14}$$

We can consider fractional (co)-frame bases on $\overset{\alpha}{\underline{T}}M$. For instance, a fractional frame basis $\overset{\alpha}{\underline{e}}_\beta = e_\beta^{\beta'}(u^\beta)\overset{\alpha}{\underline{\partial}}_{\beta'}$ is connected via a vierlbein transform $e_\beta^{\beta'}(u^\beta)$ with a fractional local coordinate basis

$$\overset{\alpha}{\underline{\partial}}_{\beta'} = \left(\overset{\alpha}{\underline{\partial}}_{j'} = \,_{1x^{j'}}\overset{\alpha}{\underline{\partial}}_{j'}, \overset{\alpha}{\underline{\partial}}_{b'} = \,_{1y^{b'}}\overset{\alpha}{\underline{\partial}}_{b'}\right), \tag{19.15}$$

for $j' = 1, 2, ..., n$ and $b' = n+1, n+2, ..., n+n$. The fractional cobases are related via $\overset{\alpha}{\underline{e}}^{\beta} = e_{\beta'}^{\beta}(u^\beta)\overset{\alpha}{du}^{\beta'}$, where $\overset{\alpha}{du}^{\beta'} = \left((dx^{i'})^\alpha, (dy^{a'})^\alpha\right)$.

The fractional absolute differential $\overset{\alpha}{d}$ is written as $\overset{\alpha}{d} := (dx^j)^\alpha\,_0\overset{\alpha}{\underline{\partial}}_j$, where $\overset{\alpha}{dx}^j = (dx^j)^\alpha\frac{(x^j)^{1-\alpha}}{\Gamma(2-\alpha)}$, where we consider $_1x^i = 0$.

Acknowledgements This chapter summarizes the results presented in our talk at the Conference "New Trends in Nanotechnology and Nonlinear Dynamical Systems," 25–27 July, 2010, Çankaya University, Ankara, Turkey.

References

1. Vacaru S. Fractional nonholonomic Ricci flows, arXiv: 1004.0625
2. Vacaru S. Fractional dynamics from Einstein gravity, general solutions, and black holes, arXiv: 1004.0628
3. Baleanu D, Vacaru S (2010) Fractional almost Kahler–Lagrange geometry, published online in: Nonlinear Dyn. arXiv: 1006.5535
4. Baleanu D, Vacaru S (2011) Fedosov quantization of fractional Lagrange spaces. Int J Theor Phys 50:233–243
5. Baleanu D, Vacaru S (2011) Constant curvature coefficients and exact solutions in fractional gravity and geometric mechanics. Centr Eur J Phys 9; arXiv: 1007.2864
6. Baleanu D, Vacaru S. Fractional curve flows and solitonic hierarchies in gravity and geometric mechanics. arXiv: 1007.2866
7. Vacaru S (2010) Curve flows and solitonic hierarchies generated by Einstein metrics. Acta Appl Math 110:73–107
8. Anco S, Vacaru S (2009) Curve flows in Lagrange–Finsler geometry, bi-Hamiltonian structures and solitons. J Geom Phys 59:79–103

Part IV
Fractional Order Modeling

Chapter 20
Front Propagation in an Autocatalytic Reaction-Subdiffusion System

Igor M. Sokolov and Daniela Froemberg

1 Introduction

The theory of reactions controlled by subdiffusion attracted much interest in the recent few years both because of practical needs (reactions in porous media and geological formations, and in crowded cellular environments) and because of very unusual mathematical structure of the corresponding equations. In what follows, we concentrate on the A+B→2A autocatalytic reaction leading to a propagation of the pulled front into the unstable B-domain [3]. Under normal diffusion, the reaction is described by the Fisher–Kolmogorov–Petovskii–Piskunov (FKPP) equation which is mathematically well understood. In the case when all particles were initially B and were distributed in space with the constant concentration B_0 (which will be put to unity in what follows), the equation for the concentration B of B reads

$$\frac{\partial B}{\partial t} = D\Delta B - k(B_0 - B)B. \tag{20.1}$$

Here, D is the diffusion coefficient of the particles (assumed equal for A and B particles), and k is the reaction rate. The two initial reaction-diffusion equations for A and for B particles are reduced to (20.1) by using the conservation law $A + B = B_0$ following from the stoichiometry. The initial condition corresponds to a droplet of A particles introduced at the origin. We concentrate on the reaction front propagating to the right.

Assuming the front to propagate with a constant velocity along the x-direction, one changes to a comoving frame thus obtaining an ordinary differential equation for the stable front form. This equation can be linearized close to the leading edge of the front, and the spectrum of possible velocities is obtained by requesting

I.M. Sokolov (✉) • D. Froemberg
Institut für Physik, Humboldt-Universität zu Berlin, Newtonstraße 15, D-12489 Berlin, Germany
e-mail: igor.sokolov@physik.hu-berlin.de; daniela@physik.hu-berlin.de

D. Baleanu et al. (eds.), *Fractional Dynamics and Control*,
DOI 10.1007/978-1-4614-0457-6_20, © Springer Science+Business Media, LLC 2012

the concentration to be nonnegative everywhere. This condition defines then the minimal possible propagation velocity. The fact that this minimal propagation velocity is the one really attained under a sharp initial condition (marginal stability principle) follows from the stability analysis of perturbations.

The minimal propagation velocity in FKPP front is $v = 2\sqrt{DkB_0}$, and the characteristic width of the propagating front is $w \simeq \sqrt{D/kB_0}$. In what follows, we confine ourselves to a one-dimensional situation, where the concentrations A, B, and B_0 have the dimension of the inverse length L^{-1}, and the reaction rate coefficient k has a dimension of $[k] = LT^{-1}$.

The case of subdiffusion is much more complicated. First, different types of subdiffusive behaviors are possible, corresponding either to disordered systems (percolation, energetic disorder, etc.) or to systems with slow modes (polymers). Second, even if the model of the subdiffusion is fixed, for example, to be continuous time random walk (CTRW) with the power-law probability density function (PDF) of waiting times

$$\psi(t) \simeq \frac{\tau^\alpha}{t^{1+\alpha}} \tag{20.2}$$

for $t > \tau$, different reaction-subdiffusion equations emerge when considering situations when the "internal clock" of the particle is reset after the reaction or not [1]. In what follows, we consider the situation when the diffusion on the large scales is hindered, but the small-scale reactions follow the mass action law. The reaction does not reset waiting times, since the last one describes only the large-scale behavior of the system. The simplest case of this situation (the isomerization reaction A→B) was discussed in [5] and leads to equations where the reaction term is not simply added to a subdiffusion equation like it is the case in (20.1) but enters as well the transport operator. The corresponding equation for the A+B→2A reaction was derived in [2], where the analysis of the corresponding reaction-subdiffusion equation showed that the minimal propagation velocity is zero, which fact was interpreted as propagation arrest. The situation was clarified in numerical simulations of [4], where two propagation regimes were identified, both corresponding to propagation with the velocity which decays with time. Thus, for small reaction rates and low concentrations the front velocity behaved as $v(t) \propto t^{(\alpha-1)/2}$, where α is the subdiffusion exponent (governing the mean squared displacement of a particle $\langle x^2(t) \rangle \propto t^\alpha$ in the reaction-free case), while for larger concentrations and high reaction rates (a fluctuation-dominated regime which cannot be described within continuous reaction-diffusion or reaction-subdiffusion equations), the propagation velocity behaves as $v \propto t^{\alpha-1}$.

The longer numerical simulations of [1] identified the second, but not the first propagation regime, claiming that the first one "is not robust." In what follows, we discuss this situation in some detail and present new results on the front propagation in such a system. We show that the continuous description of the reaction-subdiffusion reaction gives hints in favor of propagation of a front with a velocity $v(t) \propto t^{(\alpha-1)/2}$. The intermediate asymptotic behavior $v(t) \propto t^{(\alpha-1)/2}$ holds as long as the continuous description of the reaction is possible. The front width, however, decays with time as $w \propto t^{\frac{\alpha-1}{2}}$, and gets of the order of one of the microscopic scales of the problem, so that the continuous description inevitably breaks down at longer time. Then another, final asymptotic behavior ($v \propto t^{\alpha-1}$ for a one-dimensional case) sets in. The present contribution reports on a work in progress; however, we are highly confident that the overall physical understanding is reached, although a large amount of computations still has to be performed.

2   Model and Equations

Let us consider the medium as consisting of compartments of size a. The transport of particles between these compartments is governed by CTRW: the waiting time of a particle in a compartment is given by the PDF (20.2). Within a compartment i the particles react according to classical kinetic law, i.e. the transformation from B to A follows at the rate kA_iB_i, where A_i and B_i are the numbers of the particles within the compartment, and k is the properly renormalized reaction rate constant.

Following the same procedure as in [2,5] we start from the balance equation for B-particles:

$$\dot{B}_i(t) = \frac{1}{2}j_{i-1}^-(t) + \frac{1}{2}j_{i+1}^-(t) - j_i^-(t) - \kappa A_i(t)B_i(t),$$

where $j_i^-(t)$ is the loss flux from the compartment i given by

$$j_i^-(t) = \psi(t)P_s(t,0)B_i(0) + \int_0^t \psi(t-t')P_s(t,t')\left[\dot{B}_i(t') + j_i^-(t') + \kappa A_i(t')B_i(t')\right]dt'$$

and

$$P_s(t,t') = \exp\left[-\kappa \int_{t'}^t A_i(t'')dt''\right]$$

is the survival probability of a B particle. Since the equation for the loss flux only involves the concentrations at one site, it can be easily solved by means of Laplace transform, and the solution can be inserted into the first equation for the concentration. The equation for A follows in a similar way. Afterwards, the transition to the continuous limit in space is performed leading to the following equation for B

$$\frac{\partial}{\partial t}B(x,t) = -k[1 - B(x,t)]B(x,t) + \frac{a^2}{2}\Delta \int_0^t M(t-t')$$

$$\times B(x,t')\exp\left[-\int_{t'}^t k[1 - B(x,t'')]dt''\right]dt', \qquad (20.3)$$

where we used the conservation law $A(x,t) = 1 - B(x,t)$, and with $M(t)$ given by the inverse Laplace transform of $M(u) = u\psi(u)/[1 - \psi(u)]$. The equation for A follows from the one for B by using the conservation law. For the Markovian process with $\psi(t) = \tau^{-1}\exp(-t/\tau)$ one obtains $M(t)$ to be a δ-function, $M(t) = \tau^{-1}\delta(t)$ and the equations for the concentrations reduce to partial differential equations, the FKPP case. For subdiffusion with waiting time density following (20.2), the integral operator with the $M(t)$-kernel is proportional to the Riemann–Liouville fractional derivative of order $1 - \alpha$. Thus, in a subdiffusive case the equations for A and B are nonlinear fractional partial differential equations of quite a complex structure.

3   Absence of the Constant Front Velocity

Let us give a short sketch of the calculations done in [2] and leading to the conclusions that no front propagation at a constant velocity is possible. Assuming that $A(x,t)$ is small at the very far edge of the front and linearizing the reaction-subdiffusion equation for $A(x,t)$, one can look for an exponential solution of the form $A = A_0 \exp[-\lambda(x - vt)]$. This solution has to satisfy the equation

$$\lambda v\left(A_0 \exp[-\lambda(x - vt)]\right) = -kA_0 \exp[-\lambda(x - vt)]$$
$$+ \frac{a^2}{2}A_0\left[-\lambda^2 + \frac{k\lambda}{v}\right]\int_0^t M(t - t')\exp[-\lambda(x - vt')]\,dt'$$
$$- \frac{a^2}{2}\frac{k\lambda}{v}A_0 \exp[-\lambda(x - vt)]\int_0^t M(t - t')\,dt'. \qquad (20.4)$$

For the Markovian case, the standard expression for the minimal velocity of the stable propagation is reproduced: Taking $M(t) = \tau^{-1}\delta(t)$, introducing a new variable $z = x - vt$, and concentrating on the leading edge of the front ($z \to \infty$), we find the dispersion relation:

$$\frac{a^2}{2\tau}\lambda^2 - v\lambda + k = 0. \qquad (20.5)$$

The quadratic equation (20.5) has two complex conjugated roots. Since the roots λ corresponding to the propagating front need to stay real (to prevent concentration from taking negative values which are inevitable if the solutions oscillate), the condition $v \geq v_{\min} = 2\sqrt{a^2 k/2\tau} \equiv 2\sqrt{Dk}$ follows for the propagation velocity, with $D = a^2/2\tau$ being the diffusion coefficient. In this case the two roots $v = \pm v(\lambda)$ correspond to the two possible directions of the front propagation.

For waiting time PDFs decaying as a power law, $\psi(t) \propto t^{-1-\alpha}$, $0 < \alpha < 1$ for large t, we find with $\hat{t} = t - t'$ that $\int_0^t M(t - t')\exp[\lambda vt']\,dt' = \exp[\lambda vt]\tilde{M}(\lambda v)$ and

$$R(t) = \int_0^t M(t - t')\,dt' = \frac{const}{\tau^\alpha}t^{\alpha-1}, \qquad (20.6)$$

so that the last term in (20.4) vanishes for large t. Note that the integral $R(t)$, (20.6), gives the rate of jumps of a particle performing CTRW. Finally, with $z = x - vt$ we get

$$-\lambda v A_0 \exp[-\lambda z] = -k A_0 \exp[-\lambda z] + \frac{a^2}{2} A_0 \exp[-\lambda z] \left[\left(-\lambda^2 + \frac{k\lambda}{v} \right) \tilde{M}(\lambda v) \right],$$

from which the dispersion relation

$$0 = -\lambda v + k + \frac{a^2}{2} \left(\lambda^2 - \frac{k\lambda}{v} \right) \tilde{M}(\lambda v)$$

follows. Taking $\tilde{M}(u) = \tau^{-\alpha} u^{1-\alpha}$ this last one can be put into the form

$$(v\lambda - k) \left(\frac{a^2}{2\tau^\alpha} \lambda^{2-\alpha} v^{-\alpha} - 1 \right) = 0$$

and possesses two nonnegative roots for any $v \geq 0$, at variance with the Markovian case, where such roots exist only for $v > v_{min}$. This finding means that the minimal propagation velocity in this case is zero, so that the front velocity tends to zero in the course of time.

4 Front Moving at a Decaying Velocity

Numerical simulations of [4] suggest that the front does propagate, but its propagation velocity decays in the course of time. Assuming the constant front form, one could imagine that the asymptotic solution of the linearized equation could, for example, follow the pattern

$$A(x,t) = A_0 \exp\left[-\lambda_0 \left(x - v_0 t^{\frac{\alpha-1}{2}} \right) \right] = A_0 \exp(-\lambda_0 z) \tag{20.7}$$

with the constant v_0 indicating now the *subvelocity* of the front. Here, $z = x - v_0 t^{\frac{1+\alpha}{2}}$ is the variable defining the comoving frame of the front whose velocity decays as $v \propto t^{(\alpha-1)/2}$. However, substitution of (20.7) into (20.4) shows that (20.7) is not an appropriate ansatz at all. Thus, the subdiffusion analog of the FKPP equation does not possess a front solution of constant form with the velocity decaying as $v \propto t^{(\alpha-1)/2}$.

Interestingly enough, a different form of the solution is possible, the one with decaying width:

$$A(x,t) = A_0 \exp\left[-\lambda_0 t^{\frac{1-\alpha}{2}} \left(x - v_0 t^{\frac{1+\alpha}{2}} \right) \right], \tag{20.8}$$

where $\lambda(t) = \lambda_0 t^{\frac{1-\alpha}{2}}$ gives the time-dependent width of the front. Proceeding as in [2] we have in first order for the A particles:

$$
\frac{\partial A(x,t)}{\partial t} \approx kA(x,t) + \frac{a^2}{2} \int_0^t M(t-t')\Delta \exp\left[-\lambda_0 t'^{\frac{1-\alpha}{2}}\left(x - v_0 t'^{\frac{\alpha+1}{2}}\right)\right] dt'
$$

$$
+ \frac{a^2}{2} \int_0^t M(t-t')k \times \int_{t'}^t \Delta \exp\left[-\lambda_0 t''^{\frac{1-\alpha}{2}}\left(x - v_0 t''^{\frac{1+\alpha}{2}}\right)\right] dt'' \, dt'
$$

Evaluating the integrals we get for both $z = x - v_0 t^{\frac{1+\alpha}{2}}$ and t large:

$$
\lambda_0 v_0 \exp\left[-\lambda_0 t^{\frac{1-\alpha}{2}}\left(x - v_0 t^{\frac{1+\alpha}{2}}\right)\right] = \exp\left[-\lambda_0 t^{\frac{1-\alpha}{2}}\left(x - v_0 t^{\frac{1+\alpha}{2}}\right)\right]
$$

$$
\times \left(\frac{a^2}{2\Gamma(\alpha)\Gamma(1-\alpha)\tau^\alpha}\left(C\lambda_0^2 + \frac{k\lambda_0}{v_0}(1-C)\right) + k\right),
$$

where C is a constant (depending on the parameters of the model) for which the inequality $B(\alpha, 2 - \alpha) \geq C \geq 0$ holds. The upper bound $B(\alpha, 2 - \alpha)$ is the β function. This yields the dispersion relation for λ_0:

$$
0 = \lambda_0^2 - \frac{kK_\alpha^*(1-C)/v_0 - v_0}{K_\alpha^* C}\lambda_0 + \frac{k}{K_\alpha^* C},
$$

with $K_\alpha^* = a^2/2\Gamma(\alpha)\Gamma(1-\alpha)\tau^\alpha = K_\alpha/\Gamma(\alpha)$, where K_α is the generalized diffusion constant. Solving this equation for λ, we find a restriction on the values of v_0 for which this λ is real: $(kK_\alpha^*/v_0(1-C) - v_0)^2 \geq 4kK_\alpha^* C$, a quartic equation in v_0 which yields in general four symmetric roots

$$
v_0^2 = K_\alpha^* k \left[1 + C \pm 2\sqrt{C}\right]. \tag{20.9}
$$

In the FKPP case pertinent to normal diffusion, the value of C is $C = 1$, the minimal front velocity $v_{\min} = \pm 2\sqrt{Dk}$ is reproduced; the other solution is a double root $v = 0$, which is a nonphysical one and appears due to the overall higher order of the dispersion relation obtained by this method. For any C other than $C = 1$, there exists bounded domain of real roots around zero, $-v_- \leq v_0 \leq v_-$ (the subscript "$-$" corresponds to the minus sign in (20.9)) separated by gaps from another domain of real roots $|v_0| > v_+$. The existence of the gap and of the corresponding minimal velocity can be interpreted in favor of propagation of the corresponding front. Of course, such an analysis is still incomplete without looking at the stability of corresponding perturbations. There exists, however, a strong physical argument in favor of the existence of the propagation mode described above.

5 The Crossover Argument

In order to gain intuition about the front's behavior, we make use of the following idea: for any waiting time PDF $\psi(t)$ with finite mean waiting time $\langle t \rangle$, the classical (FKPP) behavior is recovered if only the time t is large enough, $t \gg \langle t \rangle$. We, therefore, consider a truncated power-law waiting time distribution

$$\psi_T(t) = \frac{\tau^\alpha(\tau+T)^\alpha}{(\tau+T)^\alpha - \tau^\alpha} \frac{\alpha}{(\tau+t)^{1+\alpha}} \Theta(T-t)$$

with $T \gg \tau$ which possesses a mean

$$\langle t \rangle = \frac{\alpha T \tau + \tau(\tau^\alpha - (T-\tau)^\alpha)}{(\alpha-1)(\tau^\alpha - (T-\tau)^\alpha)}.$$

For $T \gg \tau$, $\langle t \rangle \approx \frac{\alpha}{1-\alpha}\tau^\alpha T^{1-\alpha}$. For short times $\tau < t \ll T$, when particles cannot yet feel the cutoff, this distribution is practically a power law (20.2), and the behavior of the front velocity will be similar to that in subdiffusion, whereas for large times the behavior is the classical one with a constant velocity given by the minimal propagation velocity in FKPP. There has to be a crossover at a time t_{cr} between these two regimes. Thus, we assume that in the anomalous domain the velocity is time-dependent, $v_{SD} \propto t^\beta$, and that after a crossover to normal behavior $v_D = const \sim \sqrt{kD(T)}$ sets on. Here, $D(T)$ is the final diffusion coefficient, $D(T) = a^2/2\langle t \rangle$ depending on the cutoff time T. The subscripts SD and D indicate the regimes of subdiffusion and of normal diffusion, respectively. The number of performed steps, a measure of mobility, is $n_D(t) = t/\langle t \rangle$ in the normal regime $t \gg t_{cr}$, and $n_{SD}(t) = (\Gamma[1+\alpha]\tau^\alpha)^{-1}t^\alpha$ in the subdiffusive regime $t \ll t_{cr}$. By equating $n_{SD}(t_{cr}) = n_D(t_{cr})$ at the two sides of the crossover, we find $t_{cr} \propto T$. The velocities on both sides at the crossover time have to be of the same order of magnitude and, therefore, $v_{SD}(t_{cr}) \propto t_{cr}^\beta \propto \sqrt{kD(T)}$. Since $t_{cr} \sim T$ and $D(T) \sim T^{\alpha-1}$, we get $\beta = (1-\alpha)/2$ and thus

$$v(t) \propto t^{\frac{\alpha-1}{2}}$$

in the subdiffusive regime $t \ll t_{cr}$.

The same argument applies to the front's width. The width of the front in the normal FKPP regime is of the order of $w \simeq D/v = \sqrt{D/k}$. Taking the width of the front to behave as $w(t) \propto t^\gamma$ for $t < t_{cr}$ and matching this width with the width of the front in FKPP at t_{cr} we get $w \propto t^{\frac{\alpha-1}{2}}$, in accordance with the previous section.

6 Breakdown of the Continuous Description and Final Asymptotics

Since the subdiffusive front is not only slowing down but also becomes steeper in the course of time, a Monte-Carlo simulation, if performed long enough, enters a regime where the width of the front is comparable to the one of microscopic scales of the problem, the compartment size a or the interparticle distance $B_0^{-1} = 1$. The first happens at high concentrations, where there are many particles per compartment, the second at low concentrations. Both situations lead to similar behavior of the front's velocity.

Since in continuous time random walks the rate of the particle's jumps $R(t)$ decays in the course of time, at longer time one enters the regime, when the mean time between the two jumps of the particles within the front region gets large compared to the time of the order of kB_0 necessary for full conversion of all particles from B to A in a compartment where at least one A particle is present. Within this picture, the front can be considered as "atomically sharp," and is placed exactly between the last compartment containing A particles and the first A-free compartment. This front moves exactly one a-step forward, when an A particle from the compartment left of the front makes its jump to the right. Since the rate of these jumps is proportional to the number of the particles in the compartment aB_0 and to $R(t)$, the velocity goes as

$$v \propto a^2 B_0 \left(\frac{t^{\alpha-1}}{\tau^\alpha} \right) \propto B_0 K_\alpha t^{1-\alpha}.$$

This is the situation pertinent to high concentration of particles.

The case of low concentrations (much less then one particle per compartment) needs for a slightly different discussion, parallel to the one in [6]. The front is again "atomically sharp": If the A and the B particles meet in the same compartment, they have enough time to react before making a jump, and therefore there are no B particles to the left of the front position and no A particles to the right of it. The front position can be associated with the one of the rightmost A. This one does not change at the average as long as there is a single particle in a compartment (the jumps to the left and to the right are equally probable), but does increase by a with probability 1/2 if another particle is present in the same compartment (the probability of which is of the order of aB_0) since in this case the front cannot jump back. The mean velocity differs from the previous one only in prefactor: $v \propto (a/2)aB_0 t^{\alpha-1}/\tau^\alpha \propto B_0 K_\alpha t^{1-\alpha}$.

The result $v \propto B_0 K_\alpha t^{1-\alpha}$ is exactly what follows immediately from the dimensional analysis, if one assumes that the reaction is infinitely fast on the time scale of jumps and therefore the reaction rate coefficient k cannot play any role.

7   Conclusions

We discussed the motion of a reaction front in the A+B→2A reaction under subdiffusion in a system where the transport of the particles is described by continuous time random walks but the reaction between them locally follows the mass action law. We show, that the reaction front in such a system moves at intermediate times at a decaying velocity $v \propto t^{(\alpha-1)/2}$, and that this velocity has to cross over to a faster decay $v \propto t^{\alpha-1}$ in the asymptotics of very long times.

Acknowledgements This work was supported by DFG within TP A10 of joint collaboration grant SFB 555.

References

1. Campos D, Mendez V (2009) Nonuniversality and the role of tails in reaction-subdiffusion fronts. Phys Rev E 80:021133
2. Froemberg D, Schmidt-Martens H, Sokolov IM, Sagués F (2008) Front Propagation in A+B→2A reaction under subdiffusion. Phys Rev E 78:011128
3. Panja D (2004) Effects of fluctuations on propagating fronts. Phys Rep 393:87
4. Schmidt-Martens HH, Froemberg D, Sokolov IM, Sagués F (2009) Front propagation in a one-dimensional autocatalytic reaction-subdiffusion system. Phys Rev E 79:041135
5. Sokolov IM, Schmidt MGW, Sagués F (2006) Reaction-subdiffusion equations. Phys Rev E 73:031102
6. Warren CP, Mikus G, Somfai E, Sander LM (2001) Fluctuation effects in an epidemic model. Phys Rev E 63:056103

Chapter 21
Numerical Solution of a Two-Dimensional Anomalous Diffusion Problem

Necati Özdemir and Derya Avcı

1 Introduction

In the last decade, there has been a considerable interest to the applications of fractional calculus such that many processes in the nature have been successfully modeled by a set of axioms, definitions, and methods of fractional calculus (see [1–4]). One of these processes is anomalous diffusion which is a phenomenon occurs in complex and nonhomogeneous mediums. The phenomenon of anomalous diffusion may be based on generalized diffusion equation which contains fractional order space and/or time derivatives [5]. Turski et al. [6] presented the occurrence of the anomalous diffusion from the physical point of view and also explained the effects of fractional derivatives in space and/or time to diffusion propagation. Agrawal [7] represented an analytical technique using eigenfunctions for a fractional diffusion-wave system and therefore provided that this formulation could be applied to all those systems for which the existence of eigenmodes is guaranteed. Agrawal [8] also formulated a general solution using finite sine transform technique for a fractional diffusion-wave equation in a bounded domain whose fractional term was described in sense of Caputo. Herzallah et al. [9] researched the solution of a fractional diffusion wave model which is more accurate and provides the existence, uniqueness, and continuation of the solution. Huang and Liu [10] considered a sort of generalized diffusion equation which is defined as a space-time fractional diffusion equation in sense of Caputo and Riemann-Liouville operators. In addition, Huang and Liu [11] found the fundamental solution of the space-time fractional advection-dispersion equation with Riesz–Feller derivative. Langlands [12] proposed a modified fractional diffusion equation on an infinite domain and therefore found the solution as an infinite series of Fox functions. Sokolov et al. [13] analyzed different types of distributed-order fractional diffusion

N. Özdemir (✉) • D. Avcı
Department of Mathematics, Balıkesir University, Balıkesir, Turkey
e-mail: nozdemir@balikesir.edu.tr; dkaradeniz@balikesir.edu.tr

D. Baleanu et al. (eds.), *Fractional Dynamics and Control*,
DOI 10.1007/978-1-4614-0457-6_21, © Springer Science+Business Media, LLC 2012

equations and investigated the effects of different classes of such equations. Saichev and Zaslavsky [14] presented the solutions of a symmetrized fractional diffusion equation with a source term applying a method similar to separation of variables. Mainardi et al. [15] researched the fundamental solution of a Cauchy problem for the space-time fractional diffusion equation obtained from the standard diffusion equation by replacing the second-order space derivative by a fractional Riesz or Riesz–Feller derivative, and the first-order time derivative by a fractional Caputo derivative. Gorenflo and Mainardi [16, 17] analyzed a space-fractional (or Levy–Feller) diffusion process governed by a generalized diffusion equation which generates all Levy stable probability distributions and also approximated these processes by random walk models, discreted space and time based on Gr ünwald-Letnikov (GL) approximation. Özdemir et al. [18] presented the numerical solution of a diffusion-wave problem in polar coordinates using GL approximation. Özdemir and Karadeniz [19] also applied GL formula to find the numerical results for a diffusion problem in cylindrical coordinates. Povstenko [20–23] researched the solutions of axial-symmetric fractional diffusion-wave equations in cylindrical and spherical coordinates.

In addition, numerical schemes are fine research topics in fractional calculus. Because the analytical solutions of the fractional differential equations are usually obtained in terms of Green and Fox functions which are difficult to calculate explicitly. For this reason, there are many research related to numerical approximation of space or space-time fractional diffusion equations. Shen and Liu [24] investigated the error analysis of the numerical solution of a space fractional diffusion equation obtained using an explicit finite difference method. Liu et al. [25] formulated the numerical solution of a space-time fractional advection-dispersion equation in terms of Caputo and RL derivatives using an implicit and an explicit difference methods. Lin et al. [26] considered a nonlinear fractional diffusion equation in terms of generalized Riesz fractional derivative and applied an explicit finite-difference method to find numerical solutions. Özdemir et al. [27] researched the numerical solutions of a two-dimensional space-time fractional diffusion equation in terms of Caputo and Riesz derivatives. Ciesielski and Leszczynski [28] proposed a new numerical method for the spatial derivative called Riesz–Feller operator, and hence found the numerical solutions to a fractional partial differential equation which describe an initial-boundary value problem in one-dimensional space. Ciesielski and Leszczynski [29] also presented the numerical solutions of a boundary value problem for an equation with the Riesz–Feller derivative. Liu et al. [30] presented a random walk model for approximating a Levy–Feller advection-dispersion process and proposed an explicit finite difference approximation for Levy–Feller advection-dispersion process, resulting from the GL discretization of fractional derivatives. Zhang et al. [31] considered the Levy–Feller diffusion equation and investigated their probabilistic interpretation and numerical analysis in a bounded spatial domain. Moreover, Machado [32] presented a probabilistic interpretation to the fractional-order derivatives.

The plan of this work as follows. In this work, we consider a two-dimensional anomalous diffusion problem in terms of Caputo and Riesz–Feller derivatives.

For this purpose, we give some basic definitions necessary for our formulations in Sect. 2. In Sect. 3, we formulate our considerations and find the analytical solution of the problem. We apply GL definition to find the numerical solution in Sect. 4. In Sect. 5, we choose an example and therefore show the effectiveness of the numerical approximation for our problem. Finally, we conclude our work in Sect. 6.

2 Mathematical Background

In this work, we consider an anomalous diffusion equation in two-dimensional space. We define our problem in terms of Caputo time and Riesz–Feller fractional derivatives. Therefore, let we remind the well-known definitions and origins of these operators.

Originally, Riesz introduced the pseudo-differential operator $_xI_0^\alpha$ whose symbol is $|\kappa|^{-\alpha}$, well defined for any positive α with the exclusion of odd integer numbers, then was called Riesz Potential. The Riesz fractional derivative $_xD_0^\alpha = -\,_xI_0^\alpha$ defined by analytical continuation can be represented as follows:

$$
\begin{aligned}
_xD_0^\alpha &= -|\kappa|^\alpha \\
&= -\left(\kappa^2\right)^{\frac{\alpha}{2}} \\
&= -\left(-\frac{d^2}{dx^2}\right)^{\frac{\alpha}{2}}.
\end{aligned}
\tag{21.1}
$$

In addition, Feller [33] generalized the Riesz fractional derivative to include the skewness parameter θ of the strictly stable densities. Feller showed that the pseudo-differential operator D_θ^α is as the inverse to the Feller potential, which is a linear combination of two Riemann–Liouville (or Weyl) integrals:

$$
_xI_+^\alpha f(x) = \frac{1}{\Gamma(\alpha)} \int_{-\infty}^{x} (x-\xi)^{\alpha-1} f(\xi)\,d\xi,
\tag{21.2}
$$

$$
_xI_-^\alpha f(x) = \frac{1}{\Gamma(\alpha)} \int_{x}^{+\infty} (\xi-x)^{\alpha-1} f(\xi)\,d\xi,
\tag{21.3}
$$

where $\alpha > 0$. By these definitions, the Feller potential can be defined as follows:

$$
_xI_\theta^\alpha f(x) = c_+(\alpha,\theta)\,_xI_+^\alpha f(x) + c_-(\alpha,\theta)\,_xI_-^\alpha f(x),
\tag{21.4}
$$

where the real parameters α and θ are always restricted as follows:

$$
0 < \alpha \le 2,\ \alpha \ne 1,
$$

$$
|\theta| \le \min\{\alpha, 2-\alpha\},
$$

and also the coefficients are

$$c_+ (\alpha, \theta) = \frac{\sin \left(\frac{(\alpha - \theta)\pi}{2} \right)}{\sin (\alpha \pi)},$$

$$c_- (\alpha, \theta) = \frac{\sin \left(\frac{(\alpha + \theta)\pi}{2} \right)}{\sin (\alpha \pi)}. \tag{21.5}$$

Using the Feller potential, Mainardi and Gorenflo [16] defined the Riesz–Feller derivative

$$\frac{\partial^\alpha f(x)}{\partial |x|_\theta^\alpha} = -_x I_\theta^{-\alpha} f(x) = - \left[c_+ (\alpha, \theta)_x D_+^\alpha f(x) + c_- (\alpha, \theta)_x D_-^\alpha f(x) \right],$$

where $_x D_\pm^\alpha f(x)$ are Weyl fractional derivatives defined as follows:

$$_x D_\pm^\alpha f(x) = \begin{cases} \pm \frac{d}{dx} \left[_x I_\pm^{1-\alpha} f(x) \right], & 0 < \alpha < 1, \\ \frac{d^2}{dx^2} \left[_x I_\pm^{2-\alpha} f(x) \right], & 1 < \alpha \le 2. \end{cases} \tag{21.6}$$

The Caputo fractional derivative is defined as follows:

$$\frac{\partial^\beta u(t)}{\partial t^\beta} = \frac{1}{\Gamma(n-\beta)} \int_0^t (t - \tau)^{n-\beta-1} \left(\frac{d}{d\tau} \right)^n u(\tau) \, d\tau, \tag{21.7}$$

where $0 < \beta \le n, n \in \mathbb{Z}$. Now, we can formulate our problem after these preliminaries.

3   Formulation of the Main Problem

Let us consider the following space-time fractional anomalous diffusion problem:

$$\frac{\partial^\beta u(x,y,t)}{\partial t^\beta} = \frac{\partial^\alpha u(x,y,t)}{\partial |x|_{\theta_1}^\alpha} + \frac{\partial^\mu u(x,y,t)}{\partial |y|_{\theta_2}^\mu}, \tag{21.8}$$

$$u(x,y,0) = u_0(x,y), \tag{21.9}$$

$$\lim_{x,y \to \pm\infty} u(x,y,t) = 0, \tag{21.10}$$

where $x, y \in \mathbb{R}$; β, α, μ are real parameters restricted as $0 < \beta \le 1$, $0 < \alpha < 1$, $1 < \mu \le 2$; the skewness parameters θ_1 ($\theta_1 \le \min\{\alpha, 1 - \alpha\}$) and θ_2 ($\theta_2 \le \min\{\mu, 2 - \mu\}$) are measures of the asymmetry of the probability distribution of a real-valued random variable among the x and y coordinate axes. Note that many simplistic mathematical models are defined under the Gaussian (normal) distribution; i.e., the skewness parameter is zero. However, in reality, random variables may not distribute symmetrically. Therefore, the behavior of such anomalous diffusion problem differs with the changing of θ_1 and θ_2 parameters. We first assume that the solution and the initial condition functions can be expanded into the complex Fourier series, respectively:

$$u(x,y,t) = \sum_{n=1}^{\infty} \sum_{m=1}^{\infty} u_{nm}(t) e^{inx} e^{imy}, \tag{21.11}$$

$$u_0(x,y) = \sum_{n=1}^{\infty} \sum_{m=1}^{\infty} u_{0nm} e^{inx} e^{imy}, \tag{21.12}$$

where $i^2 = -1$. Under these assumptions, we calculate the fractional derivative terms in the right-hand side of (21.8), respectively, as follows: We start with the calculation of $\frac{\partial^{\alpha} u(x,y,t)}{\partial |x|_{\theta_1}^{\alpha}}$ term which dependent on x variable and $0 < \alpha < 1$. Let us remind the definition:

$$\frac{\partial^{\alpha} u(x,y,t)}{\partial |x|_{\theta_1}^{\alpha}} = -\left[c_+(\alpha,\theta_1)_{-\infty}D_x^{\alpha} u(x,y,t) + c_-(\alpha,\theta_1)_x D_{+\infty}^{\alpha} u(x,y,t)\right], \tag{21.13}$$

where

$$_{-\infty}D_x^{\alpha} u(x,y,t) = \frac{\partial}{\partial x}\left(\frac{1}{\Gamma(1-\alpha)}\int_{-\infty}^{x}\frac{u(\xi,y,t)}{(x-\xi)^{\alpha}}d\xi\right) \tag{21.14}$$

and

$$_x D_{+\infty}^{\alpha} u(x,y,t) = -\frac{\partial}{\partial x}\left(\frac{1}{\Gamma(1-\alpha)}\int_{x}^{\infty}\frac{u(\xi,y,t)}{(\xi-x)^{\alpha}}d\xi\right) \tag{21.15}$$

are the left- and the right-side Weyl fractional derivatives. Now, substituting (21.11) into (21.14), we have

$$_{-\infty}D_x^{\alpha} u(x,y,t) = \frac{\partial}{\partial x}\left(\frac{1}{\Gamma(1-\alpha)}\sum_{n=1}^{\infty}\sum_{m=1}^{\infty} u_{nm}(t) e^{imy}\int_{-\infty}^{x}\frac{e^{in\xi}}{(x-\xi)^{\alpha}}d\xi\right)$$

$$= \frac{1}{\Gamma(1-\alpha)}\sum_{n=1}^{\infty}\sum_{m=1}^{\infty} u_{nm}(t) e^{imy}\frac{d}{dx}\left(e^{inx}\int_{0}^{\infty}\frac{e^{-inr}}{r^{\alpha}}dr\right)$$

$$= \frac{1}{\Gamma(1-\alpha)}\sum_{n=1}^{\infty}\sum_{m=1}^{\infty} u_{nm}(t) e^{imy}\frac{d}{dx}\left(e^{inx}(in)^{\alpha-1}\Gamma(1-\alpha)\right)$$

$$= \sum_{n=1}^{\infty}\sum_{m=1}^{\infty}(in)^{\alpha} u_{nm}(t) e^{imy} e^{inx}$$

and with the similar manipulations,

$$_x D_{+\infty}^{\alpha} u(x,y,t) = \sum_{n=1}^{\infty}\sum_{m=1}^{\infty}(-in)^{\alpha} u_{nm}(t) e^{imy} e^{inx}.$$

Hence, for $0 < \alpha < 1$,

$$\frac{\partial^\alpha u(x,y,t)}{\partial |x|_{\theta_1}^\alpha} = -\sum_{n=1}^\infty \sum_{m=1}^\infty n^\alpha \left\{ c_+(\alpha,\theta_1)(i)^\alpha + c_-(\alpha,\theta_1)(-i)^\alpha \right\} u_{nm}(t) e^{imy} e^{inx}.$$

(21.16)

Now, we obtain a similar computation of $\frac{\partial^\mu u(x,y,t)}{\partial |y|_{\theta_2}^\mu}$ for the case of $1 < \mu \le 2$.
Therefore, we get

$$-\infty D_y^\mu u(x,y,t) = \frac{\partial^2}{\partial y^2} \left(\frac{1}{\Gamma(2-\mu)} \int_{-\infty}^y \frac{u(x,\eta,t)}{(y-\eta)^{\mu-1}} d\eta \right)$$

$$= \frac{\partial^2}{\partial y^2} \left(\frac{1}{\Gamma(2-\mu)} \sum_{n=1}^\infty \sum_{m=1}^\infty u_{nm}(t) e^{inx} \int_{-\infty}^y \frac{e^{im\eta}}{(y-\eta)^{\mu-1}} d\eta \right)$$

$$= \frac{1}{\Gamma(2-\mu)} \sum_{n=1}^\infty \sum_{m=1}^\infty u_{nm}(t) e^{inx} \frac{d^2}{dy^2} \left(e^{imy} \int_0^\infty \frac{e^{-imk}}{k^{\mu-1}} dk \right)$$

$$= \frac{1}{\Gamma(2-\mu)} \sum_{n=1}^\infty \sum_{m=1}^\infty u_{nm}(t) e^{inx} \frac{d^2}{dy^2} \left(e^{imy}(im)^{\mu-2} \Gamma(2-\mu) \right)$$

$$= \sum_{n=1}^\infty \sum_{m=1}^\infty (im)^\mu u_{nm}(t) e^{inx} e^{imy}$$

and

$$_y D_{+\infty}^\mu u(x,y,t) = \sum_{n=1}^\infty \sum_{m=1}^\infty (-im)^\mu u_{nm}(t) e^{inx} e^{imy}.$$

Hence, we obtain

$$\frac{\partial^\mu u(x,y,t)}{\partial |y|_{\theta_2}^\mu} = -\sum_{n=1}^\infty \sum_{m=1}^\infty \left\{ c_+(\mu,\theta_2)(im)^\mu + c_-(\mu,\theta_2)(-im)^\mu \right\} u_{nm}(t) e^{imy} e^{inx}.$$

(21.17)

Consequently, substituting (21.16) and (21.17) into (21.8) we take the following time fractional differential equation

$$\frac{\partial^\beta u_{nm}(t)}{\partial t^\beta} = - \left\{ n^\alpha \left[c_+(\alpha,\theta_1)(i)^\alpha + c_-(\alpha,\theta_1)(-i)^\alpha \right] \right.$$

$$\left. + m^\mu \left[c_+(\mu,\theta_2)(i)^\mu + c_-(\mu,\theta_2)(-i)^\mu \right] \right\} u_{nm}(t).$$

(21.18)

Therefore, we reduce the (21.8) to a fractional differential equation with one fractional term. To find the $u_{nm}(t)$, we apply Laplace transform to (21.18) and obtain

$$s^\beta u_{nm}(s) - s^{\beta-1} u_{nm}(0) + A u_{nm}(s) = 0$$

(21.19)

where

$$A = n^{\alpha} \left[c_+ (\alpha, \theta_1) (\mathrm{i})^{\alpha} + c_- (\alpha, \theta_1) (-\mathrm{i})^{\alpha} \right] + m^{\mu} \left[c_+ (\mu, \theta_2) (\mathrm{i})^{\mu} + c_- (\mu, \theta_2) (-\mathrm{i})^{\mu} \right].$$

$$(21.20)$$

Using inverse Laplace transform, (21.19) reduces to

$$u_{nm} (t) = u_{nm} (0) E_{\beta,1} \left(-At^{\beta} \right), \qquad (21.21)$$

where $E_{\beta,1}(.)$ is a well-known Mittag–Leffler function. The Fourier coefficients of the (21.12) can be found by

$$u_{0nm} = \frac{1}{(2\pi)^2} \int_{-\pi}^{\pi} \int_{-\pi}^{\pi} u_0 (x,y) e^{-\mathrm{i}nx} e^{-\mathrm{i}my} \mathrm{d}x \mathrm{d}y. \qquad (21.22)$$

After some manipulations, we take $u_{nm} (0) = u_{0nm}$ and also $u_{nm} (t) = u_{0nm} E_{\beta,1} \left(-At^{\beta} \right)$. Now, we can rewrite the solution series after these computations:

$$u (x,y,t) = \sum_{n=1}^{\infty} \sum_{m=1}^{\infty} u_{0nm} (0) E_{\beta,1} \left(-At^{\beta} \right) e^{\mathrm{i}nx} e^{\mathrm{i}my}. \qquad (21.23)$$

4 Grünwald–Letnikov Approximation for Numerical Solution

In this section, we show the numerical solution of the problem by applying GL approximation for Caputo derivative. Let us first give the relation between the left RL and Caputo definitions:

$$_aD_t^{\beta} u (t) = {}_a^C D_t^{\beta} u (t) + \sum_{r=0}^{m-1} \frac{\mathrm{d}^r}{\mathrm{d}t^r} u (t) \Big|_{t=a} \frac{(t-a)^{r-\beta}}{\Gamma (r-\beta+1)},$$

where $m \in \mathbb{N}$, $m - 1 < \beta \leq m$, $a \in \mathbb{R}$. Note that under the assumption $\left| \lim_{a \to -\infty} \frac{\mathrm{d}^r}{\mathrm{d}t^r} u (t) \Big|_{x=a} \right| < \infty$ for $r = 0, 1, ..., m - 1$, we have

$$_{-\infty}D_x^{\beta} u (t) = {}_{-\infty}^C D_x^{\beta} u (t).$$

It is also valid for the upper limit case and similar assumption as follows:

$$_xD_{+\infty}^{\beta} u (t) = {}_x^C D_{+\infty}^{\beta} u (t).$$

We remind that the order of Caputo derivative is $0 < \beta \leq 1$, the lower limit of derivative $a = 0$, and so we obtain

$$_0^C D_x^{\beta} u (t) = {}_0 D_x^{\beta} u (t) - u (0) \frac{t^{-\beta}}{\Gamma (1-\beta)}.$$

It is well known that if a function has suitable properties, i.e., it has first-order continuous derivatives and its second-order derivative is integrable, the β-order derivatives of function in both RL and GL senses are the same. By this property, we discretize the RL operator applying GL definition to (21.18), and therefore we take the approximation of Caputo derivative as

$$
{}_0^C D_t^\beta u_{nm}(t) \approx \frac{1}{h^\beta} \sum_{r=0}^M w_r^{(\beta)} u_{nm}(hM - rh) - u_{nm}(0) \frac{(hM)^{-\beta}}{\Gamma(1-\beta)}, \qquad (21.24)
$$

where $M = \frac{t}{h}$ represents the number of sub-time intervals, h is step size, and $w_r^{(\beta)}$ are the coefficients of GL formula:

$$
w_0^\beta = 1, \ w_r^\beta = \left(1 - \frac{\beta+1}{r}\right) w_{(r-1)}^\beta . \qquad (21.25)
$$

Substituting (21.24) into (21.18) and after some arranging, we get

$$
u_{nm}(hM) = \frac{1}{\left(\frac{1}{h^\beta} w_0^{(\beta)} + A\right)} \left\{ u_{nm}(0) \frac{(hM)^{-\beta}}{\Gamma(1-\beta)} - \frac{1}{h^\beta} \sum_{r=1}^M w_r^{(\beta)} u_{nm}(hM - rh) \right\},
$$

$$
\qquad (21.26)
$$

where A is given by (21.20).

5 Numerical Example

In this section, we consider the following initial condition:

$$
u(x,y,0) = \sinh(x+y).
$$

In Fig. 21.1, we first validate the efficiency of numerical method by comparison of analytical and numerical solutions for $x = \frac{\pi}{5}, y = \frac{\pi}{4}, t = 5, h = 0.01$ and $n = m = 10$. It is clear from the figure that the analytical solution is in a good agreement with the numerical solution. Figure 21.2 shows the behavior of problem under the variations of μ values for $x = \frac{\pi}{5}, y = \frac{\pi}{4}, t = 5, h = 0.01, \beta = 1, \alpha = 0.3$ and $\theta_1 = 0.3$. Similarly, Fig. 21.3 shows the response of the problem for variable order of α for $t = 5, \beta = 0.5, \mu = 1.5$ and $\theta_2 = 0.5$. Figure 21.4 indicates changing behaviors of problem with respect to the variations of α, β, and μ parameters for $x = \frac{\pi}{5}, y = \frac{\pi}{4}, t = 5$. In Fig. 21.5, we get the three-dimensional surface of the problem (21.8) with respect to x and t for $y = \frac{\pi}{4}, \beta = 0.7, \alpha = 0.5, \theta_1 = 0.5$ and $\mu = 1.8, \theta_2 = 0.1$. Finally, we obtain the surface of the problem (21.8) with respect to x and y for $\beta = 0.7, \alpha = 0.5, \theta_1 = 0.5$, and $\mu = 1.8, \theta_2 = 0.1$ and $h = 0.01$ in Fig. 21.6.

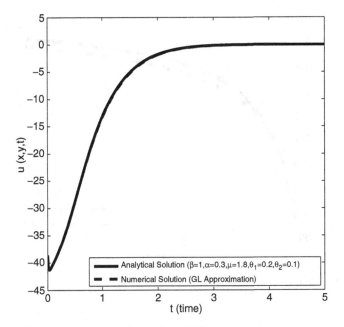

Fig. 21.1 Comparison of analytical and numerical solutions for $\beta = 1$

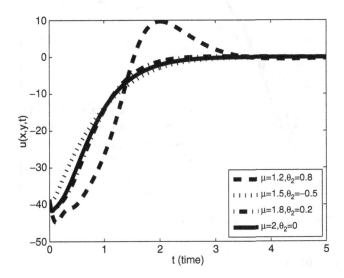

Fig. 21.2 Variations of μ parameter

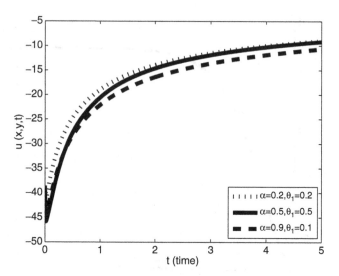

Fig. 21.3 Variations of α parameter

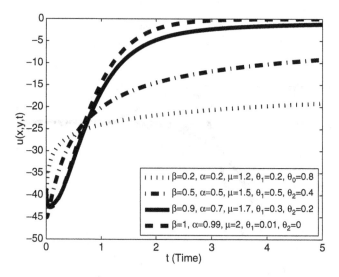

Fig. 21.4 Variations of β, α, and μ parameters

6 Conclusions

In this chapter, we have defined a two-dimensional anomalous diffusion problem with time and space fractional derivative terms. These have been described in the sense of Caputo and Riesz–Feller operators, respectively. We have purposed to find

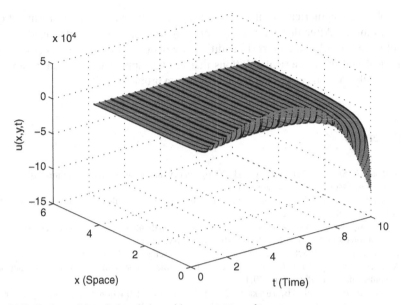

Fig. 21.5 Surface of the whole solution with respect to x and t

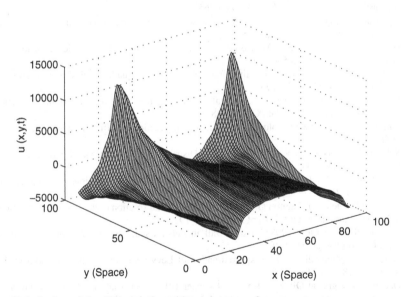

Fig. 21.6 Surface of the whole solution with respect to x and y

the exact and the numerical solutions of the problem under some assumptions. Therefore, we use Laplace and Fourier transforms for analytical solution and also prefer to apply GL definition. However, we first reduce the main problem to a fractional differential equation with time fractional term. By this way, we have obtained numerical results more easily. Finally, we apply the formulations to an example. After that we present some figures under different considerations about variations of parameters. In addition, we deduce from the comparison of the analytical and the numerical solutions that the GL approximation can be applied successfully to such type of anomalous diffusion problems.

References

1. Kilbas AA, Srivastava HM, Trujillo JJ (2006) Theory and applications of fractional differential equations. Elsevier B.V., Amsterdam
2. Miller KS, Ross B (1993) An introduction to the fractional calculus and fractional differential equations. Wiley, New York
3. Samko SG, Kilbas AA, Marichev OI (1993) Fractional integrals and derivatives-theory and applications. Gordon and Breach, Longhorne Pennsylvania
4. Podlubny I (1999) Fractional differential equations. Academic, New York
5. Metzler R, Klafter J (2000) The random walk's guide to anomalous diffusion: A fractional dynamic approach. Phys Rep 339:1–77
6. Turski AJ, Atamaniuk TB, Turska E (2007) Application of fractional derivative operators to anomalous diffusion and propagation problems. arXiv:math-ph/0701068v2
7. Agrawal OP (2001) Response of a diffusion-wave system subjected to deterministic and stochastic fields. Z Angew Math Mech 83:265–274
8. Agrawal OP (2002) Solution for a fractional diffusion-wave equation defined in a bounded domain. Nonlinear Dyn 29:145–155
9. Herzallah MAE, El-Sayed AMA, Baleanu D (2010) On the fractional-order diffusion-wave process. Rom Journ Phys 55(3–4): 274–284
10. Huang F, Liu F (2005) The space-time fractional diffusion equation with Caputo derivatives. Appl Math Comput 19:179–190
11. Huang F, Liu F (2005) The fundamental solution of the space-time fractional advection-dispersion equation. Appl Math Comput 18:339–350
12. Langlands TAM (2006) Solution of a modified fractional diffusion equation. Phys A 367:136–144
13. Sokolov IM, Chechkin AV, Klafter AJ (2004) Distributed-order fractional kinetics. Acta Phys Polon B 35:1323–1341
14. Saichev AI, Zaslavsky GM (1997) Fractional kinetic equations: Solutions and applications. Chaos 7:753–764
15. Mainardi F, Luchko Y, Pagnini G (2001) The fundamental solution of the space-time fractional diffusion equations. Fract Cal Appl Anal 4:153–192
16. Gorenflo R, Mainardi F (1998) Random walk models for space-fractional diffusion processes. Fract Cal Appl Anal 1:167–191
17. Gorenflo R, Mainardi F (1999) Approximation of Levy-Feller diffusion by random walk. J Anal Appl 18:231–146
18. Özdemir N, Agrawal OP, Karadeniz D, İskender BB (2009) Analysis of an axis-symmetric fractional diffusion-wave equation. J Phys A Math Theor 42:355208
19. Özdemir N, Karadeniz D (2008) Fractional diffusion-wave problem in cylindrical coordinates. Phys Lett A 372:5968–5972
20. Povstenko YZ (2008) Fractional radial diffusion in a cylinder. J Mol Liq 137:46–50
21. Povstenko YZ (2008) Time Ffactional radial diffusion in a sphere. Nonlinear Dyn 53:55–65
22. Povstenko YZ (2008) Fundamental solutions to three-dimensional diffusion-wave equation and associated diffusive stresses. Chaos Solitons Fractals 36:961–972

23. Povstenko YZ (2010) Signaling problem for time-fractional diffusion-wave equation in a half-space in the case of angular symmetry. Nonlinear Dyn 59:593–605
24. Shen S, Liu F (2005) Error analysis of an explicit finite difference approximation for the space fractional diffusion equation with insulated ends. ANZIAM J 46:871–887
25. Liu F, Zhuang P, Anh V, Turner I, Burrage K (2007) Stability and convergence of the difference methods for the space-time fractional advection-diffusion equation. Appl Math Comput 191:12–20
26. Lin R, Liu F, Anh V, Turner I (2009) Stability and convergence of a new explicit finite-difference approximation for the variable-order nonlinear fractional diffusion equation. ANZIAM J 212:435–445
27. Özdemir N, AvcıD, İskender BB (2011) The Numerical Solutions of a Two-Dimensional Space-Time Riesz-Caputo Fractional Diffusion Equation. An International Journal of Optimization and Control: Theories and Applications. 1(1):17–26
28. Ciesielski M, Leszczynski J (2006) Numerical treatment of an initial-boundary value problem for fractional partial differential equations. Signal Proces 86:2619–2631
29. Ciesielski M, Leszczynski J (2006) Numerical solutions to boundary value problem for anomalous diffusion equation with Riesz-Feller fractional operator. J Theoret Appl Mech 44:393-403
30. Liu Q, Liu F, Turner I, Anh V (2007) Approximation of the Levy-Feller advection-dispersion process by random walk and finite difference method. Comput Phys 222:57–70
31. Zhang H, Liu F, Anh V (2007) Numerical approximation of Levy-Feller diffusion equation and its probability interpretation. J Comput Appl Math 206:1098–1115
32. Machado JAT (2003) A probabilistic interpretation of the fractional-order differentiation. Fract Cal Appl Anal 6:73–80
33. Feller W (1952) On a generalization of Marcel Riesz' potentials and the semi-groups generated by them. Meddeladen Lund Universitets Matematiska Seminarium, Tome suppl.dedie a M. Riesz, Lund, 73–81

Chapter 22
Analyzing Anomalous Diffusion in NMR Using a Distribution of Rate Constants

R.L. Magin, Y.Z. Rawash, and M.N. Berberan-Santos

1 Introduction

There is a growing realization that relaxation and diffusion phenomena in complex materials cannot be expressed simply in terms of sum of exponential decays, each characterized by a single relaxation time or rate [8]. In nuclear magnetic resonance (NMR) and in optical luminescence studies, single-exponential models fail to describe the wide variety of relaxation times observed in synthetic and biological materials [6, 11]. In luminescence decay, for example, the observed relaxation spans many orders of magnitude – from nanoseconds to milliseconds – and a wide distribution of rate constants is needed to describe the phenomena [5]. In NMR relaxation experiments, the direct measurement of a distribution of T2 relaxation times from 1 to 1,000 ms is possible, but only for high signal-to-noise data collected from bulk samples [10]. The same is true for so-called q-space NMR diffusion measurements [1]. Our attention has been drawn recently to the success of simple empirical fits to the NMR diffusion attenuation in pulsed field gradient (PFG) studies [7], where both magnetic resonance imaging (MRI) and diffusion distribution analysis are possible. In PFG, the signal attenuation caused by diffusion can be measured for applied magnetic field gradients and the data fit to a simple-exponential decay law:

$$S(b) = S_0 \exp(-bD), \tag{22.1}$$

R.L. Magin (✉) • Y.Z. Rawash
Department of Bioengineering, University of Illinois at Chicago, 851 South Morgan Street, Chicago, IL, 60607–7052, USA
e-mail: rmagin@uic.edu; yrawas2@uic.edu

M.N. Berberan-Santos
Centro de Quimica-Fisica Molecular, Instituto Superior Tecnico, 1049–001, Lisboa, Portugal
e-mail: mbs@ist.utl.pt

D. Baleanu et al. (eds.), *Fractional Dynamics and Control*,
DOI 10.1007/978-1-4614-0457-6_22, © Springer Science+Business Media, LLC 2012

where b is a measure of the applied gradient strength and duration, and D is the effective diffusion constant, with units of mm^2/s. For high b-value experiments ($b > 1{,}000s/mm^2$) in the brain, however, the data collected from both gray and white matter do not follow a single exponential and have been simply fit using the so-called, stretched-exponential decay law:

$$S(b) = S_0 \exp\left[-(bD_\alpha)^\alpha\right], \tag{22.2}$$

where $0 < \alpha < 1$ and D_α is the fractional diffusion constant. A rationale for the empirical function was recently provided using fractional calculus [6,11,14,18]. The purpose of this chapter is to show that the stretching parameter, α, directly reflects a "distribution" of many single-exponential relaxation rates. This analysis follows closely the approach used by [3] in applying the stretched exponential and similar functions to the time decay of optical luminescence signals.

2 Underlying Distributions

The exponential function (22.1) can also be written in a more general form. Let $I(b) = S(b)/S_0$, so we can write:

$$I(b) = \exp\left(-\int_0^b w(u)\,du\right), \tag{22.3}$$

where $w(b)$ is a b-value-dependent rate coefficient, defined by:

$$w(b) = -\frac{d\ln I(b)}{db} = -\frac{1}{I(b)}\frac{dI(b)}{db}. \tag{22.4}$$

In the simplest case, $w(b)$ can be b-value independent with, for example, $w(b) = D$ and the decay is an exponential function as in (22.1). Equation (22.3) can be easily generalized to fractional order as follows:

$$I(b) = \exp\left(-[1 * w(b)]\right) = \exp\left(-[k_\alpha(b) * w(b)]\right). \tag{22.5}$$

Here, the $*$ symbol indicates the convolution operator and $k_\alpha(b)$ is a monotonically decreasing function that provides fading memory of earlier b-values. If we assume that we can write $k_\alpha(b)$ as a power law, $k_\alpha(b) = b^{\alpha-1}/\Gamma(\alpha)$, where $\Gamma(\alpha)$ is the γ function, then (22.5) takes the form of a classical Riemann–Liouville fractional integral [9], and we have:

$$I(b) = \exp\left(-I_0^\alpha[w(b)]\right). \tag{22.6}$$

Now in the case of $w(b) = D$, we find using either the properties of fractional calculus or the Laplace transformation, that:

$$I(b) = \exp\left[-b^{\alpha} D / \Gamma(\alpha)\right] = \exp\left[-\left(bD_{\alpha}\right)^{\alpha}\right], \tag{22.7}$$

where $D_{\alpha} = D / \Gamma(1 + \alpha)$ is the effective diffusion constant and the diffusion decay curve is a stretched-exponential function of b.

The relaxation function $I(b)$ used to describe diffusion signal attenuation is thus seen to span the range from a single exponential to a stretched exponential depending upon the amount of memory included in the convolution; with longer memory, additional, slower components of the response could be expected to become more evident. Another way of spanning both long and short diffusion components is to assume two or more single exponentials are present. This is the usual approach to incorporate multi-compartmental and multi-component behavior. Here we will take this idea to the extreme and assume a distribution of exponential terms that can also be written as:

$$I(b) = L\left[\rho\left(k_{D}\right)\right] = \int_{0}^{\infty} \rho\left(k_{D}\right) e^{-bk_{D}} dk_{D}. \tag{22.8}$$

This relation is valid as long as the integral exists (i.e. $\rho(k_{D})$ grows no faster than an exponential) and shows $\rho(k_{D})$ to be the inverse Laplace transform of $I(b)$. The function $\rho(k_{D})$ is normalized, as $I(0) = 1$ implies:

$$\int_{0}^{\infty} \rho\left(k_{D}\right) dk_{D} = 1. \tag{22.9}$$

The integrand of (22.8), $\rho(k_{D})\exp(-bk_{D})$, can be seen as a particular diffusion term in the distribution with k_{D} viewed either as the corresponding diffusion coefficient or as the rate constant for the process that proceeds with advancing b.

In previous work, we found that the stretched-exponential function is well suited to describe signal attenuation caused by diffusion phenomena in gels, cartilage, and brain tissue [11]. We now will show that the distribution of rate constants – the probability density function – outlined above for a stretched-exponential function is well suited to describe and characterize intrinsic properties of the complex environment that water inhabits in each of these cases.

Recovery of the distribution $\rho(k_{D})$ from experimental data is an ill-conditioned problem [12]. In other words, a small change in $I(b)$ can cause an arbitrarily large change in $\rho(k_{D})$. In general, high signal-to-noise is needed so that $\rho(k_{D})$ can be recovered from the experimental relaxation decay results. A simple form of the inverse Laplace transform of a relaxation function can be obtained by the method outlined by [3]. Briefly, the three following equations can be used for the computation of $\rho(k_D)$ from $I(b)$:

$$\rho(k_D) = \frac{2}{\pi} \int_0^\infty \mathrm{Re}\,[I(i\omega)] \cos(k_D\omega)\,d\omega \tag{22.10}$$

$$\rho(k_D) = -\frac{2}{\pi} \int_0^\infty \mathrm{Im}\,[I(i\omega)] \sin(k_D\omega)\,d\omega \tag{22.11}$$

$$\rho(k_D) = \frac{1}{\pi} \int_0^\infty \mathrm{Re}\,[I(i\omega)] \sin(k_D\omega)\,d\omega - \mathrm{Im}\,[I(i\omega)] \sin(k_D\omega)\,d\omega. \tag{22.12}$$

3 Anomalous Diffusion Model

The stretched-exponential model follows from a fundamental extension of the Bloch–Torrey equation through application of the operators of fractional calculus [11]. This promising model needs further investigation and study; however, like models of [2] and of [7], it describes the anomalous diffusion behavior of signal attenuation using a stretched-exponential function. The specific formula derived by Magin et al. is:

$$I(b) = \exp\left[-D\mu^{2(\beta-1)} (\gamma G_z\delta)^{2\beta} \left(\Delta - \frac{2\beta-1}{2\beta+1}\delta\right)\right]. \tag{22.13}$$

In this equation, γ is the gyromagnetic ratio for the water proton (42.58 MHz/T), G_z is the spatial gradient in z direction, Δ is the time between the two gradient pulses, δ is the gradient pulse duration, and D is the diffusion coefficient. In this model there are also two fractional order-related parameters: μ, a space constant needed to maintain consistent units and β, the order of the fractional calculus operator, $0.5 < \beta \leq 1$, which can be related to the complexity of the material.

For the case of $\beta = 1$, the classic expression for diffusion is recovered, that is:

$$I(b) = \exp\left[-D(\gamma G_z\delta)^2 (\Delta - \delta/3)\right]. \tag{22.14}$$

Using (22.13) the theoretical curves are plotted in Fig. 22.1, as $I(b)$ versus the gradient parameter b (where for $\Delta \gg \Delta$, $b = (\gamma G_z\delta)^2\delta$) for selected values of μ and β. In this figure a gradient pulse sequence G_z, Δ, and δ is assumed with G_z varying from 0 to 1,500 mT/m. We observe that as β decreases from 1.0 to 0.6, the attenuation curves change from a simple exponential (a straight line on the semi-log

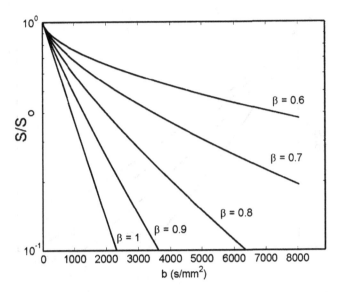

Fig. 22.1 Normalized signal attenuation decay (22.13), plotted versus b, where $b = (\gamma G_z \delta)^2 \Delta$, for selected values of β. In each curve, G_z increases from 0 to 1,500 mT/m while all other parameters are fixed: $D(1 \times 10^{-3} \text{mm}^2/\text{s})$, Δ (50 ms), δ (1 ms), and μ (35μm)

graph) to heavy-tailed curved shape that strongly resembles the behavior recorded in restricted diffusion – particularly at high b-values.

In Fig. 22.2, (22.13) is plotted for a series of μ-values ranging from 20 to 80μm with $\beta = 0.8$. The G_z, Δ, and δ-values in Fig. 22.2 are the same as those used in Fig. 22.1. Here we see that increasing the value of μ appears to increase the contribution of restricted diffusion in the diffusion attenuation curve for a fixed value of β. This behavior is evident when (22.13) is expressed either in terms of a single-exponential decay, $\exp\{-bD_{apl}\}$, where the apparent diffusion coefficient is expressed as:

$$D_{apl} = D/((\gamma G_z \delta)\mu)^{2(1-\beta)}, \qquad (22.15)$$

or when (22.13) is written as a stretched exponential, $\exp\{-(bD_F)^\beta\}$, where:

$$D_F^\beta = D \left(\Delta/\mu^2\right)^{1-\beta}. \qquad (22.16)$$

In addition, when $\mu = \sqrt{D\Delta}$, (22.13) corresponds with the "stretched exponential" result, $\exp\{-(bD)^\beta\}$, considered by [2]. In the example plotted in Fig. 22.2 this correspondence occurs for $\mu = 7.07$μm. Overall, Figs. 22.1 and 22.2 show for (22.13) a decrease in the apparent diffusion coefficient as the values of β decrease and values of μ increase.

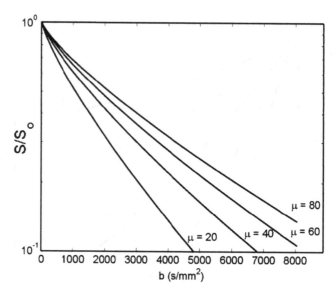

Fig. 22.2 Normalized signal attenuation decay (22.13), plotted versus b, where $b = (\gamma G_z \delta)^2 \Delta$, with selected values of μ. In each curve, G_z increases from 0 to 1,500 mT/m while all other parameters are fixed: $D(1 \times 10^{-3} \, \text{mm}^2/\text{s})$, Δ (50 ms), δ (1 ms), and β (0.8)

4 Anomalous Distribution Model

The stretched-exponential decay function was also recently used in the analysis of single-molecule fluorescence, quantum dot luminescence, and in the fluorescence lifetime imaging of biological tissues [6]. The determination of $\rho(k_D)$ for a given $I(b)$ amounts to finding the inverse Laplace transform of the stretched-exponential decay function.

The result, first obtained by [15], and an equivalent integral representation found by [3], is:

$$\rho(k_D) = \frac{1}{\pi D_F} \int_0^\infty e^{-u^\beta \cos\left(\frac{\beta\pi}{2}\right)} \cos\left(u^\beta \sin\left(\frac{\beta\pi}{2}\right) - \frac{k_D u}{D_F}\right) du. \qquad (22.17)$$

In addition, a convergent power series for $\rho(k_D)$ is known:

$$\rho(k_D) = \frac{1}{\pi D_F} \sum_{n=1}^\infty \frac{(-1)^{n+1}}{n!} \frac{\sin(n\beta\pi)}{(k_D/D_F)^{(1+n\beta)}} \Gamma(1+n\beta). \qquad (22.18)$$

It can be seen that the asymptotic form of $\rho(k_D)$ is:

$$\rho(k_D) = \frac{\Gamma(1+\beta)}{\pi D_F} \frac{\sin(\beta\pi)}{(k_D/D_F)^{1+\beta}}. \qquad (22.19)$$

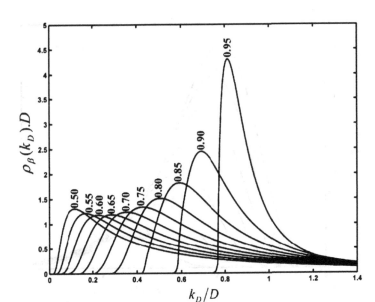

Fig. 22.3 A series of plots of the distribution of diffusional decay rates calculated using (22.18). The β-values span the range from 0.5 to 0.95

We have calculated the distribution of diffusional rate constants $\rho(k_D)$ using (22.18) for a range of β-values from 0.5 to 0.95, and the results are plotted in Fig. 22.3. As the value of β approaches 1, the distribution appears to converge toward a Dirac Delta function corresponding to a single-exponential function with $k_D = D$, whereas, as β approaches the value of 0.5, the distribution of diffusional relaxation rates broadens and becomes more uniform. A similar broadening of the distribution occurs for a fixed value of β as μ is increased, as is shown in Fig. 22.4.

The stretched-exponential decay function is necessarily of an approximate nature, as it implies an infinitely fast initial rate of decay. Super-exponential short-time behavior results from the fat tail of the stretched-exponential distribution [4]. Usually, one is, however, more interested in the long-time behavior, which is governed by the shape of the distribution of rate constants near the origin. With this aim in view, an exponentially attenuated modified form of the stretched-exponential distribution was proposed [4]. The resulting decay law retains the original long-time behavior but no longer suffers from the short-time problems of the original distribution. Other well-behaved decay functions that encompass the exponential function as a special case such as the compressed hyperbola have been proposed and successfully applied to several problems where a continuous distribution of rate constants was found to exist [16, 17].

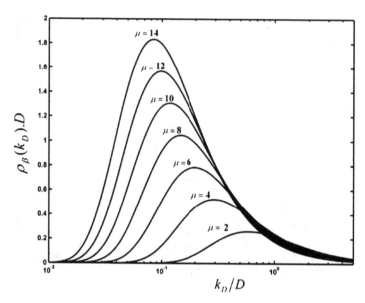

Fig. 22.4 A series of plots of the distribution of diffusional decay rates calculated using (22.18) with β of 0.5. The μ-values span the range from 2 to 14 μm

5 Experimental Results

Two diffusion-weighted MRI (DWI) experiments were carried out at 11.74 T (500 MHz for protons) to illustrate applications of the stretched-exponential model to characterized biological complexity. In first experiment, glass capillary tubes were filled with Sephadex (Sigma-Aldrich, St. Louis, MO) gels type G-25, G-50, and G-100, and placed in a 5 mm NMR tube filled with distilled water. Diffusion-weighted images were acquired using a Stejskal–Tanner diffusion-weighted spin–echo pulse sequence with the following parameters: TR = 1,000 ms, TE = 60 ms, slice thickness = 1.5 mm, Δ = 45 ms, δ = 1 ms, and 4 averages. The FOV was 0.6 cm × 0.6 cm, which for a matrix size of 64 × 64 corresponds to an in-plane resolution of 94 μm × 94 μm.

In the second experiment, diffusion-weighted brain imaging was carried out on a healthy human volunteer at the University of Illinois Medical Center. Axial images were acquired with multiple b-values using a customized single-shot EPI. The key data acquisition parameters were: TR = 4,000 ms, TE = 96.6 ms, slice thickness of 4 mm, slice gap of 3 mm, Δ = 42.6 ms, δ = 32.2 ms, and 4 averages. Diffusion-weighted images were acquired with a maximum b-value of 3,300 s/mm^2.

The signal attenuation results from DWI for the two experiments (Figs. 22.5 and 22.6) was measured for selected regions of interest (ROI) at increasing b-values,

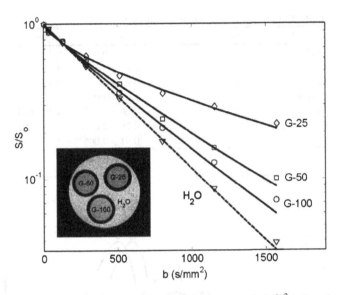

Fig. 22.5 Normalized signal intensity plotted versus b, where $b = (\gamma G_z \delta)^2 \Delta$, for selected ROI in samples of distilled water and Sephadex G-25, G-50, and G-100. The experimental data were fit to the fractional-order model (22.13) to determine D_F, β, and μ

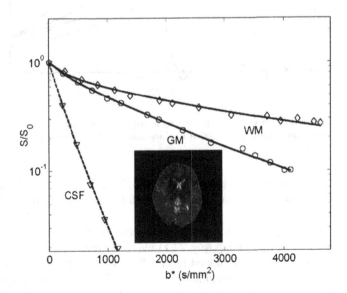

Fig. 22.6 Normalized signal intensity plotted versus b^*, from (22.13) for selected ROI in white matter, gray matter and cerebrospinal fluid for a human brain. The experimental data were fit to the fractional-order stretched-exponential model. (22.13) for D, β, and μ, where $b^* = (\gamma G_z (\delta)^2 (\Delta - ((2\beta - 1)/(2\beta + 1))\delta)$

Table 22.1 Diffusion and complexity measurement

	$D_F \times 10^{-3}\ mm^2/s$	β (a.u.)	μ (μm)
G-25	1.1 ± 0.04	0.71 ± 0.06	6.4 ± 0.1
G-50	1.5 ± 0.03	0.8 ± 0.05	5.7 ± 0.1
G-100	1.8 ± 0.06	0.91 ± 0.08	4.4 ± 1.6
Distilled water	2.1 ± 0.02	1.0 ± 0.003	2.9 ± 0.3
White matter	0.41 ± 0.006	0.60 ± 0.008	4.3 ± 0.04
Gray matter	0.75 ± 0.08	0.78 ± 0.03	4.9 ± 0.02
CSF	2.8 ± 0.18	0.91 ± 0.005	3.0 ± 1.3

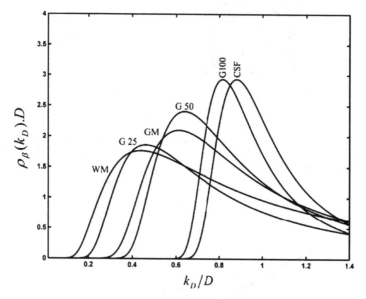

Fig. 22.7 Distribution of rate constants (probability density function) calculated using (22.18) for the stretched-exponential relaxation function for (Sephadex G-25, G-50 and G-100, CSF, gray matter, and white matter)

and the data fit to the stretched-exponential model (22.13) using the Levenberg–Marquardt nonlinear least squares algorithm. A summary of the results is listed in Table 22.1.

We have used (22.18) to calculate the distribution of diffusional rate constants $\rho(k_D)$ for the data on Sepadex gels (Fig. 22.5) and for human brain tissue (Fig. 22.6). The resulting distributions are plotted in Fig. 22.7. The results provide a graphic depiction of the effect of the fractional-order parameter β to embed a wide range of diffusion rates. As expected, the water and water-like cerebral spinal fluid (CSF) have very narrow distributions while the Sephadex gels, beginning with the G-100 (largest pore size) fan out with decreasing G-number (effectively the gel's molecular exclusion value in kilodaltons). Interestingly, the gray matter distribution peaks between the G-25 and G-50 Sephadex samples, while the white matter shows an almost uniform distribution of diffusion rates.

6 Discussion and Conclusions

In this work, we demonstrated that a stretched-exponential model describing anomalous diffusion in disordered system can be employed with success to characterize the diffusion results from the MR signal attenuation curves obtained from Sephadex gels and biological tissue during DWI experiments. Moreover, the probability density function of diffusional rate constants was found to describe and characterize the complexity of gels and biological tissue in a manner consistent with the water distribution inside the biological environments.

The general properties and conditions of the MR signal attenuation curves were outlined for the MR signal in order to have a probability density function of rate constants; it was concluded that the decay must be either exponential or sub-exponential for all b-values, if $\rho(k_D)$ is to be a probability density function. The use of the stretched-exponential model for analyzing MR signal attenuation curves has been shown to provide excellent tissue discrimination [11, 18]. Even more importantly, the stretched-exponential function not only describes the decay profiles almost exactly but also derives from the more realistic decay model of continuous distributions in biological tissue, rather than from an arbitrary assumption of single or multiple discrete exponential decay components.

The heterogeneity parameter β is important because it enables the study of mechanisms that cause a continuous distribution to broaden or narrow. Future work on a variety of different tissue types and in different environmental conditions will hopefully provide more insight in this matter. It is possible that the heterogeneity parameter β is sensitive to the gradient direction, particularly in highly anisotropic structures in the brain. This is expected because diffusion imaging of the brain has revealed considerable anisotropy. In the present studies, diffusion gradients were applied in only one direction to avoid averaging the data in more than one direction. Results showed reduced μ in regions of human brain corresponding to the superior sagittal sinus. This is thought to occur because of partial volume effects between the relatively fast-diffusing CSF in the sinus and the slow-diffusing gray matter parenchymal protons and white matter in between, so it is likely that μ can be related to the porosity and tortuosity in biological tissue, and more specifically to the heterogeneity of the biological tissue.

A stretched-exponential model that describes diffusion in a random disordered system and fractal spaces was used to parameterize the diffusion and biological complexity from diffusion-weighted MR signals. The model performed well on data obtained from Sephadex gel type G-100, G-50, and G-25 as well as on data from human brain tissues (white and gray matter) with different predicted diffusion characteristics. This approach has potential to be applied in clinical studies and may aid in monitoring the developmental as well as pathological changes to biological tissues.

Acknowledgment The authors R.L. Magin and Y.Z. Rawash would like to acknowledge the support of NIH grant R01 EB007537.

References

1. Assaf Y, Cohen Y (2000) Assignment of the water slow-diffusing component in the central nervous system using q-space diffusion MRS: implications for fiber track imaging. Magn Reson Med 43: 191–199
2. Bennett KM, Schmainda KM, Bennett RT, Rowe DB, Lu H, Hyde JS (2003) Characterization of continuously distributed cortical water diffusion rates with a stretched exponential model. Magn Reson Med 50:727–734
3. Berberan-Santos MN (2005) Analytical inversion of the Laplace transform without contour integration: Application to luminescence decay laws and other relaxation functions. J Math Chem 38:165–173
4. Berberan-Santos MN, Bodunov EN, Valeur B (2005) Mathematical functions for the analysis of luminescence decays with underlying distributions 1. Kohlrausch decay function (stretched exponential). J Lumin 126:263–272
5. Berberan-Santos MN, Valeur B (2007) Luminescence decays with underlying distributions: General properties and analysis with mathematical functions. 1. Kohlrausch decay function (stretched exponential). Chem Phys 315:171–182
6. Berberan-Santos MN, Bodunov EN, Valeur B (2008) Luminescence decays with underlying distributions of rate constants: General properties and selected cases. In Bereran-Santos MN (ed) Fluorescence of supermolecules, polymers, and nanosystems. Springer, Berlin, pp 105–116
7. Hall MG, Barrick TR (2008) From diffusion-weighted MRI to anomalous diffusion imaging. Magn Reson Med 59:447–455
8. Hilfer R (2002) Applications of fractional calculus in physics. World Scientific, Singapore
9. Kilbas AA, Srivastava HM, Trujillo JJ (2006) Theory and applications of fractional differential equations. Elsevier, Amsterdam
10. Lin PC, Reiter DA, Spencer RG (2009) Classification of degraded cartilage through multi-parametric MRI analysis. J Magn Reson 201:61–71
11. Magin RL, Abdullah O, Baleanu D, Zhou XJ (2008) Anomalous diffusion expressed through fractional order differential operators in the Bloch–Torrey equation. J Magn Reson 190: 255–270
12. Mollay B, Kauffman HF (1994) Dynamics of energy transfer in aromatic polymers. In: Richert R, Blumen A (eds) Disorder effects on relaxational processes: glasses, polymers, proteins. Springer, Berlin, pp 509–541, 509
13. Narayanan A, Hartman JS, Bain AD (1995) Characterizing nonexponential spin–lattice relaxation in solid-state NMR by fitting to the stretched exponential. J Magn Reson Ser A 112:58–65
14. Peled S, Cory DG, Raymond SA, Kirschner DA, Jolesz FA (1999) Water diffusion, T_2, and compartmentation in frog sciatic nerve. Magn Reson Med 42:911–918
15. Pollard K (1946) The representation of $\exp(-x^\alpha)$ as a Laplace integral. Bull Amer Math Soc 52:908–910
16. Souchon V, Leray I, Berberan-Santos MN, Valeur B (2009) Multichromophoric supramolecular systems. Recovery of the distributions of decay times from the fluorescence decays. Dalton Trans 20:3988–3992
17. Whitehead L, Whitehead R, Valeur B, Berberan-Santos MN (2009) A simple function for the description of near-exponential decays: The stretched or compressed hyperbola. Am J Phys 77:173–179
18. Zhou XJ, Gao Q, Abdullah O, Magin RL (2010) Studies of anomalous diffusion in the human brain using fractional order calculus. Magn Reson Med 63:562–569

Chapter 23
Using Fractional Derivatives to Generalize the Hodgkin–Huxley Model

Hany H. Sherief, A.M.A. El-Sayed, S.H. Behiry, and W.E. Raslan

1 Introduction

In recent years fractional differential equations have gained a considerable amount of interest due to their many applications in the fields of physics [16], signal processing [1], fluid mechanics [2], mathematical biology [14], and bioengineering [15].

Fractional calculus has been used successfully to modify many existing models of physical processes. The first application of fractional derivatives was given by Abel who applied fractional calculus in the solution of an integral equation that arises in the formulation of the tautochrone problem [17]. Subsequently, Caputo and Mainardi [3] applied fractional calculus for the description of viscoelastic materials and found good agreement with experimental results and established a connection between fractional derivatives and the theory of linear viscoelasticity [4,5].

Recent monographs and symposia proceedings have highlighted the application of fractional calculus in physics, continuum mechanics, signal processing, and electromagnetism, but with few examples of applications in bioengineering. This is surprising because the methods of fractional calculus, when defined as a Laplace or Fourier convolution product, are suitable for solving many problems in biomedical

H.H. Sherief • A.M.A. El-Sayed (✉)
Department of Mathematics, Faculty of Science, University of Alexandria, Alexandria, Egypt
e-mail: HanySherief@Gmail.com; amasayed@hotmail.com

S.H. Behiry
General Required Courses Department, Jeddah *Community College*, King Abdulaziz
University, Jeddah 21589, Kingdom of Saudi Arabia
e-mail: salahb@mans.edu.eg

W.E. Raslan
Mathematics and Engineering Physics Department, Faculty of Engineering,
University of Mansoura, Mansoura, Egypt
e-mail: w_raslan@yahoo.com

D. Baleanu et al. (eds.), *Fractional Dynamics and Control*,
DOI 10.1007/978-1-4614-0457-6_23, © Springer Science+Business Media, LLC 2012

research. For example, early studies by Cole and Hodgkin [6] of the electrical properties of nerve cell membranes and the propagation of electrical signals are well characterized by differential equations of fractional order. The solution involves a generalization of the exponential function to the Mittage-Leffler function, which provides a better fit to the observed cell membrane data [15].

Since action potentials are the most important phenomenon of the nervous system, a great deal of scientific research was conducted in deciphering their proprieties. Alan Hodgkin and Andrew Huxley (HH) established a model system in 1952 that is until today one of the key achievements in cellular biophysics [10–13]. Their work on the squid giant axon was awarded a Nobel prize in physiology and medicine together with John Eccles in 1963. Hodgkin and Huxley conducted a series of experiments that allowed the determination of the time and voltage dependence of the sodium and potassium conductance. Their set of differential equations is able to model all that was known about action potentials at that time with remarkable accuracy. Their model is an empirical approach to the kinetic description of excitable membranes and their goal to predict the properties of action potentials was met with astonishing accuracy. They were able to fit their research on the squid giant axon into a set of four differential equations and various supporting functions. The main equation, describing the membrane in terms of an electric circuit, is a nonlinear differential equation, whereas the remaining three describe the properties of the ion channels using ordinary first order differential equations.

2 Fractional HH Model

As in the original HH model, we assume that the electrical properties of a segment of nerve membrane can be modeled by an equivalent circuit. In this circuit, current flow across the membrane has two major components one, I_C ($\mu A\,cm^{-2}$) associated with charging the membrane capacitance C ($\mu F\,cm^{-2}$) and another, I_{ion} associated with the movement of specific types of ions across the membrane. In squid giant axon, the ionic current is further subdivided into three distinct components, a sodium current I_{Na}, a potassium current I_K, and a small leakage current I_L that is primarily carried by chloride ions. We thus have:

$$I_C + I_{ion} = I_{ext},\qquad (23.1)$$

where I_{ext} is the external applied current.

For the fractional model, we suggest that:

$$I_C = C^c D_t^\mu v,\qquad (23.2)$$

where v is the membrane voltage (mV) and $^c D_t^\mu$ is the Caputo fractional derivative of order μ, $0 < \mu \leq 1$, defined by [17]:

$$^c D_t^\mu f(t) = \int_0^t \frac{(t-s)^{-\mu}}{\Gamma(1-\mu)} f'(s)ds,$$

where $\Gamma(.)$ is the gamma function.

Substituting (23.2) into (23.1), we obtain:

$$C^c D_t^\mu v + I_{ion} = I_{ext}. \tag{23.3}$$

Each individual ionic component has an associated conductance value denoted by G $(mS\,cm^{-2})$ and an equilibrium potential denoted by E (the potential for which the net ionic current flowing across the membrane is zero). Each current component is assumed to be proportional to the conductance times the driving force. Thus, the ionic current in (23.3) can be written as [18]:

$$I_{ion} = G_{Na}(v - E_{Na}) + G_K(v - E_K) + G_L(v - E_L). \tag{23.4}$$

Let the gating variables m, n, and h represent the sodium activation, potassium activation and sodium de-inactivation, respectively. If we consider a large number of channels, rather than an individual channel, we can also define a variable p to be the fraction of gates in that population that is in the permissive state and $(1 - p)$ to be the fraction in the nonpermissive state. In (23.4), sodium conductance is a function of both m and h, while the potassium conductance is a function of n only [13]. Transitions between permissive and nonpermissive states are assumed to obey the relation:

$$^c D_t^{\mu_p} p = \alpha_p(v)(1 - p) - \beta_p(v)p,$$
$$p = m, n, h, \quad 0 < \mu_p \le 1, \tag{23.5}$$

where α_p and β_p are voltage-dependent rate constants describing the nonpermissive to permissive and permissive to nonpermissive transition rates, respectively.

Equation (23.5) can be rewritten in the form:

$$^c D_t^{\mu_p} p = \frac{p_\infty - p}{\tau_p}, \tag{23.6}$$

where $p_\infty(v) = \frac{\alpha_p(v)}{\alpha_p(v) + \beta_p(v)}$, $\tau_p(v) = \frac{1}{\alpha_p(v) + \beta_p(v)}$.

The above model is a generalization of the HH model [12] given by:

$$C\frac{dv}{dt} = -G_{Na}(v - E_{Na}) - G_K(v - E_K) - G_L(v - E_L) + I_{ext} \tag{23.7}$$

$$\frac{dp}{dt} = \alpha_p(v)(1 - p) - \beta_p(v)p = \frac{p_\infty - p}{\tau_p}, \quad p = m, n, h \tag{23.8}$$

with the initial conditions, $v(0) = v_0$ and $p(0) = p_0 = p_\infty(v_0)$ [13].

Equations (23.7) and (23.8) are special cases of (23.3), (23.4), and (23.6). They can be obtained by passing to the limit as μ, $\mu_p \to 1$.

3   Solution of the Fractional Model

Taking the Laplace transform, denoted by an over bar, of both sides of (23.6), we obtain after some manipulations:

$$\bar{p}(s) = \frac{p_0}{s^{\mu_p} + 1/\tau_p} + p_\infty \left[\frac{1}{s} - \frac{s^{\mu_p - 1}}{\left(s^{\mu_p} + 1\tau_p \right)} \right]. \tag{23.9}$$

Inverting the Laplace transform in (23.9), we obtain:

$$p(t) = p_\infty - \left[p_\infty E_{\mu_p, 1} \left(-\frac{t^{\mu_p}}{\tau_p} \right) - p_0 t^{\mu_p - 1} E_{\mu_p, \mu_p} \left(-\frac{t^{\mu_p}}{\tau_p} \right) \right], \tag{23.10}$$

where $E_{\alpha, \beta}(z)$ is the Mittage function defined by [9]:

$$E_{\alpha, \beta}(z) = \sum_{k=0}^{\infty} \frac{z^k}{\Gamma(\beta + \alpha k)}, \quad \alpha > 0, \beta \in R, z \in C, \tag{23.11}$$

from which, we can find:

$$n(t) = n_\infty - \left[n_\infty E_{\mu_n, 1} \left(-\frac{t^{\mu_n}}{\tau_n} \right) - n_0 t^{\mu_n - 1} E_{\mu_n, \mu_n} \left(-\frac{t^{\mu_n}}{\tau_n} \right) \right], \tag{23.12}$$

$$m(t) = m_\infty - \left[m_\infty E_{\mu_m, 1} \left(-\frac{t^{\mu_m}}{\tau_m} \right) - m_0 t^{\mu_m - 1} E_{\mu_m, \mu_m} \left(-\frac{t^{\mu_m}}{\tau_m} \right) \right], \tag{23.13}$$

and

$$h(t) = h_\infty - \left[h_\infty E_{\mu_h, 1} \left(-\frac{t^{\mu_h}}{\tau_h} \right) - n_0 t^{\mu_h - 1} E_{\mu_h, \mu_h} \left(-\frac{t^{\mu_h}}{\tau_h} \right) \right]. \tag{23.14}$$

4   Numerical Results

The conductance for each current component can be written in the form [10],

$$G_{Na} = g_{Na} m^3 h, \quad G_k = -g_K n^3, \tag{23.15}$$

where g_{Na} and g_K are the maximal conductance of sodium and potassium, respectively, and G_L is a constant.

Substituting (23.15) into (23.4), we obtain:

$$I_{ion} = -g_{Na} m^3 h(v - E_{Na}) - g_K n^3 (v - E_K) - g_L(v - E_L). \tag{23.16}$$

We take the applied external current and the parameters of the problem as [13]:

$$I_{ext} = \begin{cases} 40 \text{ when } 0.1 \le t \le 0.3 \\ 0 \text{ otherwise} \end{cases}, \qquad (23.17)$$

and $g_{Na} = 120\,\text{mS/cm}^2$, $g_K = 36\,\text{mS/cm}^2$, $g_L = 0.3\,\text{mS/cm}^2$, $E_{Na} = 50\,\text{mV}$, $E_k = -77\,\text{mV}$, $E_L = 50\,\text{mV}$, and $v_0 = -65\,\text{mV}$

Note that: As μ_m, μ_n, and μ_h tend to one, we recover the HH model. In this case, the rates α_p and β_p will be approximated by [13]:

$$\alpha_n(v) = \frac{0.01(v+55)}{1-e^{-\frac{(v+55)}{10}}}, \quad \beta_n(v) = 0.125e^{-\frac{v}{80}}, \qquad (23.18)$$

$$\alpha_m(v) = \frac{0.1(v+40)}{1-e^{-\frac{(v+40)}{10}}}, \quad \beta_m(v) = 4e^{-\frac{(v+65)}{18}}, \qquad (23.19)$$

and

$$\alpha_h(v) = 0.07e^{-\frac{(v+40)}{10}}, \quad \beta_h(v) = \frac{1}{1+e^{-\frac{(v+35)}{10}}}. \qquad (23.20)$$

Using (23.18)–(23.20), (23.3) takes the form:

$$
^c D_t^\alpha v = -g_{Na}\left(m_\infty - \left[m_\infty E_{\mu_m,1}\left(-\frac{t^{\mu_m}}{\tau_m}\right) - m_0 t^{\mu_m-1} E_{\mu_m,\mu_m}\left(-\frac{t^{\mu_m}}{\tau_m}\right)\right]\right)^3
$$
$$
\left(h_\infty - \left[h_\infty E_{\mu_h,1}\left(-\frac{t^{\mu_h}}{\tau_h}\right) - n_0 t^{\mu_h-1} E_{\mu_h,\mu_h}\left(-\frac{t^{\mu_h}}{\tau_h}\right)\right]\right)(v - E_{Na})
$$
$$
-g_K\left(n_\infty - \left[n_\infty E_{\mu_n,1}\left(-\frac{t^{\mu_n}}{\tau_n}\right) - n_0 t^{\mu_n-1} E_{\mu_n,\mu_n}\left(-\frac{t^{\mu_n}}{\tau_n}\right)\right]\right)^3 (v - E_K)
$$
$$
-g_L(v - E_L) + I_{ext}. \qquad (23.21)
$$

To solve the above nonlinear fractional differential equation, we shall use a numerical predictor-corrector algorithm given by Diethelm for solving initial value problems with Caputo derivatives [7,8]. We have used the algorithm of Gorenflo [9] to evaluate the Mittage function.

The following figures illustrate the response of the membrane voltage for different values of μ.

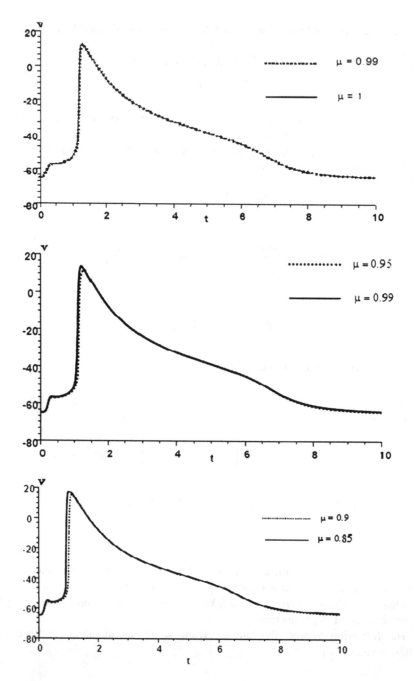

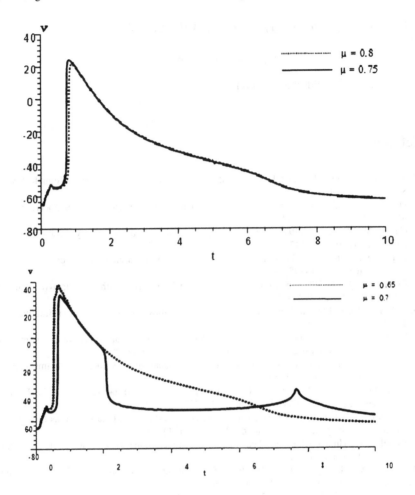

5   Discussion

The fractional HH model is a generalization of the original HH model. We believe that it may be applicable to a wider range of organisms. In this model four new parameters are introduced, we can obtain the action potential characteristics of a different organism by choosing the appropriate parameters in each case.

It can be noted that as these parameters tend to unity, the classical HH model is obtained. Also in our proposed model, the calculations of voltage-dependent rate constants α_p and β_p are dependent on the values of μ_p which reflect the generality, we can obtain different forms for α_p and β_p by changing the parameters.

In future work, we think that a specific data set should be used to describe how each parameter affects the shape of the action potential response.

Note that: as μ_m, μ_n, and μ_h tend to 1, (23.10) becomes:

$$p(t) = p_\infty - (p_\infty - p_0)e^{-t/\tau_p},$$

which is in accord with the HH model.

References

1. Baleanu D (2006) Fractional Hamiltonian analysis of irregular systems. Signal Process 86(10):2632–2636
2. Debnath L (2003) Recent applications of fractional calculus to science and engineering. Int J Mathematics Mathematical Sci 54:3413–3442
3. Caputo M, Mainardi F (1971) A new dissipation model based on memory mechanism. Pure Appl Geophys 91:134–147
4. Caputo M, Mainardi F (1971) Linear model of dissipation in anelastic solids. Rivis ta del Nuovo Cimento 1:161–198
5. Caputo M (1974) Vibrations on an infinite viscoelastic layer with a dissipative memory. J Acoust Soc Am 56:897–904
6. Cole KS (1933) Electrical excitation in nerve. Cold Spring Harbor Symp Quantitative Biol 1:131–137
7. Diethelm K, Ford NJ, Freed AD (2002) A predictor-corrector approach for the numerical solution of fractional differential equations. Nonlinear Dynamics 29:3–22
8. Diethelm K, Ford NJ, Freed AD (2004). Detailed error analysis for a fractional Adams method. Numer Algorithms 36(1):31–52
9. Gorenflo R, Loutchko I, Luchko Yu (2002) Computation of the Mittag-Leffler function $E_{\alpha,\beta}(z)$ and its derivatives. Fract Calculus Appl Anal 5:491–518
10. Hodgkin AL, Huxley AF (1952) Currents carried by sodium and potassium ions through the membrane of the giant axon of Loligo. J Physiol 116:449–472
11. Hodgkin AL, Huxley AF (1952) The components of membrane conductance in the giant axon of Loligo. J Physiol 116:473–496
12. Hodgkin AL, Huxley AF (1952) The dual effect of membrane potential on sodium conductance in the giant axon of Loligo. J Physiol 116:497–506
13. Hodgkin AL, Huxley AF (1952) A quantitative description of membrane current and its application to conduction and excitation in nerve. J Physiol 117:500–544
14. Kilbas A, Srivastava HM, Trujillo JJ (2006) Theory and applications of fractional differential equations. North-Holland Mathematics Studies
15. Magin RL (2006) Fractional calculus in bioengineering. Begell House Publisher, Connecticut
16. Oldham K, Spanier J (1974) The fractional calculus. Academic, New York
17. Podlubny I (1999) Fractional differential equations. Academic, New York
18. Westerlund S, Ekstam L (1994) Capacitor theory. IEEE Trans Diel Elect Insul 1:826–839

Chapter 24
An Application of Fractional Calculus to Dielectric Relaxation Processes

M.S. Çavuş and S. Bozdemir

1 Introduction

Fractional calculus, which is the field of mathematical analysis dealing with the investigation and applications of integrals and derivatives of arbitrary order, has attracted in recent years a considerable interest in many disciplines. It has been found that the behavior of many physical systems can be more properly defined by using the fractional theory. The flexibility of degrees of freedom, which is very easily obtained in the fractional theory, is one of the most important advantages of the fractional order modeling. Moreover, in recent years, the use of the fractional calculus in the analysis of the fractional diffusion equations has been a field of increasing interest [5, 11, 13–16, 21, 22].

1.1 The Fractional Integral and Riemann–Liouville Fractional Derivative

According to the Riemann–Liouville approach, the fractional integral of order $\alpha > 0$ is defined as,

$$_aJ_t^{-\alpha}U(t) = \frac{1}{\Gamma(\alpha)} \int_a^t (t-\tau)^{\alpha-1}U(\tau)d\tau \tag{24.1}$$

$$_aJ_t^0 U(t) = U(t) \tag{24.2}$$

M.S. Çavuş (✉)
Department of Physics, Faculty of Arts and Sciences, Kastamonu University, Kastamonu, Turkey
e-mail: mserdarcavus@kastamonu.edu.tr

S. Bozdemir
Department of Physics, Faculty of Arts and Sciences, Çukurova University, Adana, Turkey
e-mail: sbozdemir@cu.edu.tr

D. Baleanu et al. (eds.), *Fractional Dynamics and Control*,
DOI 10.1007/978-1-4614-0457-6_24, © Springer Science+Business Media, LLC 2012

Moreover, for $\alpha, \beta > 0, t > 0$, and $\upsilon > -1$ (24.1) has the following properties:

$$J_t^{-\alpha} J_t^{-\beta} U(t) = J_t^{-(\alpha+\beta)} U(t) \tag{24.3}$$

and

$$J_t^{-\alpha} t^\upsilon = \frac{\Gamma(\upsilon+1)}{\Gamma(\upsilon+1+\alpha)} t^{\upsilon+\alpha} \tag{24.4}$$

Also,

$$_a D_t^p U(t) = \left(\frac{d}{dt}\right)^{m+1} \int_a^t (t-\tau)^{m-p} U(\tau) d\tau. \tag{24.5}$$

The expression (24.5) is the most widely known definition of the fractional derivative and is usually called the Riemann–Liouville fractional derivative definition.
The most important property of the Riemann–Liouville fractional approach is
given by:

$$_a D_t^\alpha \left(_a J_t^{-\alpha} U(t)\right) = U(t) \tag{24.6}$$

The Riemann–Liouville fractional differentiation operator is a left inverse to the
Riemann–Liouville fractional integration operator of the same order α. The detailed
properties of the operator J^α and D^p can be found in [17, 19, 20].

1.2 Adomian Decomposition Method

Adomian decomposition method (ADM) introduced by Adomian in 1980 has
proved to be a very useful tool in the solution of nonlinear functional equations.
The decomposition method consists of finding a solution in the form,

$$U(x,t) = \sum_{n=0}^{\infty} U_n(x,t), \tag{24.7}$$

where the components $U_n(x,t)$ will be determined recursively. More information
about ADM can be found in [1].

2 Dielectric Relaxation Processes

Relaxation properties are generally expressed in terms of time-domain response
function $f(t)$ or of the frequency-dependent real and imaginary components of its
Fourier transform [23]:

$$\tilde{f}(i\omega) = \int_0^\infty e^{-i\omega t} f(t) dt = \phi'(\omega) - i\phi''(\omega). \tag{24.8}$$

Classically, relaxation processes are described in terms of the exponential function:

$$\varphi(t) = \exp(-t/\tau), t \geq 0 \tag{24.9}$$

that is generally referred to as Maxwell–Debye relaxation. However, in many systems the dynamical behavior shows conspicuous deviations from the ideal exponential pattern. Therefore, in general the empirical expressions, involving adjustable parameters, have been widely used in the literature.

Commonly three general relaxation laws are encountered in the experimental studies of complex systems:

(i) Stretched exponential (KWW) function [24]

$$f(t) \approx \exp[-(t/\tau)^\alpha], 0 < \alpha < 1, t > \tau \tag{24.10}$$

(ii) Exponential–logarithmic function

$$f(t) \approx \exp[-B\ln^\alpha(t/\tau)] \tag{24.11}$$

(iii) Algebraic decay function

$$f(t) \approx (t/\tau)^\alpha \tag{24.12}$$

where α, τ, and B are the appropriate fitting parameters [18].

By definition, the normalized susceptibility, $\chi(\omega)$, is connected to the normalized relaxation function through the relation:

$$\chi(\omega) = \int_0^\infty e^{-i\omega t} d(-\varphi(t)) = 1 - i\omega \int_0^\infty e^{-i\omega t} \varphi(t) dt, \tag{24.13}$$

where $\varphi(t) = \Phi(t)/\varphi(0)$. A significant amount of experimental data on disordered systems supports the following empirical expressions for dielectric loss spectra, namely, the Cole–Cole equation [4],

$$\chi(\omega) = \frac{\chi_0}{1 + (i\omega\tau)^\alpha}, 0 < \alpha \leq 1 \tag{24.14}$$

the Cole–Davidson equation [6],

$$\chi(\omega) = \frac{\chi_0}{(1 + i\omega\tau)^\beta}, 0 < \beta \leq 1 \tag{24.15}$$

and the Havriliak–Negami equation [8] considered as a general expression for the universal relaxation law [10],

$$\chi(\omega) = \frac{\chi_0}{(1 + (i\omega\tau)^\alpha)^\beta}, 0 < \alpha \text{ and } \beta \leq 1. \tag{24.16}$$

Here, we should point out that the Havriliak–Negami equation is a combination of the Cole–Cole and Cole–Davidson equations.

3   The Ising Model and Fractional Relaxation

The spin–spin time correlation functions in a one-dimensional Ising model [9] with Glauber dynamics [7] was studied by Bozdemir [3], and later by Brey and Parados [2]. The main idea in those studies is the spin time autocorrelation function obtained in the one-dimensional Ising model with Glauber dynamics which is assumed to be identical with the dipole correlation function of a molecular chain. Based on this assumption, the system can be analyzed in the following way: The energy of the system in the one-dimensional Ising model for a spin configuration σ is

$$H(\sigma) = -J \sum_i \sigma_i \sigma_{i+1}, \tag{24.17}$$

where J is a positive coupling constant. The state of the system is specified by the spin vector $\sigma = \{\sigma_i\}$, where $\sigma_i = \pm 1$ is the spin at site i. The evolution of the system is described by a Markov process with Glauber dynamics. So, the conditional probability $P_{1/1}(\sigma, t/\sigma', t')$ of finding the system in the state σ at a time t, provided that it was given in the state σ' at a time t', obeys the master equation:

$$\frac{\partial P_{1/1}(\sigma, t/\sigma', t')}{\partial t} = \sum_{i=-\infty}^{\infty} \left[\omega_i(R_i \sigma) p_{1/1}(R_i \sigma, t/\sigma', t') \right.$$
$$\left. - \omega_i(\sigma) p_{1/1}(\sigma, t/\sigma', t') \right], \tag{24.18}$$

where $R_i \sigma$ is the configuration obtained from σ by flipping the ith spin and $\omega_i(\sigma)$ is the transition rate for the flip. Following the above procedure, the spin–spin–time correlation function, in the low temperature limit, was found by Brey and Parados as the following differential equation,

$$\frac{\partial f_n}{\partial t} = -\alpha f_n(t) + \frac{\alpha\gamma}{2}[f_{n-1} + f_{n+1}], \tag{24.19}$$

where n is an integer in the range $-\infty < n < \infty$, α is a positive constant defining the time scale of the evolution of the system and γ is a function of temperature T of the heat bath given as:

$$\gamma = \tanh\frac{2J}{k_B T}, \tag{24.20}$$

where k_B is the Boltzmann's constant. Equation (24.19), which is a function of time and position, can be expressed as:

$$\frac{\partial f(x,t)}{\partial t} = -\alpha f(x,t) + \frac{\alpha\gamma}{2}[f(x-1,t) + f(x+1,t)]. \qquad (24.21)$$

If one takes the Taylor expansion of Equation (24.21) and retains only terms up to second order, one obtains a diffusion type equation:

$$\frac{\partial f(x, t)}{\partial t} = (\alpha\gamma - \alpha) f(x, t) + \frac{\alpha\gamma}{2}\frac{\partial^2 f(x, t)}{\partial x^2} \qquad (24.22)$$

If equation (24.22) is converted to fractional differential equation form, one gets

$$D_t^\xi f(x,t) = (\alpha\gamma - \alpha) f(x,t) + \frac{\alpha\gamma}{2}\frac{\partial^2 f(x,t)}{\partial x^2}, \qquad (24.23)$$

where D_t^ξ is the Riemann–Liouville fractional differentiation operator, and the initial condition for $f(x,t)$ is

$$f(x,0) = e^{-|x|}. \qquad (24.24)$$

We adopt ADM for solving (24.23). According to this method we assume that

$$f(x,t) = \sum_{n=0}^{\infty} f_n(x,t). \qquad (24.25)$$

Now, the fractional differential equation (24.25) can be written as, for $\upsilon + \zeta = 1$,

$$D_t^\upsilon (D_t^\xi f(x,t)) = (\alpha\gamma - \alpha)D_t^\upsilon f(x,t) + \frac{\alpha\gamma}{2}D_t^\upsilon \frac{\partial^2 f(x,t)}{\partial x^2}. \qquad (24.26)$$

If we operate on both sides of this relation with integral operator Ω_t^{-1}, we reach to

$$\Omega_t^{-1} D_t^\upsilon \left(D_t^\xi f(x,t)\right) = (\alpha\gamma - \alpha)\Omega_t^{-1}(D_t^\upsilon f(x,t)) + \frac{\alpha\gamma}{2}\Omega_t^{-1}\left(D_t^\upsilon \frac{\partial^2 f(x,t)}{\partial x^2}\right)$$

$$(24.27a)$$

$$f(x,t) = (\alpha\gamma - \alpha)\Omega_t^{-1}(D_t^\upsilon f(x,t)) + \frac{\alpha\gamma}{2}\Omega_t^{-1}\left(D_t^\upsilon \frac{\partial^2 f(x,t)}{\partial x^2}\right). \qquad (24.27b)$$

Moreover, the recursive relations related to the above equation are given in the following forms:

$$f(0) = f(x,0) = e^{-|x|}$$

$$f(1) = (\alpha\gamma - \alpha)D_t^{-\xi} f(0) + \frac{\alpha\gamma}{2}D_t^{-\xi}\frac{\partial^2 f(0)}{\partial x^2} = \left(\alpha\gamma - \alpha + \frac{\alpha\gamma}{2}\right)\frac{e^{-|x|}t^\xi}{\Gamma(\xi + 1)}.$$

$$f(2) = (\alpha\gamma - \alpha)D_t^{-\xi}f(1) + \frac{\alpha\gamma}{2}D_t^{-\xi}\frac{\partial^2 f(1)}{\partial x^2} = \left(\alpha\gamma - \alpha + \frac{\alpha\gamma}{2}\right)^2 \frac{e^{-|x|}t^{2\xi}}{\Gamma(2\xi + 1)}.$$

$$f(3) = (\alpha\gamma - \alpha)D_t^{-\xi}f(2) + \frac{\alpha\gamma}{2}D_t^{-\xi}\frac{\partial^2 f(2)}{\partial x^2} = \left(\alpha\gamma - \alpha + \frac{\alpha\gamma}{2}\right)^3 \frac{e^{-|x|}t^{3\xi}}{\Gamma(3\xi + 1)}.$$

and so on. Therefore, the solution (24.27b) becomes:

$$f(x,t) = \sum_{n=0}^{\infty} \left(\alpha\gamma - \alpha + \frac{\alpha\gamma}{2}\right)^n \frac{e^{-|x|}t^{n\xi}}{\Gamma(n\xi + 1)}$$

$$= e^{-|x|}E_\xi\left\{\alpha(-1 + 3\gamma/2)t^\xi\right\}, \quad 0 < \xi < 1, \qquad (24.28)$$

where $E_\xi\{\cdot\}$ is the Mittag–Leffler function given by:

$$E_v\{Z\} = \sum_{n=0}^{\infty} \frac{Z^n}{\Gamma(vn + 1)}. \qquad (24.29)$$

If we assume that the position of dipoles located between x and $x + x_0$ have a probability density given by:

$$f(x) = \frac{1}{x_0}e^{(-x/x_0)}, \qquad (24.30)$$

and substitute it into (24.30), integrate over all the space, we can obtain the time-dependent correlation function:

$$f(t) = \int_0^\infty e^{(-x/x_0)}e^{-|x|}E_\xi\{\alpha(-1 + 3\gamma/2)t^\xi\}dx = \frac{E_\xi\{\alpha(-1 + 3\gamma/2)t^\xi\}}{1 + x_0}, \quad (24.31)$$

where x_0 is the average value of x, and $1/2x_0$ is the average number of dipoles per unit length.

If equation (24.31) is substituted into (24.13), one obtains

$$\chi(\omega) = 1 - i\omega \int_0^\infty e^{-i\omega t}\frac{E_\xi\{\alpha(-1 + 3\gamma/2)t^\xi\}}{1 + x_0}dt. \qquad (24.32)$$

From this expression, in the frequency zone, the empiric Cole–Cole type equation is obtained as:

$$\chi(\omega) = \frac{x_0}{1 + (i\omega\tau)^\xi}, \quad 0 < \xi \leq 1, \qquad (24.33)$$

where $\tau = [\alpha(-1 + 3\gamma/2)]^{-\xi}$, $\chi_0 = 1 + \lambda(i\omega\tau)^\xi$, and $\lambda = 1 - 1/(1 + x_0)$.

Moreover, for sufficiently small times, (24.33), which is a Mittag–Leffter type function, exhibits the same behavior as with the stretched exponential function [12]:

$$f(t) \approx 1 - \frac{(t/\tau)^{\xi}}{\Gamma(\xi+1)} + \ldots \approx \exp\left[-\frac{(t/\tau)^{\xi}}{\Gamma(\xi+1)}\right], \quad 0 < t \ll 1, \tag{24.34}$$

which is known as Kolraush–William–Watts (KWW) function. Also, by using the asymptotic expansions it can be written as,

$$f(t) \approx \frac{\Gamma(\xi)\sin(\pi\xi)}{\pi}(t/\tau)^{-\xi}, \quad t \to \infty, \tag{24.35}$$

which has the same form with that of empirical algebraic decay function (24.12). When (24.19) is solved by using the eigenfunctions method with appropriate boundary conditions, which was done by [2],

$$\chi(\omega)\frac{\alpha(1-\eta^2)}{(1+\eta^2)\left[(i\omega+\alpha)^2 - \alpha^2\gamma^2\right]}^{1/2} \tag{24.36}$$

is obtained, where $\eta = \tan J/K_B T$. This expression, at low temperature, converts to a special case of the Cole–Davidson equation:

$$\chi(\omega) = \frac{1}{(1+i\omega\tau)^{1/2}}, \tag{24.37}$$

where τ is the relaxation time [2].

4   Conclusion

In this study, it is shown that fractional solution of the diffusion equation obtained from the stochastic Ising model, where we used the Adomian decomposition method for solving the fractional diffusion equation, gives a non-Debye type behavior which can also be represented by the Mittag–Leffler decay function. Furthermore, we may say that fractional dynamics in polar dielectric systems are a result of fractional dipole distribution in the medium. In the fractional approaches, the variable parameter α, especially used in the forming of the fractional order modeling, exhibits that the space of physical processes has a fractional form. Therefore, the irregularity (or chaos) in the nature compels us to use the fractional theory.

We have seen that the fractional order of the differential equations, which is compatible with most of the experimental results, is generally smaller than that of the integer order of differential equations. Likely, in the medium, the nearest neighbor interactions between dipoles or charged particles have not the same behaviors as that of the linear systems in terms of times and velocities, because the time (or energy) is fractionally changing in time. The local spaces of charged particles which have different time and energy intervals should lead to be resulted to have fractional order differential equations. Moreover, it may be said that in the atomic levels (or electronic) the flow of the time is quantized. Therefore, the interactions between dipoles in questions may also be quantized in time domain. That is, quantization of the energy may be a result of the time quantization. As a result of these processes, the order of the differential equations should be changing during the interaction times.

Acknowledgements We thank our friends Prof. Dr Kerim Kıymaç and Prof. Dr Metin Özdemir for their reading and correcting the article.

References

1. Adomian G (1994) Solving frontier problems of physics: The decomposition method, Kluwer Academic, Dordrecht
2. Brey JJ, Prados A (1996) Low-temperature relaxation in the one-dimensional Ising model, Phys Rev E 53(1):458–464
3. Bozdemir S (1981) Phys Status Solidi B 103:459, Phys Status Solidi B 104 (1981) 37
4. Cole KS, Cole RN (1941) Dispersion and absorption in dielectrics: 1. Alternating current characteristics. J Chem Phys 9:341
5. Das S (2009) A note on fractional diffusion equations.Chaos Solitons Fractals 42:2074–2079
6. Davidson DW, Cole RH (1951) Dielectric relaxation in glycerol propylene and n-propanol. J Chem Phys. 19:1484–1490
7. Glauber RG (1963) Time-dependent statistics of the Ising model. J Math Phys 4:294
8. Havriliak S, Negami S (1966) A complex plane analysis of α dispersions in some polymers. J Polym Sci 14(B):99–117
9. Ising E (1925) Z Phys 31:253
10. Jonscher AK (1983) Dielectric relaxation in solids. Chelsea Dielectrics, London
11. Mainardi F, Luchko Y, Pagnini G (2001) The fundamental solution of the space–time fractional diffusion equation, Fractional Calculus Appl Anal 4(2):153–192
12. Mainardi F, Raberto M, Gorenflo R, Scalas E (2000) Fractional calculus and continuous-time finance II: The waiting-time distribution. *arXiv:cond-mat/*0006454 **v2** 11 Nov
13. Mainardi F (1997) Fractional calculus: some basic problems in continuum and statistical mechanics, In: Fractals and fractional calculus in continuum mechanics, Springer-Verlag, New York, pp 291–348
14. Metzler R, Klafter J (2002) From stretched exponential to inverse power-law: fractional dynamics, Cole–Cole relaxation processes, and beyond. J Non Cryst Solids 305:81–87
15. Metzler R, Klafter J (2000) The random walk's guide to anomalous diffusion: a fractional dynamics approach, Phys Rep 339:1–77
16. Metzler R, Barkai E, Klafter J (1999) Anomalous diffusion and relaxation close to thermal equilibrium: a fractional Fokker–Planck equation approach. Phys Rev Lett 82(18):3563–3567
17. Miller KS, Ross B (1993) An introduction to the fractional calculus and fractional differential equations. Wiley, New York
18. Novikov VV, Wojciechowski KW, Privalko VP (2000) Anomalous dielectric relaxation of inhomogeneous media with chaotic structure. J Phys Condens Matter 12:4869–4879 (printed in the UK)
19. Oldham KB, Spanier J (1974) The fractional calculus. Academic, New York
20. Podlubny I (1999) Fractional differential equations. Academic, New York

21. Ray SS (2008) A new approach for the application of Adomian decomposition method for the solution of fractional space diffusion equation with insulated ends. Appl Math Comput 202:544–549
22. Schneider WR, Wyss W (1989) Fractional diffusion and wave equations. J Math Phys 30:134–144
23. Uchaikin VV (2003) Relaxation processes and fractional differential equations. Int J Theor Phys 42(1), pp. 121–134
24. Williams G, Watts DC, Dev SB, North AM (1971) Further considerations of non-symmetrical dielectric relaxation behaviour arising from a simple empirical decay function. Trans Faraday Soc 67:1323335

Chapter 25
Fractional Wave Equation for Dielectric Medium with Havriliak–Negami Response

R.T. Sibatov, V.V. Uchaikin, and D.V. Uchaikin

1 Introduction

When a step-wise electric field is applied, polarization of a material approaches its equilibrium value not instantly but after some time. Hereditary effect of polarization is expressed through the integral relation. For electric induction in an isotropic medium, we have [1]

$$\mathbf{D}(t) = \varepsilon_\infty \, \mathbf{E}(t) + \int\limits_{-\infty}^{t} \kappa(t - t') \, \mathbf{E}(t')\mathrm{d}t'. \qquad (25.1)$$

Fourier transformation of the last expression gives

$$\widetilde{\mathbf{D}}(\omega) = \varepsilon^*(\omega)\widetilde{\mathbf{E}}(\omega), \quad \varepsilon^*(\omega) = \varepsilon_\infty + \widetilde{\kappa}(\omega).$$

Relaxation properties of different media (dielectrics, semiconductors, ferromagnetics, and so on) are normally expressed in terms of time-domain response function $\phi(t)$ which represents the current flowing under the action of a step-function electric field, or of the frequency-dependent real and imaginary components of its Fourier transform:

$$\widetilde{\phi}(\omega) = \int\limits_{0}^{\infty} \mathrm{e}^{-\mathrm{i}\omega t}\phi(t)\mathrm{d}t = \phi'(\omega) - \mathrm{i}\phi''(\omega).$$

R.T. Sibatov (✉) • V.V. Uchaikin • D.V. Uchaikin
Ulyanovsk State University, 42 Leo Tolstoy street, Ulyanovsk, Russia
e-mail: ren_sib@bk.ru; vuchaikin@gmail.com; uchaikin@gmail.com

D. Baleanu et al. (eds.), *Fractional Dynamics and Control*,
DOI 10.1007/978-1-4614-0457-6_25, © Springer Science+Business Media, LLC 2012

The Fourier transform of the response function $\phi(t)$ is related to the frequency dependence of dielectric permittivity through the following relation

$$\tilde{\phi}(\omega) = \frac{\varepsilon^*(\omega) - \varepsilon_\infty}{\varepsilon_s - \varepsilon_\infty}, \quad \tilde{\kappa}(\omega) = (\varepsilon_s - \varepsilon_\infty)\tilde{\phi}(\omega).$$

Here $\varepsilon_s = \varepsilon^*(0)$ is the stationary dielectric permittivity.

The classical Debye expression (see [2]) for a system of noninteracting randomly oriented dipoles freely floating in a neutral viscous liquid is

$$\tilde{\phi}(\omega) = \frac{1}{1 + i\omega\tau},$$

where τ is the temperature-dependent relaxation time characterizing the Debye process:

$$\phi(t) = \phi(0)\exp(-t/\tau), \quad t > 0.$$

The latter function obeys the simple differential equation

$$\frac{d\phi(t)}{dt} + \tau^{-1}\phi(t) = 0.$$

Numerous experimental data gathered, for instance, in books [3, 4] convincingly show that this theory is not able to describe relaxation processes in solids, and the relaxation behavior may deviate considerably from the exponential Debye pattern and exhibits a broad distribution of relaxation times. There exist a few other empirical response functions for solids: the Cole–Cole function [5]

$$\tilde{\phi}(\omega) = \frac{1}{1 + (i\omega/\omega_p)^\alpha}, \quad 0 < \alpha < 1, \tag{25.2}$$

the Cole–Davidson function [6]

$$\tilde{\phi}(\omega) = \frac{1}{(1 + (i\omega/\omega_p))^\beta}, \quad 0 < \beta < 1, \tag{25.3}$$

and the Havriliak–Negami function [7]

$$\tilde{\phi}(\omega) = \frac{1}{(1 + (i\omega/\omega_p)^\alpha)^\beta}, \quad 0 < \alpha, \beta < 1. \tag{25.4}$$

Here ω_p is the peak of losses, and α and β are constant parameters.

In this chapter, we obtain the fractional relaxation equation in dielectrics with response function of the Havriliak–Negami type. With the help of definitions of the functional analysis, we derive the explicit expression for fractional operator in this equation. Then we construct a Monte Carlo algorithm for calculation of action of this operator and of the inverse one. The algorithm represents a numerical way of calculation in relaxation problems with arbitrary initial and boundary conditions. Then, we obtain the equation for propagation of electromagnetic waves in such dielectric media.

2 Havriliak–Negami's Relaxation

The most general approximation for frequency dependence of response function is given by two-parameter formula proposed by [7]. The solution of the corresponding fractional differential equation

$$[1 + (\tau\,{}_0D_t)^\alpha]^\beta\,\phi(t) = \delta(t), \qquad \tau = 1/\omega_p, \tag{25.5}$$

based on the expansion of fractional power of operator sum into infinite Newton's series

$$[1 + (\tau\,{}_0D_t)^\alpha]^\beta = \sum_{n=0}^{\infty} \binom{\beta}{n} (\tau\,{}_0D_t)^{\alpha(\beta-n)},$$

has been obtained by [8]:

$$\phi(t) = -\frac{1}{\Gamma(\beta)} \sum_{n=0}^{\infty} \frac{(-1)^n \Gamma(n+\beta)}{n!\Gamma(\alpha(n+\beta))} \left(\frac{t}{\tau}\right)^{\alpha(n+\beta)}.$$

The operator $\left[1 + \omega_p^{-\alpha}({}_0D_t)^\alpha\right]^\beta$ can also be presented in the form [9]:

$$W_{\omega_p}^{\alpha,\beta} f(t) = [1 + \omega_p^{-\alpha}\,{}_0D_t^\alpha]^\beta f(t) = \omega_p^{-\alpha\beta} \exp\left(-\frac{\omega_p^\alpha t}{\alpha}\,{}_0D_t^{1-\alpha}\right)\,{}_0D_t^{\alpha\beta} \exp\left(\frac{\omega_p^\alpha t}{\alpha}\,{}_0D_t^{1-\alpha}\right) f(t).$$

The HN function is considered as a general expression for the universal relaxation law [10]. This universality is observed in dielectric relaxation in dipolar and nonpolar materials, conduction in hopping electronic semiconductors, conduction in ionic conductors, trapping in semiconductors, decay of delayed luminescence, surface conduction on insulators, chemical reaction kinetics, mechanical relaxation, and magnetic relaxation. Despite of quite different intrinsic mechanisms, the processes manifest astonishing similarity [10]. The situation seems to be similar to diffusion processes. Random movements of small pollen grain visible under a microscope, and neutrons in nuclear reactors, electrons in semiconductors are quite different processes from physical point of view, but they are the same processes of Brownian motion from stochastic point of view. This analogy stimulates search of an appropriate stochastic model for the universal relaxation law. Investigations of such kind have been carried out in the works [11–18].

Weron et al. [19] have shown how to modify the random-walk scheme underlying the classical Debye response to derive the empirical Havriliak–Negami function. Moreover, they have derived formulas for simulation of random variables with probability density function, Fourier transform of which is the Havriliak–Negami function. These relations contain stable random variables.

Coffey et al. [16] reformulated the Debye theory of dielectric relaxation of an assembly of polar molecules using a fractional noninertial Fokker–Planck equation to explain anomalous dielectric relaxation.

Déjardin [17] considered the fractional approach to the orientational motion of polar molecules acted on by an external perturbation. The problem is treated in terms of noninertial rotational diffusion (configuration space only) which leads to solving a fractional Smoluchowski equation. This model is in a good agreement with experimental data for the third-order nonlinear dielectric relaxation spectra of a ferroelectric liquid crystal.

Here with the help of relations of the functional analysis for fractional powers of operators, we derive a new numerical algorithm of solution of fractional equations corresponding to the Havriliak–Negami response. This algorithm is based on Monte Carlo simulation of one-sided stable random variables.

3 Fractional Operator Corresponding to the Havriliak–Negami Response

Let the equi-continuous semigroup $\{T_t; t \geq 0\}$ of C_0-class be defined on locally convex linear topological B-space F. The infinitesimal generating operator A of the semigroup $\{T_t\}$ is defined as

$$Af = \lim_{h\downarrow 0} h^{-1}(T_h - I) f$$

with domain

$$D(A) = \{f \in F; \lim_{h\downarrow 0} h^{-1}(T_h - I) f \text{ exists in } F\}.$$

According to S. Bochner [20] and R. S. Phillips [21], the operators

$$\widetilde{T}_{t,\alpha} x \equiv \widetilde{T}_t x = \begin{cases} \int\limits_0^\infty t^{-1/\alpha} g^{(\alpha)}(t^{-1/\alpha}s)\, T_s x\, ds, & t > 0, \\ x, & t = 0 \end{cases},$$

where $g^{(\alpha)}(t)$ is a one-sided stable density, which constitute an equi-continuous semigroup of C_0 class.

The corresponding infinitesimal operators are connected through the following relation [22]:

$$\tilde{A}_\alpha x = -(-A)^\alpha x, \qquad \forall x \in D(A).$$

The infinitesimal operator $-[1 +_{-\infty} D_t^\alpha]$ generates the semigroup

$$T_h = e^{-h} e^{-h_{-\infty} D_z^\alpha},$$

According to the Bochner–Philips relation, the semigroup generated by the infinitesimal operator $[1 +_{-\infty} D_t^\alpha]^\beta$, where $\beta < 1$, has the form

$$\widehat{T}_t f = \int_0^\infty t^{-1/\beta} g^{(\alpha)} \left(t^{-1/\beta} \tau \right) \widetilde{T}_\tau f \, d\tau$$

$$= \int_0^\infty d\tau \, e^{-\tau} t^{-1/\beta} g^{(\beta)} \left(t^{-1/\beta} \tau \right) \int_0^\infty \tau^{-1/\alpha} g^{(\alpha)} \left(\tau^{-1/\alpha} u \right) T_u f \, du.$$

Considering this integral as averaging over ensembles of stable random variables, we obtain the following relationship:

$$\widehat{T}_t f = \left\langle \exp\left(-t^{1/\beta} S_\beta \right) f \left(z - \left[t^{1/\beta} S_\beta \right]^{1/\alpha} S_\alpha \right) \right\rangle.$$

From this semigroup, we can obtain the corresponding infinitesimal operator $[1 +_{-\infty} D_t^\alpha]^\beta$.

To find the inverse operator

$$[1 +_{-\infty} D_t^\alpha]^{-\beta}, \qquad 0 < \alpha, \beta < 1,$$

we use the relation for potential operator

$$A^{-1} f = \int_0^\infty T_s f \, ds.$$

Consequently,

$$[1 +_{-\infty} D_t^\alpha]^{-\beta} f = \left\langle \int_0^\infty \exp\left(-t^{1/\beta} S_\beta \right) f \left(z - \left(t^{1/\beta} S_\beta \right)^{1/\alpha} S_\alpha \right) dt \right\rangle.$$

Here S_α and S_β are one-sided stable random variables with characteristic exponents $0 < \alpha \leq 1$ and $0 < \beta \leq 1$.

Introducing exponentially distributed random variable U, we arrive at

$$[1 + {}_{-\infty}D_t^\alpha]^{-\beta} f = \left\langle \int_0^\infty e^{-\xi} f\left(z - S_\alpha \xi^{1/\alpha}\right) \frac{\beta \xi^{\beta-1}}{S_\beta^\beta} d\xi \right\rangle$$

$$= \beta \left\langle S_\beta^{-\beta} U^{\beta-1} f\left(z - S_\alpha U^{1/\alpha}\right) \right\rangle. \qquad (25.6)$$

We use this formula to find the solution of fractional relaxation equation for arbitrary prehistories of charging–discharging process.

4   Fractional Wave Equation

Substituting the Havrilyak–Negami dependence of permittivity on frequency

$$\varepsilon^*(\omega) = \varepsilon_\infty + \frac{\varepsilon_s - \varepsilon_\infty}{[1 + (i\omega/\omega_p)^\alpha]^\beta}$$

into the Fourier transform of (25.1), we obtain

$$\mathbf{D}(t) = \varepsilon_\infty \mathbf{E}(t) + (\varepsilon_s - \varepsilon_\infty) \, \omega_p^{\alpha\beta} \times \exp\left(-\frac{\omega_p^\alpha}{\alpha} t \, {}_{-\infty}D_t^{1-\alpha}\right) \, {}_{-\infty}I_t^{\alpha\beta} \exp\left(\frac{\omega_p^\alpha}{\alpha} t \, {}_{-\infty}D_t^{1-\alpha}\right) \mathbf{E}(t).$$

Here special forms of fractional operators arise

$$\mathsf{W}_{\omega_p}^{\alpha,\beta} f(t) = [1 + \omega_p^{-\alpha} \, {}_{-\infty}D_t^\alpha]^\beta f(t)$$

$$= \omega_p^{-\alpha\beta} \exp\left(-\frac{\omega_p^\alpha t}{\alpha} \, {}_{-\infty}D_t^{1-\alpha}\right) \, {}_{-\infty}D_t^{\alpha\beta} \exp\left(\frac{\omega_p^\alpha t}{\alpha} \, {}_{-\infty}D_t^{1-\alpha}\right) f(t).$$

The inverse operator has the form

$$\mathsf{W}_{\omega_p}^{\alpha,-\beta} f(t) = [1 + \omega_p^{-\alpha} \, {}_{-\infty}D_t^\alpha]^{-\beta} f(t)$$

$$= \omega_p^{\alpha\beta} \exp\left(-\frac{\omega_p^\alpha}{\alpha} t \, {}_{-\infty}D_t^{1-\alpha}\right) \, {}_{-\infty}I_t^{\alpha\beta} \exp\left(\frac{\omega_p^\alpha}{\alpha} t \, {}_{-\infty}D_t^{1-\alpha}\right) f(t).$$

The following asymptotical relationships take place:

$$\mathsf{W}_{\omega_p}^{\alpha,\beta} f(t) \sim \begin{cases} [1 + \beta \omega_p^{-\alpha} \, {}_{-\infty}D_t^\alpha] f(t), & t \gg 1/\omega_p, \\ \omega_p^{-\alpha\beta} \, {}_{-\infty}D_t^{\alpha\beta} f(t), & t \ll 1/\omega_p, \end{cases} \qquad (25.7)$$

$$W_{\omega_p}^{\alpha,-\beta} f(t) \sim \begin{cases} [1 - \beta \omega_p^{-\alpha} \, _{-\infty}D_t^{\alpha}]f(t), & t \gg 1/\omega_p, \\ \omega_p^{\alpha\beta} \, _{-\infty}I_t^{\alpha\beta} f(t), & t \ll 1/\omega_p. \end{cases} \tag{25.8}$$

Maxwell's equations,

$$\text{rot } \mathbf{H} = \frac{4\pi}{c}\mathbf{j} + \frac{1}{c}\frac{\partial \mathbf{D}}{\partial t},$$

$$\text{rot } \mathbf{E} = -\frac{1}{c}\frac{\partial \mathbf{B}}{\partial t},$$

in combination with the material relations

$$\mathbf{D} = \varepsilon_\infty \mathbf{E} + (\varepsilon_s - \varepsilon_\infty)[1 + \omega_p^{-1} \, _{-\infty}D_t^{\alpha}]^{-\beta}\, \mathbf{E}, \quad \mathbf{B} = \mu\mathbf{H}$$

lead to the following wave equation

$$\frac{\mu\varepsilon_\infty}{c^2}\frac{\partial^2 \mathbf{E}}{\partial t^2} + \frac{\mu(\varepsilon_s - \varepsilon_\infty)}{c^2}[1 + \omega_p^{-1} \, _{-\infty}D_t^{\alpha}]^{-\beta}\frac{\partial^2 \mathbf{E}}{\partial t^2} + \nabla(\text{div } \mathbf{E}) - \nabla^2\mathbf{E} = \frac{4\pi\mu}{c^2}\frac{\partial \mathbf{j}}{\partial t}. \tag{25.9}$$

At small times, we have $(t \ll 1/\omega_p)$

$$\frac{\mu\varepsilon_\infty}{c^2}\frac{\partial^2 \mathbf{E}}{\partial t^2} + \frac{\mu(\varepsilon_s - \varepsilon_\infty)}{c^2}\omega_p^{\alpha\beta} \, _{-\infty}D_t^{2-\alpha\beta}\, \mathbf{E} + \nabla(\text{div } \mathbf{E}) - \nabla^2\mathbf{E} = \frac{4\pi\mu}{c^2}\frac{\partial \mathbf{j}}{\partial t}.$$

At large times, we have $(t \gg 1/\omega_p)$

$$\frac{\mu\varepsilon_s}{c^2}\frac{\partial^2 \mathbf{E}}{\partial t^2} - \frac{\mu(\varepsilon_s - \varepsilon_\infty)}{c^2}\beta\omega_p^{-\alpha} \, _{-\infty}D_t^{2+\alpha}\, \mathbf{E} + \nabla(\text{div } \mathbf{E}) - \nabla^2\mathbf{E} = \frac{4\pi\mu}{c^2}\frac{\partial \mathbf{j}}{\partial t}.$$

The wave equation presented above is concordant with the equation obtained by Tarasov [23] from Jonscher's universal law.

5 Conclusion

Let us summarize the results of this chapter. The fractional relaxation equation (25.5) and the fractional wave equation (25.9) for dielectrics with the response function of the Havriliak–Negami type are considered. The explicit expression for the fractional operator in these equations is obtained. The Monte Carlo algorithm for calculation of action of this operator and of the inverse one is constructed. The algorithm is derived from the Bochner–Phillips relation for a semigroup generated by a fractional power of initial infinitesimal operator. The method is based on averaging procedure over ensembles of one-sided stable variables.

Relaxation functions calculated numerically according to this scheme coincide with analytical functions obtained earlier by other authors. The algorithm represents a numerical way of calculation in relaxation problems with arbitrary initial and boundary conditions.

Acknowledgment The authors are grateful to the Russian Foundation for Basic Research (grant 10-01-00618) for financial support.

References

1. Fröhlich H (1958) Theory of dielectrics, 2nd edn. Oxford University Press, Oxford
2. Debye P (1954) Polar molecules. Dover, New York
3. Jonscher AK (1977) The "universal" dielectric response. Nature 267:673
4. Ramakrishnan TV, Raj Lakshmi M (1987) Non-debye relaxation in condensed matter. World Scientific, Singapore
5. Cole KS, Cole RH (1941) Dispersion and absorption in dielectrics. J Chem Phys 9:341
6. Davidson DW, Cole RH (1951) Dielectric relaxation in glycerol, propylene glycol, and n-propanol. J Chem Phys 19:1484
7. Havriliak S, Negami S (1966) A complex plane analysis of α-dispersions in some polymer systems. J Polymer Sci 14:99
8. Novikov VV, Wojciechowski KW, Komkova OA, Thiel T (2005) Anomalous relaxation in dielectrics. Equations with fractional derivatives. Mater Sci Poland 23:977
9. Nigmatullin RR, Ryabov YaE (1997) Cole-Davidson dielectric relaxation as a self-similar relaxation process. Phys Solid State 39:87
10. Jonscher AK (1996) Universal relaxation law. Chelsea-Dielectrics Press, London
11. Weron K (1991) How to obtain the universal response law in the Jonscher screened hopping model for dielectric relaxation. Phys Condens Matter 3:221
12. Weron K, Kotulski M (1996) On the equivalence of the parallel channel and the correlated cluster relaxation models. J Stat Phys 88:1241
13. Nigmatullin RR (1984) To the theoretical explanation of the "universal response". Phys Stat Sol b 123:739–745
14. Glöckle WG, Nonnenmacher TF (1993) Fox function representation of non-Debye relaxation processes. J Stat Phys 71:741
15. Jurlewicz A, Weron K (2000) Relaxation dynamics of the fastest channel in multichannel parallel relaxation mechanism. Chaos Solitons Fractals 11:303
16. Coffey WT, Kalmykov YuP, Titov SV (2002) Anomalous dielectric relaxation in the context of the Debye model of noninertial rotational diffusion. J Chem Phys 116:6422
17. Déjardin J-L (2003) Fractional dynamics and nonlinear harmonic responses in dielectric relaxation of disordered liquids. Phys Rev E 68:031108
18. Aydiner E (2005) Anomalous rotational relaxation: A fractional Fokker-Planck equation approach. Phys Rev E 71:046103
19. Weron K, Jurlewicz A, Magdziarz M (2005) Havriliak–Negami response in the framework of the continuous-time random walk. Acta Physica Polonica B 36:1855–1868
20. Bochner S (1949) Diffusion equation and stochastic processes. Proc Nat Acad Sci USA 35:368–370

21. Phillips RS (1952) On the generation of semigroups of linear operators. Pacific J Math 2:343–369
22. Yosida K (1980) Functional analysis. Springer, New York
23. Tarasov VE (2008) Universal electromagnetic waves in dielectric. J Phys Condens Matter 20:175223

Index

D. Baleanu et al. (eds.), *Fractional Dynamics and Control*,
DOI 10.1007/978-1-4614-0457-6, © Springer Science+Business Media, LLC 2012